AF361764

Impulse-Based Manufacturing Technologies

Impulse-Based Manufacturing Technologies

Editor

Verena Psyk

MDPI • Basel • Beijing • Wuhan • Barcelona • Belgrade • Manchester • Tokyo • Cluj • Tianjin

Editor
Verena Psyk
Fraunhofer Institute for Machine Tools and
Forming Technology IWU
Germany

Editorial Office
MDPI
St. Alban-Anlage 66
4052 Basel, Switzerland

This is a reprint of articles from the Special Issue published online in the open access journal
Journal of Manufacturing and Materials Processing (ISSN 2504-4494) (available at: https://www.mdpi.
com/journal/jmmp/special_issues/impulse_based_manufacturing).

For citation purposes, cite each article independently as indicated on the article page online and as
indicated below:

LastName, A.A.; LastName, B.B.; LastName, C.C. Article Title. *Journal Name* **Year**, *Volume Number*,
Page Range.

ISBN 978-3-0365-2890-8 (Hbk)
ISBN 978-3-0365-2891-5 (PDF)

Contents

About the Editor

Verena Psyk (Dr.-Ing.), head of the department of Profile-Based Processes and High-Velocity Technologies at the Fraunhofer Institute for Machine Tools and Forming Technologies IWU, started working on impulse-based technologies as a student. She received a doctorate with a distinction, focusing on the topic of *Process Chain Curving - Electromagnetic Compression - Hydroforming for Tubes and Hollow Profile-shaped Parts*. In parallel, she worked as a scientist and head of the department of High-Speed Forming and Joining at the Institute of Forming Technology and Lightweight Construction, TU Dortmund. Since 2011, she has been working in different positions at the Fraunhofer IWU. During this time, she coordinated, among other things, the European project Join'EM and developed solutions for different companies. She is a member of the International Impulse Forming Group and served as Vice Chairperson from 2014 to 2019. In 2019/2020, she received the Best Reviewer Award from the *Journal of Materials Processing Technology*.

Journal of
*Manufacturing and
Materials Processing*

Editorial

Impulse-Based Manufacturing Technologies

Verena Psyk

Fraunhofer Institute for Machine Tools and Forming Technology IWU, 09126 Chemnitz, Germany;
verena.psyk@iwu.fraunhofer.de

Citation: Psyk, V. Impulse-Based
Manufacturing Technologies. *J.
Manuf. Mater. Process.* **2021**, *5*, 133.
https://doi.org/10.3390/
jmmp5040133

Received: 4 December 2021
Accepted: 7 December 2021
Published: 9 December 2021

Modern manufacturing faces extensive technological and economic challenges to remain competitive under the current political and social conditions. Complex component geometries must be formed with high precision from materials that often provide limited formability. Joining technology must enable connections between different material combinations so that the resulting joints meet high requirements in terms of, e.g., mechanical strength, electrical conductivity, or tightness. Cutting operations must ensure a high-quality cut surface for materials of different strengths. In all areas, cost effectiveness, resource efficiency, and flexibility must be considered, and the manufacturing industry must be able to react quickly to changing boundary conditions and adapt their production in the sense of a resilient system. The clever use of velocity effects can help to meet these demands. This especially concerns impulse-based manufacturing technologies, in which the required energy acts within a very short period of time and suddenly accelerates workpiece areas to very high velocities. The correspondingly high strain rates, together with inertia and adiabatic effects, affect the behavior of many materials, resulting in technological benefits such as improved formability, reduced localizing, springback, extended possibilities regarding the production of high-quality multi material joints, and burr-free cutting. Furthermore, it is often possible to apply smaller and simpler presses.

This Special Issue of *JMMP* presents the current research findings, which focus on exploiting the full potential of impulse-based processes by providing a deep understanding of the technology and material behavior and detailed knowledge of sophisticated process and equipment design. The range of considered processes covers electromagnetic forming and joining by magnetic pulse welding, electrohydraulic forming, the use of vaporizing foil actuators, and adiabatic cutting with hydraulically driven presses. The presented work spans from basic investigations via application-oriented research to the development of industrial series processes.

In recent years, magnetic pulse welding—an application of electromagnetic forming that allows for the impact welding of various material combinations, including ones that are not compatible in conventional fusion-welding processes—has been a topic of interest in research and among industrial users. Corresponding with this trend, several papers in this Special Issue are dedicated to this technology: Faes et al. investigate the magnetic pulse welding of copper and steel tubes with and without an internal support. They use a coupled numerical simulation to quantify the collision parameters (impact angle and velocity) and predict the resulting shape of the joint. This shape is verified experimentally and fundamental correlations between influencing process parameters, collision properties and joint characteristics are identified [1]. Complementarily to this, Zhang and Kinsey study the tube-to-rod welding of aluminum to titanium and copper. They analyze the weld morphology via microstructural analysis and multiscale simulation. The research shows that yield strength and the density of the joining partners influence the wave formation and shear stresses are decisive, while shear velocity and contact pressure seem to have a minor influence [2]. Drehmann et al. focus on sheet metal magnetic pulse welding of aluminum to press-hardening steel and prove, via lap shear tests, that high quality is achievable. Similar to [1], they determine the collision parameters via coupled numerical simulation and deduce a quantitative collision-parameter-based process window for the material combination. They also perform an in-depth microstructural analysis and correlate the results with the mechanical quality of the welds [3]. Khalil et al. consider a special variant of magnetic pulse welding, which is suitable for the spot welding of metal sheets. They focus on tool design and assess different geometries and materials with regard to their efficiency and lifetime performance. In this context, again, the impact velocity is calculated, and the resulting aluminum steel joints are evaluated in quasistatic shear tests and fatigue tests [4]. These contributions reflect the significance of the collision properties in magnetic pulse welding. However, the technology does not allow for the direct adjustment of these important parameters, and even their measurement is hardly possible. To overcome these restrictions, Groche and Niessen present a model experiment that allows for the direct setting of the collision velocity and angle, independently of each other. They quantify a process window for aluminum/copper welds and compare this to the corresponding process windows identified for aluminum/aluminum and copper/copper welds. Microstructural investigations help to evaluate the weld quality and identify bonding mechanisms [5]. Similarly, Prasad et al. present a model setup that allows for an investigation of high-speed collision processes. They use a vaporizing foil actuator to accelerate a projectile to velocities in the range of 300–400 m/s. The setup serves for impulse-based manufacturing, such as collision welding or powder compaction, as well as for characterizing, e.g., the spall strength of materials in shock physics [6].

In addition to magnetic pulse welding, shaping by electromagnetic forming is considered in some contributions to this Special Issue. While earlier work dedicated to this technology focused on modelling and understanding the general deformation course, the work presented here focusses on equipment design and the development of complex industrial parts. Goyal et al. present fundamental results regarding the mechanical loads acting on the coil. They consider skin and proximity effects to numerically calculate the current density distribution and the resulting deformation of the coil winding as a function of the arrangement of the individual winding turns relative to each other and the workpiece [7]. As a supplement, Beckschwarte et al. examined the molding tools. They point out that, when forming thin foils, both the mechanical and electrical properties of the dies must be considered because, for this type of workpiece, the shielding of the magnetic field is not very effective. To avoid significant reductions in the process efficiency, the die materials should feature low electrical conductivity and high permeability [8]. Avrillaud et al. comprehensively reviewed how high strain rates can improve the formability of materials and applied this knowledge to industrial applications from the automotive sector, luxury products, or hadron collider components. The application examples consider the formation of deep or complex geometries, embossing, and post-forming to shape sharp edges. They show that, depending on the manufacturing task, electromagnetic or electrohydraulic forming could be suitable [9]. The contribution of Mamutov et al. also deals with the implementation of electrohydraulic forming in industrial manufacturing. They present a simple stepwise process design method: first, critical areas requiring a high forming pressure are identified; then, a hydrodynamic simulation serves to optimize the pressure chamber and, finally, a simple preform design method allows for the strains to be redistributed in the workpiece [10].

Compared to the technologies considered so far, high-speed impact cutting with hydraulically driven presses is a relatively slow process. However, the punch speed can still reach up to 10 m/s, and corresponding strain rates are in the range of 10^3/s, so the technology can still benefit from strain rate, inertia, and adiabatic heating. Winter et al. investigate the formation of adiabatic shear bands and the corresponding geometric accuracy of the cut surface under different process parameters. Specifically, they consider the influences of the relative clearance and the corresponding stress state, impact velocity, and impact energy [11].

All contributions to this Special Issue use numerical simulation to visualize and quantify process parameters, which can hardly be measured. Obviously, the quality of such simulations significantly depends on the implemented material model and the accuracy of the material parameters under the specific technological circumstances, such as strain rate and triaxiality. Therefore, technology-adapted material characterization and modelling is highly relevant in impulse-based manufacturing. Silva et al. suggest a me-thodology for the mechanical characterization and modelling of materials under extreme load conditions considering strain, strain rate, temperature, and state of stress [12]. Al-though the validation in this paper is based on metal cutting, these load conditions show comprehensive similarities with the situation in impulse-based manufacturing, so the approach is relevant and could benefit these technologies, too. The consequent improvement in the simulation via establishing efficient coupled and multiscale tools, as well as well-adapted material models, provides a prerequisite for the future implementation of digitalization and artificial intelligence, and thus contributes to the development of smart, impulse-based manufacturing processes.

Funding: This research received no external funding.

Conflicts of Interest: The author declares no conflict of interest.

References

1. Faes, K.; Shotri, R.; De, A. Probing Magnetic Pulse Welding of Thin-Walled Tubes. *J. Manuf. Mater. Process.* **2020**, *4*, 118. [CrossRef]
2. Zhang, S.; Kinsey, B.L. Influence of Material Properties on Interfacial Morphology during Magnetic Pulse Welding of Al1100 to Copper Alloys and Commercially Pure Titanium. *J. Manuf. Mater. Process.* **2021**, *5*, 64. [CrossRef]
3. Drehmann, R.; Scheffler, C.; Winter, S.; Psyk, V.; Kräusel, V.; Lampke, T. Experimental and Numerical Investigations into Magnetic Pulse Welding of Aluminum Alloy 6016 to Hardened Steel 22MnB5. *J. Manuf. Mater. Process.* **2021**, *5*, 66. [CrossRef]
4. Khalil, C.; Marya, S.; Racineux, G. Magnetic Pulse Welding and Spot Welding with Improved Coil Efficiency—Application for Dissimilar Welding of Automotive Metal Alloys. *J. Manuf. Mater. Process.* **2020**, *4*, 69. [CrossRef]
5. Groche, P.; Niessen, B. The Energy Balance in Aluminum–Copper High-Speed Collision Welding. *J. Manuf. Mater. Process.* **2021**, *5*, 62. [CrossRef]
6. Prasad, K.S.; Mao, Y.; Vivek, A.; Niezgoda, S.R.; Daehn, G.S. A Rapid Throughput System for Shock and Impact Characterization: Design and Examples in Compaction, Spallation, and Impact Welding. *J. Manuf. Mater. Process.* **2020**, *4*, 116. [CrossRef]
7. Goyal, S.P.; Lashkari, M.; Elsayed, A.; Hahn, M.; Tekkaya, A.E. Analysis of Proximity Consequences of Coil Windings in Electromagnetic Forming. *J. Manuf. Mater. Process.* **2021**, *5*, 45. [CrossRef]
8. Beckschwarte, B.; Langstädtler, L.; Schenck, C.; Herrmann, M.; Kuhfuss, B. Numerical and Experimental Investigation of the Impact of the Electromagnetic Properties of the Die Materials in Electromagnetic Forming of Thin Sheet Metal. *J. Manuf. Mater. Process.* **2021**, *5*, 18. [CrossRef]
9. Avrillaud, G.; Mazars, G.; Cantergiani, E.; Beguet, F.; Cuq-Lelandais, J.-P.; Deroy, J. Examples of How Increased Formability through High Strain Rates Can Be Used in Electro-Hydraulic Forming and Electromagnetic Forming Industrial Applications. *J. Manuf. Mater. Process.* **2021**, *5*, 96. [CrossRef]
10. Mamutov, A.V.; Golovashchenko, S.F.; Bessonov, N.M.; Mamutov, V.S. Electrohydraulic Forming of Low Volume and Prototype Parts: Process Design and Practical Examples. *J. Manuf. Mater. Process.* **2021**, *5*, 47. [CrossRef]
11. Winter, S.; Nestler, M.; Galiev, E.; Hartmann, F.; Psyk, V.; Kräusel, V.; Dix, M. Adiabatic Blanking: Influence of Clearance, Impact Energy, and Velocity on the Blanked Surface. *J. Manuf. Mater. Process.* **2021**, *5*, 35. [CrossRef]
12. Silva, T.E.F.; Gregório, A.V.L.; de Jesus, A.M.P.; Rosa, P.A.R. An Efficient Methodology towards Mechanical Characterization and Modelling of 18Ni300 AMed Steel in Extreme Loading and Temperature Conditions for Metal Cutting Applications. *J. Manuf. Mater. Process.* **2021**, *5*, 83. [CrossRef]

Journal of
*Manufacturing and
Materials Processing*

Article

Probing Magnetic Pulse Welding of Thin-Walled Tubes

Koen Faes [1,*]**, Rishabh Shotri** [2] **and Amitava De** [2]

[1]   Belgian Welding Institute, Technologiepark Zwijnaarde 935, B-9052 Ghent, Belgium
[2]   Indian Institute of Technology, Bombay, Mumbai 400076, India; rishabhshotri@gmail.com (R.S.);
     amit@iitb.ac.in (A.D.)
*   Correspondence: koen.faes@bil-ibs.be

Received: 29 October 2020; Accepted: 9 December 2020; Published: 11 December 2020

Abstract: Magnetic pulse welding is a solid-state joining technology, based on the use of electromagnetic forces to deform and to weld workpieces. Since no external heat sources are used during the magnetic pulse welding process, it offers important advantages for the joining of dissimilar material combinations. Although magnetic pulse welding has emerged as a novel technique to join metallic tubes, the dimensional consistency of the joint assembly due to the strong impact of the flyer tube onto the target tube and the resulting plastic deformation is a major concern. Often, an internal support inside the target tube is considered as a solution to improve the stiffness of the joint assembly. A detailed investigation of magnetic pulse welding of Cu-DHP flyer tubes and 11SMnPb30 steel target tubes is performed, with and without an internal support inside the target tubes, and using a range of experimental conditions. The influence of the key process conditions on the evolution of the joint between the tubes with progress in time has been determined using experimental investigations and numerical modelling. As the process is extremely fast, real-time monitoring of the process conditions and evolution of important responses such as impact velocity and angle, and collision velocity, which determine the formation of a metallic bond, is impossible. Therefore, an integrated approach using a computational model using a finite-element method is developed to predict the progress of the impact of the flyer onto the target, the resulting flyer impact velocity and angle, the collision velocity between the flyer and the target, and the evolution of the welded joint, which are usually impossible to measure using experimental observations.

Keywords: magnetic pulse welding; dissimilar material combinations; impact velocity; impact angle; collision velocity

---

1. Introduction

Magnetic pulse welding (MPW) is a solid-state impact welding technology able to create joints between two overlapping parts by a progressive collision, which is generated by an intense electromagnetic (EM) impulse [1]. Figure 1 shows a schematic layout for the MPW of overlapping tubes. A capacitor bank is charged by a power supply to store the required amount of energy, which is released instantaneously into a coil by using a high-current switch. The resulting discharge current of high magnitude and high frequency induces an intense transient EM field inside the coil, which induces eddy current in the outer tube [2]. The induced eddy current causes a differential EM field on both sides of the outer tube, resulting in an EM pressure that, in turn, causes the outer tube to impact onto the internal tube with high velocity [3]. As a result of the collision between the outer and the inner tubes at a certain angle, the tubes experience intense localized plastic deformation and a jet is generated along the surfaces of the materials before they make contact, which is able to remove the surface impurities and promote consolidation between the clean mating surfaces under EM pressure [2,3]. The consolidation between the two parts occurs without any bulk melting [4], although local melting at scattered locations along the joint interface is reported for MPW of AA6060 flyer tubes and copper target rods [5], AA1050 flyer and target sheets [6] and AA6060 flyer tubes and AlSi10Mg target rods [7]. For the overlapping assembly, the outer part is referred to as the flyer and the inner part is referred to as the target. Since the bulk melting of materials is avoided, MPW is increasingly considered for the joining of dissimilar materials [8].

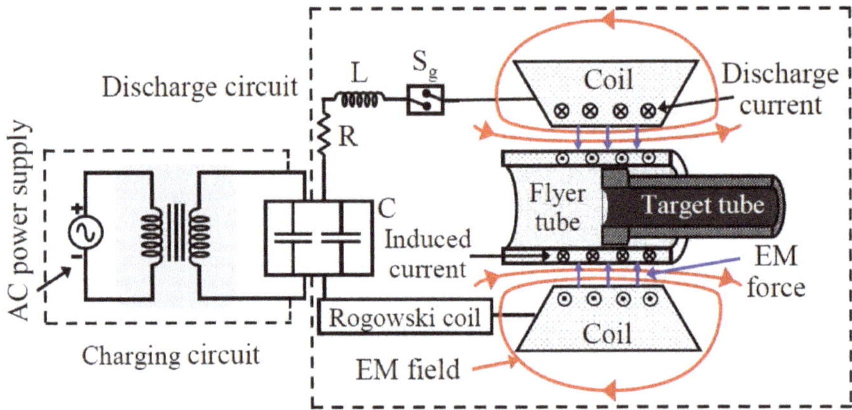

Figure 1. Schematic layout of magnetic pulse welding (MPW) set-up for joining of tubes.

The joining of tubular parts with MPW involves the use of a thick circular coil, with the assembly of the overlapping flyer and target tubes placed inside the coil [9]. The coil imposes a radial EM impulse pressure on the flyer tube for a very short duration, that forces the flyer to impact onto the target tube in a progressive manner, resulting in jetting of surface impurities from the original interface, a controlled plastic deformation and bonding between two tubes [10]. The welded interface in MPW is characteristically similar to that in explosion welding, which is created as a result of a traverse of collision points, followed by plastic deformation at a high strain rate and jetting out of surface impurities, and consolidation between clean metallic layers along the interface between the impacting flyer and target [5]. The phenomena of jetting due to the high velocity impact between the flyer and target and the consequent shearing of very thin layers from the metallic surfaces, the composition of the metal jets and the possible effect of jetting on the profile of the welded interface are examined by several researchers for MPW of similar and dissimilar materials [11–13].

Often, an internal support is employed to enhance the stiffness and avoid excessive plastic deformation and fracture of the tube assembly as a result of the high velocity impact of the flyer tube onto the target tube [14]. Especially, tubes with a small wall thickness need to be supported, because they can hardly resist radial forces. A study comparing the weld performance in terms of the contour deformation, microstructure and tensile strength of tubular joints, achieved with and without internal support, was found in [15]. An aluminium AA6060-O flyer tube (outer diameter: 76 mm × 2.5 mm) was joined onto a steel 34 (St34 mod) target tube (outer diameter: 76 mm × 2.5 mm). Other sources report the use of an internal support placed inside the target tube for MPW of AA5052 flyer tubes of wall thickness of 1 mm and SS304 target tubes of wall thickness of 1 mm [16], and for MPW of AA6060 flyer tubes of wall thickness of 1.5 mm and Cu-ETP target tubes of wall thickness of 1 mm [17]. Although these studies showed the need for an internal support during MPW of tubular parts, detailed quantitative analyses of MPW with and without an internal support are not provided.

The plastic deformation of the flyer and target tube during MPW is influenced by the nature and magnitude of the applied discharge current, the original joint configuration, the mechanical properties of the tube materials, the wall thickness of the tubes and the initial standoff distance between them, and the impact velocity of the flyer. For example, typical AA6061 flyer tubes (diameter: 40 mm × 2 mm) experienced a radial plastic deformation in the range of 1 to 2 mm during the joining with AISI1045 target rods [18]. The impact velocity of the flyer when impacting with the target rods was measured and was in the range of 245 to 305 m/s [18]. MPW of AA3003 flyer tubes (diameter: 20 mm × 1 mm) and steel target tubes (diameter: 15.2 mm × 1.7 mm) resulted in an inward plastic deformation of 1.5 mm of the joint assembly [8]. Likewise, MPW of AA6060 flyer tubes (diameter: 20 mm × 1 mm) and steel target tubes (diameter: 14.4 mm × 2.7 mm) led to an inward plastic deformation of 1.4 mm of the joint assembly [19]. An inward plastic deformation of 4.2 mm of the joint assembly was reported by Guigliemetti et al. [20] during MPW of aluminium tubes with a wall thickness of 2 mm. For MPW of Cu-DHP flyer tubes (diameter: 22.22 mm × 0.89 mm) and 11SMnPb30 steel target tubes (diameter: 16.44 mm × 2 mm), Shotri et al. [10] reported an inward deformation of approximately 5.8 mm of the joint assembly [10]. The aforementioned studies showed that the MPW of tubes would lead to an inward plastic deformation of the joining partners, and to a distortion of the final joint geometry, which may often be unacceptable and unwarranted for the intended final purpose.

A solution to restrict the inward plastic deformation of tubes during MPW when the flyer impacts on the target is to use a solid mandrel inside the target tube to enhance the stiffness of the original joint assembly. Shotri et al. [21] reported the use of a steel insert for MPW of AA2017 flyer tubes (diameter: 20 mm × 1 mm) and SS304 target tubes (diameter: 20 mm × 2 mm). Cui et al. [22] used an AA6061 supporting insert for the joining of AA5052 flyer tubes (diameter: 30 mm × 1 mm) and carbon fibre composite target tubes (diameter: 28 mm × 1.5 mm). Faes et al. [23] reported a decrease of the inward plastic deformation of the final joint assembly from 2.9 to 1 mm in MPW of Cu-DHP flyer tubes with 11SMnPb30 steel target tubes using polyurethane internal supports. The use of polyurethane internal support was also reported for MPW of AA6061 flyer tubes with a diameter of 40 mm and Cu-ETP target tubes, in order to minimize the maximum plastic deformation of the joint assembly [17]. Although these studies have showed that the use of an additional insert inside the target tube can avoid excessive plastic deformation, a detailed analysis of MPW of tubes with and without such inserts is rarely conducted.

In the present work, an effort is made to investigate the relative influence of the discharge energy, the initial standoff distance, and the use of an internal support for MPW of thin-walled Cu-DHP flyer and 11SMnPb30 steel target tubes using a bitter plate coil with a field shaper. The welded joints were produced at two different discharge energies (14 and 16 kJ), using variable standoff distances between the flyer and the target tubes, and with and without a polyurethane internal support inside the target tube. The nature of the progressive plastic deformation and evolution of the joint profile was examined in detail experimentally and using a numerical process model.

2. Materials and Methods

Figure 2 shows schematically the MPW coil, which is connected in series to an electrical discharge circuit. A five turn AA6082 bitter plate coil with a CuCrZr field shaper was used. The bitter plate coil consists of five hollow circular plates, which are made of AA6082 and held together by a set of insulated steel bolts in staggered positions. The circular coil plates with a diameter (m_o) of 280 mm and a thickness (m_b) equal to 12 mm have *i/o* (in/out) terminals at 10°, through radial cut-out sections. An insert of Cu-ETP with a thickness of 3 mm is located between the start and end of the radial sections and connects the adjacent plates (Figure 2), for a uniform circumferential flow of the discharge current. Table 1 presents the details of the discharge energy circuit and its characteristics.

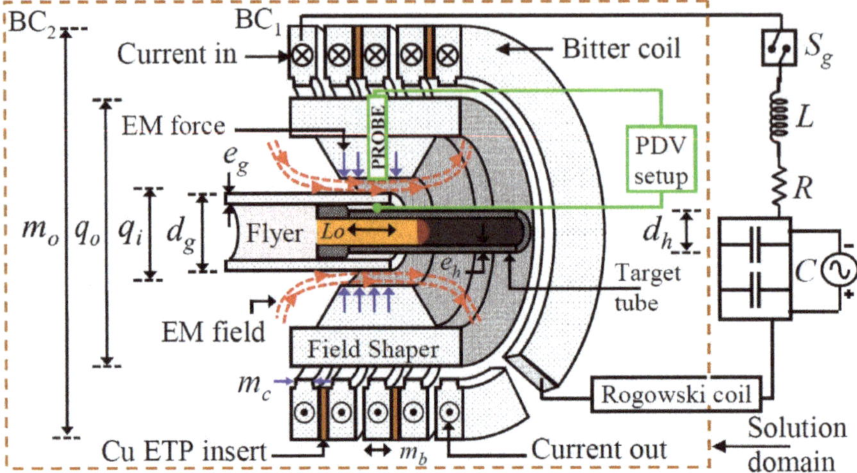

Figure 2. Schematic representation of the coil-tube assembly and solution domain.

Table 1. Details of discharge circuit elements and energy characteristics.

	Capacitance (μF)	Resistance (mΩ)	Inductance (μH)
Discharge circuit	C = 160	R = 14.3	L = 0.55
Generator capacity		E = 50 kJ, U = 25 kV	
Characteristics of the discharge energy			
Pulse frequency	$f = 1/(2\pi \sqrt{LC}) = 17$ kHz; $\omega = 2\pi f = 106.76 \times 10^3$ rad/s		
Damping coefficient	$\tau = (2L)/R = 7.69 \times 10^5$ s		
Discharge current	$I = I_m e^{(-t/\tau)} \sin \omega t$, where $I_m = U \sqrt{C/L} e^{(-t/4)/\tau}$		

A high frequency damped sinusoidal current flows through the coil when the high current switch (S_g) is activated and creates a transient EM field that induces a secondary eddy current in the inner coaxial CuCrZr field shaper. The field shaper has a tapered geometry that allows a concentrated inward flow of the surface current towards its internal face, using a radial slit across its thickness. The inner and the outer width of the field shaper are 80 and 15 mm, respectively, and its tapered angle θ is 30°. A constant initial radial gap (a) of 1.14 mm is maintained between the field shaper internal surface and the flyer tube. The process conditions, and the dimensions of the coil, field shaper and tubes assembly are presented in Table A1 (Appendix A).

A Rogowski coil was used to measure the magnitude and nature of the discharge current [10] for the applied energy of 14 and 16 kJ. A photon Doppler velocimetry (PDV) setup was used to measure the velocity of the flyer tube during the course of its acceleration, impact on the target and complete deceleration. The collimator probe of the PDV set-up has an outer diameter of 2.5 mm and is integrated into the field shaper, as shown in Figure 2. The actual arrangement of the coil and the field shaper is illustrated in Figure 3. Several bore holes and pockets were made in the field shaper in order to allow the collimator probe of the PDV set-up to access the outer surface of the flyer tube. This setup allows radial velocity measurements at the centre of the field shaper. This corresponds with a velocity measurement location at 0.5 mm from the tube extremity. The measured velocity of the flyer tube when it impacts onto the target tube is used for the comparison with the computed impact velocity of the flyer tube.

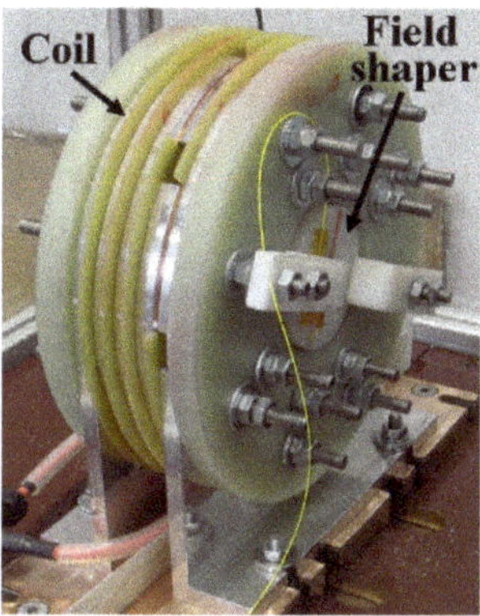

Figure 3. Actual arrangement of multi-turn bitter coil and field shaper with the collimator probe (yellow) of the photon Doppler velocimetry (PDV) setup coming out of the field shaper.

The tube assembly is comprised of a flyer tube made of the Cu-DHP alloy with a diameter (d_g) equal to 22.22 mm and a wall thickness (e_g) equal to 0.89 mm, and a target tube made of 11SMnPb30 steel with a wall thickness (e_h) equal to 1 mm. The overlap between the flyer and the target tubes was equal to 25 mm, creating a free length of the flyer tube of 15 mm. The initial standoff distance (s) between the flyer and the target tubes was varied from 1 to 2 mm, by employing target tubes with different external diameters (d_h) ranging from 18.44 to 16.44 mm.

For the purpose of minimizing the deformation of the parent part, a 50 mm-long tube made of polyurethane with an inserted solid steel bolt was used as an internal support for the target tube (Figure 4). The polyurethane had a hardness of 92 Shore A. The tubular internal supports were pre-stressed via the inserted bolt, a washer and nuts. After the MPW process, the inserts could be removed manually by releasing the bolt and could be re-used. Tables 2 and 3 show the material properties of the coil, field shaper, flyer and target tubes, the internal polyurethane tube and the steel bolt support, respectively. These materials properties are used further in computational modelling of the MPW process in the present work.

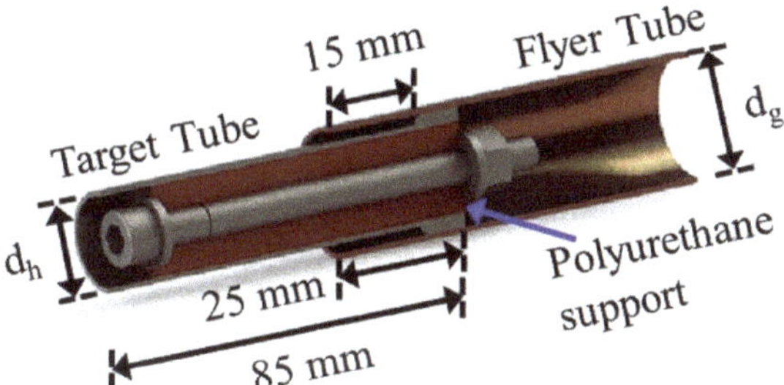

Figure 4. Schematic representation of the tubular internal support used in the experiments with flyer tube diameter d_g = 22.22 mm and target tube diameter d_h = 16.44 to 18.44 mm.

Table 2. Material properties of coil, field shaper and tubes (flyer-target) [10].

Property	Symbol	Coil AA6082	Field Shaper CuCrZr	Flyer Cu-DHP	Target 11SMnPb30
Relative permeability	μ	1	0.99	0.99	B-H curve *
Electrical conductivity (S/m)	σ	26.32×10^6	58×10^6	46.9×10^6	5.75×10^6
Density (kg/m^3)	ρ	2700	8933	8900	7800
Specific heat (J/kg/K)	c_p	900	385	386	472
Shear modulus (GPa)	G	26.3	45	47.8	129
Poisson ratio	ν	0.33	0.38	0.38	0.29

* B-H curve for 11SMnPb30 steel is presented as Figure A1 in Appendix C.

Table 3. Material properties of the polyurethane [24] and steel bolt [10].

Property	Symbol	Internal Support Polyurethane	AISI 1006 Bolt
Density (kg/m^3)	ρ	1150	7896
Shear modulus (GPa)	G	0.7	81.8
Poisson ratio	ν	0.4	0.29

Figure 5 shows a sample welded tubular assembly, together with a typical cross-section of the joined zone. The final cross-section of each welded sample was examined along the original flyer-target overlap length, to measure the joint length and the overall deformation of the assembly. The longitudinal cross-section of each welded joint was examined using optical microscopy to obtain a macroscopic view of the entire flyer-target contact and using scanning electron microscopy (SEM) to investigate the microscopic nature of the contact interface at multiple locations. The welded length is judged based on the intimate contact along the flyer-target interface as viewed under the optical microscope. Subsequently, the microscopic nature of the interface was examined at a few locations along the welded interface using SEM. For a sample weld prepared with an internal polyurethane support, the macroscopic view of the longitudinal cross-section of the entire tubular joint and the SEM backscattered images at two random locations along the welded interface are shown together in Figure 6. Similar SEM backscattered images at six to ten locations along the welded length are examined for each welded sample to observe the interface nature. In most cases, a slightly wavy nature of the interface has been observed along the welded length, which is anticipated from the macroscopic view, based on a close scrutiny using optical microscopy.

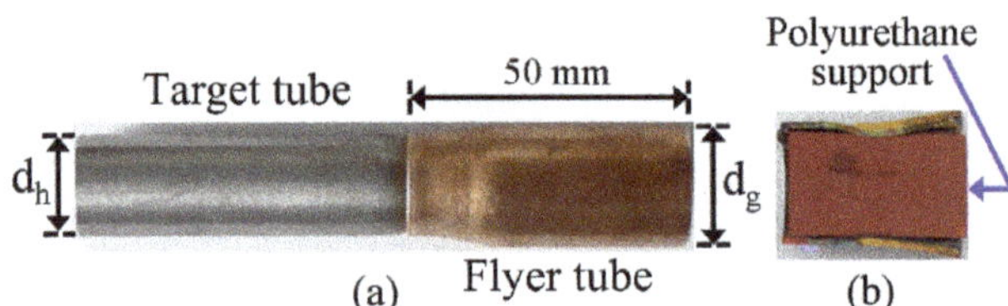

Figure 5. (**a**) Typical welded assembly and (**b**) cross-section in the as-welded condition with flyer tube diameter d_g = 22.22 mm and target tube diameter d_h = 16.44 to 18.44 mm.

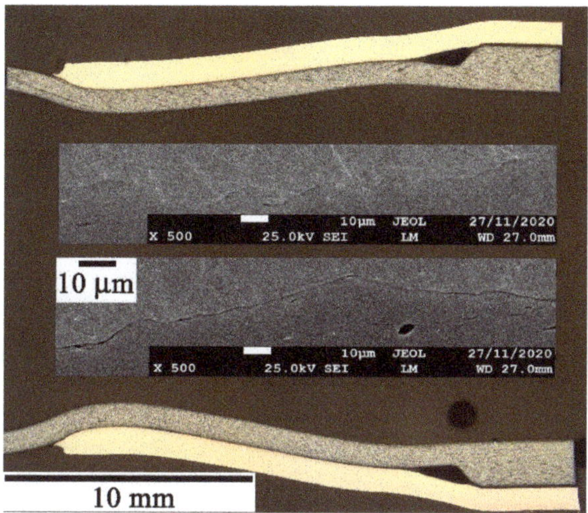

Figure 6. Macroscopic view of the longitudinal joint cross-section with SEM backscattered images at two random locations along the welded length, showing a slightly wavy interface profile for a weld made with a polyurethane internal support. Conditions for this welded sample: discharge energy = 14 kJ, standoff distance = 1 mm, flyer tube diameter and wall thickness = 22.22 and 0.89 mm, target tube diameter and wall thickness = 18.44 and 1 mm.

3. Theoretical Formulation

The EM field, **H**, around the coil and tube assembly in the MPW process, is obtained by solving the Maxwell's governing equation of the magnetic diffusion as [25,26]:

$$\frac{1}{\mu\sigma}\nabla^2\mathbf{H} = \frac{\partial\mathbf{H}}{\partial t} \tag{1}$$

where, σ and μ represent the electrical conductivity and magnetic permeability. The resultant EM force, **F**, over the flyer tube can be estimated as:

$$\mathbf{F} = \mathbf{J} \times \mathbf{B} \tag{2}$$

where $\mathbf{J} = \nabla \times \mathbf{H} = \nabla \times (\mathbf{B}/\mu)$, **B** and **J** refer to the magnetic flux density vector and the eddy current density vector, respectively.

The dynamic impact analysis requires the application of a transient EM pressure on the flyer tube. The transient EM pressure, p, on the flyer tube is therefore estimated using the EM force, **F**, as,

$$p = \int_0^\delta \mathbf{F}dt = \int_0^\delta -\mu_0\mu_r\mathbf{H}\frac{\partial\mathbf{H}}{\partial r}dt = \frac{1}{2}\mu_0\mu_r\left(H_s^2 - H_p^2\right) \tag{3}$$

where μ_0 and μ_r refer to the relative magnetic permeability in the air and in the flyer tube. H_s and H_p represent the magnetic field over the surface and at a depth δ, referred to as the skin depth of the flyer tube [21]. The field intensity at a depth δ, where $\delta = 1/\sqrt{\mu\pi f\sigma}$, is very small and hence neglected in the present analysis.

Figure 2 shows schematically the solution domain, which is considered for the EM field analysis with the density, relative permeability and electrical conductivity of air equal to 1.1614 kg/m^3, 1 and zero, respectively. The solution domain is extended to 800 mm at all sides of the coil-tube assembly. The EM field remains continuous within the solution domain and is represented as [27]:

$$\hat{a}_n \cdot (\mathbf{B_1} - \mathbf{B_2}) = 0, \; \hat{a}_n \times (\mathbf{H_1} - \mathbf{H_2}) = \mathbf{J_s} \tag{4}$$

where $\hat{a}_n$ is the unit vector normal to the surface, $\mathbf{J_s}$ is the surface current density vector, $\mathbf{B_1}$ and $\mathbf{B_2}$ are the magnetic flux density vectors at the coil cross-sections. $\mathbf{H_1}$ represents the magnetic field intensity at the coil inner surface, interacting with the field shaper. $\mathbf{H_2}$ is the magnetic field intensity at the coil outer surface. At the domain boundary, BC$_2$ [Figure 2], the EM field intensity becomes negligible and is represented as [27]:

$$\hat{a}_n \cdot (\mathbf{H_1} - \mathbf{H_2}) = 0 \tag{5}$$

The numerical model for the EM field and the mechanical analysis is undertaken using the finite-element software ANSYS (ver. 14.5). For the EM field analysis, a total of 174,000 three-dimensional tetrahedral elements were used to discretize the solution domain (Figure 2), with the current density vector as nodal input along the coil boundary and the EM field vector as the nodal degree of freedom. The estimated EM pressure distribution is employed as boundary condition for the dynamic impact analysis using a MATLAB-based code. The dynamic impact analysis includes the overlapping tubes assembly, which was discretized with 125,000 solid hexahedral elements that can consider non-linear constitutive models, which are presented in Appendix B. The computed results of the EM field and dynamic impact analyses, along with the experimentally measured results, are illustrated in the subsequent section.

4. Results and Discussions

Figure 7a,b represents the measured and estimated currents for two typical values of the discharge energy of 14 and 16 kJ, which were utilized in the present work to prepare the welded samples. The peak current for a discharge energy of 14 kJ was approximately equal to 197 kA at the time instant of 13 μs. Likewise, the peak current for the discharge energy of 16 kJ was measured to be around 202 kA at the same time instant of 13 μs. The frequency of the discharge current was observed to be approximately 17 kHz. The corresponding maximum values of the EM force were calculated to be around 47.2 and 51.2 kN for a discharge energy of 14 and 16 kJ, respectively.

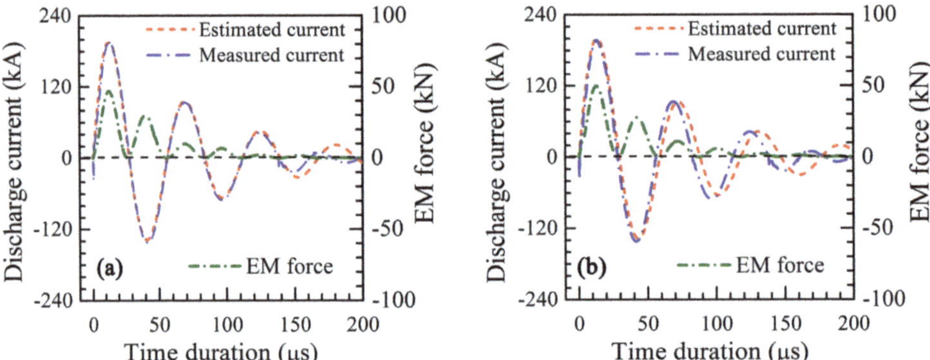

Figure 7. Experimentally measured and analytically estimated nature of the discharge current, and numerically computed EM force for an applied energy of (**a**) 14 kJ and (**b**) 16 kJ.

Figure 8 shows the computed results of the EM field and pressure for an applied discharge energy of 14 kJ at the time instant of 13 μs. The EM field vectors are distributed around the coil-field shaper and the flyer-target tubular assembly, as shown in Figure 8a. The EM fields are concentrated around the flyer tube portion in the proximity of the field shaper internal surface, due to its shaped geometry. The maximum values of the EM field and the eddy current density over the flyer tube surface were calculated as 39.8 T and 5.8×10^{10} A/m^2 at the time instant of 13 μs.

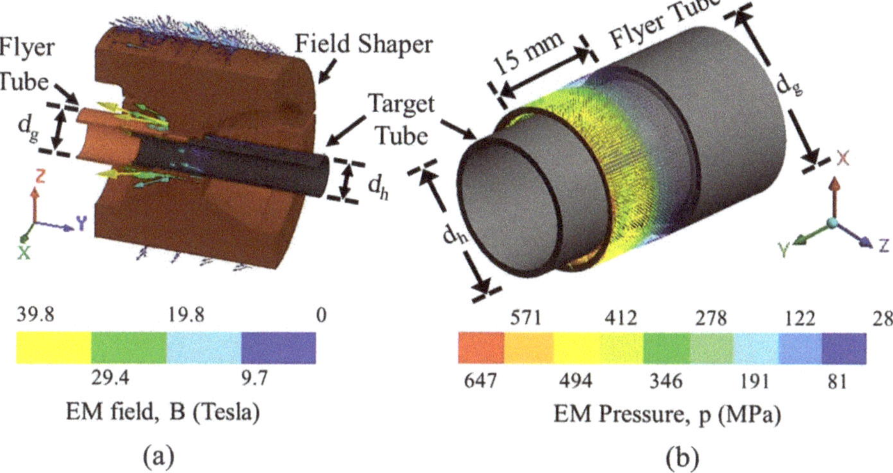

Figure 8. Numerically computed distribution of the maximum (**a**) EM field over the field shaper-tube assembly, and (**b**) EM pressure over the flyer tube for an applied energy of 14 kJ.

Figure 8b shows the calculated EM pressure over the flyer tube at the time instant of 13 μs. The estimated EM pressure remains circumferentially uniform, with the maximum value equal to 647 MPa, which corresponds with the maximum EM force of 47.2 kN for the discharge energy of 14 kJ. The EM pressure on the flyer tube is maximal at its free end and decreases along its length beneath the field shaper, resulting in an inward bending and oblique impact of the flyer on the target tube. As the portion of the flyer bends and moves away from the field shaper, the EM field and pressure distribution change. This results in a progressive oblique impact of the flyer onto the target tube along their overlapping length. While Figure 8b shows the computed EM pressure distribution over the flyer tube at the time instant of 13 μs, similar EM pressure distributions are extracted for multiple discrete time-steps and used as input for the dynamic impact analysis of the tube assembly. The results of this are presented in the text below.

Figure 9 shows the computed results of the progressive impact and deformation of the tube assembly and, the flyer-target contact length at multiple consecutive time instants for an applied discharge energy of 14 kJ. The colour bars under each figure represent the total plastic deformation in mm, which is the resultant of the computed plastic deformations in the x-, y- and z- directions. The calculated flyer-target contact length is determined by examining the deformed portion of the flyer, which has impacted and remained in contact with the target tube under the influence of the EM pressure. The actual contact is deemed to have established for the length of the continuous segment, along which the vertical distance between the flyer and the target (i.e., the internal surface of the flyer and the outer surface of the target) tubes is equal to zero. This calculated flyer-target contact length is subsequently compared with the experimentally measured welded length.

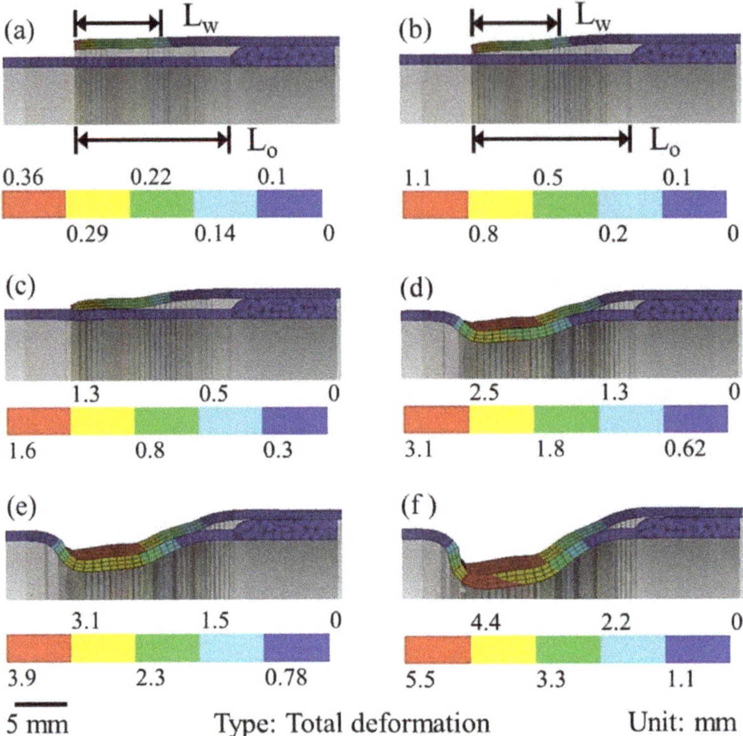

Figure 9. Computed results of the progressive impact and plastic deformation of the tube assembly at the time instants of (**a**) 7 µs, (**b**) 9 µs, (**c**) 11 µs, (**d**) 17 µs, (**e**) 21 µs and (**f**) 26 µs for a discharge energy of 14 kJ and with no internal support inside the target tube. (Initial conditions: wall thicknesses of flyer = 0.89 mm and of target = 1 mm; standoff distance between flyer and target = 1 mm; overlapping length between field shaper and flyer, L_w = 8 mm and free length of the flyer tube, L_o = 15 mm).

Figure 9a shows that there is very little deformation of the flyer tube after 7 µs for a peak EM field and pressure of 33.2 T and 441 MPa. At 9 µs, the typical inward bending of the flyer can be observed, resulting in the calculated flyer-target contact length of around 1 mm for a peak EM field and pressure of 36.1 T and 520 MPa (Figure 9b). The increase of the EM field and pressure is attributed to the increase of the peak current at the time instants of 7 and 9 µs. The inward bending of the flyer continues and the calculated flyer-target contact length increases further to 8 mm at the time instant of 11 µs (Figure 9c). However, as the flyer moves away from the coil (field shaper), the peak EM field and pressure are reduced to 35.6 T and 505 MPa. With further progress in time, at 17 µs, the calculated flyer-target contact length increases to 11 mm. The inward deformation of the joint assembly also increases to 2.5 mm, while the peak EM field and pressure are reduced to 32.6 T and 430 MPa, respectively (Figure 9d).

After 21 µs, the flyer-target assembly showed an increased inward deformation of 3.1 mm, while the calculated flyer-target contact length remains the same as that at 17 µs, which is attributed to a significantly reduced peak EM field intensity and pressure of 23.3 T and 216 MPa, respectively (Figure 9e). As the time increases further, the target tube also experiences more deformation, resulting in further inward distortion of the overall tubular assembly, as shown in Figure 9f. A greater inward deformation of the tubular assembly especially at longer time durations can be attributed to the thinning of the tube walls and the consequent loss of stiffness of the tubular assembly. The computed results of the deformed tubular assembly depict a maximum thinning of both the flyer and the target tube walls by approximately 0.2 mm at the end of the flyer-target overlapping length.

The excessive inward plastic deformation of the tube assembly as shown in Figure 9f indicates that MPW of tubes can lead to a significant distortion of the tube assembly. Lueg-Althoff et al. [17] reported a similar range of plastic deformation for MPW of AA6060 flyer and AISI 1045 target tubes, with the wall thickness of the target tube equal to 1 mm. These authors suggested the use of an internal support to constrain the inward deformation of the joint assembly, while a further examination of the influence of the internal support to improve the joint profile and quality was not undertaken.

Figure 10 shows the computed results of the progressive impact and deformation of the tube assembly at multiple consecutive time instants when using a polyurethane internal support placed inside the target tube. Figure 10a confirms a calculated contact length of 1 mm between the flyer and the target tube at 9 µs. The peak EM field and pressure at 9 µs were calculated as 36.5 T and 529 MPa, respectively. After 11 µs, the calculated flyer-target contact length increases to 8.5 mm (Figure 10b) for a higher peak EM field and pressure of 37.9 T and 571 MPa, respectively. It is noteworthy in Figure 10a,b that only the flyer experiences plastic deformation and that the target tube has suffered no inward distortion. The calculated contact length between the flyer and the target tube increases further to 10.5 mm after 15 µs (Figure 10c). The computed values of the peak EM field and pressure at 15 µs showed a drop to 33.6 T and 451 MPa, respectively.

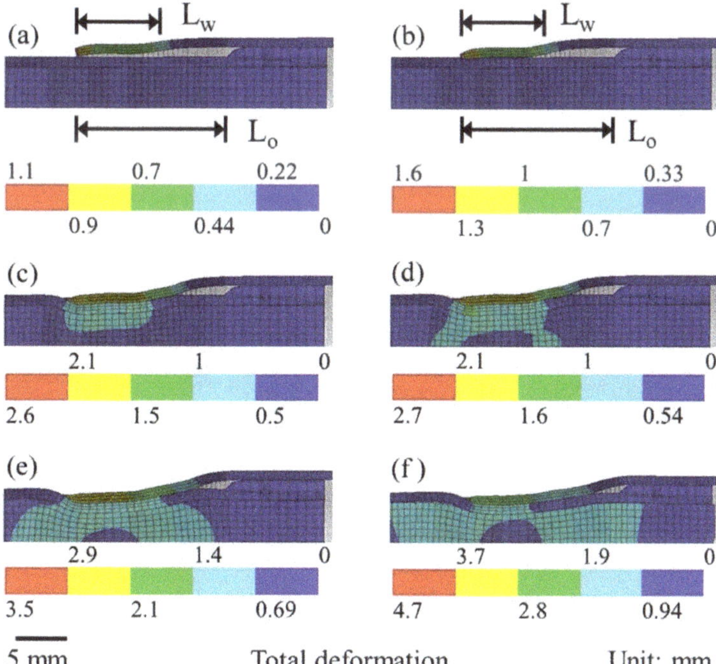

Figure 10. Computed results of the progressive impact and plastic deformation of the tube assembly at the time instants of (**a**) 9 µs, (**b**) 11 µs, (**c**) 15 µs, (**d**) 17 µs, (**e**) 21 µs and (**f**) 26 µs for an applied discharge energy of 14 kJ and with a polyurethane internal support for the target tube. (Initial conditions: wall thicknesses of the flyer = 0.89 mm and of the target = 1 mm; standoff distance between the flyer and the target = 1 mm; overlapping width between the field shaper and the flyer, L_w = 8 mm and free length of the flyer tube, L_o = 15 mm).

After 17 μs, the computed flyer-target contact length shows a small increase to approximately 11.5 mm (Figure 10d), and the calculated values of the peak EM field and pressure show a further drop to 32.2 T and 415 MPa, respectively. The calculated contact length between the flyer and the target shows no further increase in the subsequent time instants of 21 μs (Figure 10e) and 26 μs (Figure 10f). This is consistent with the continuing decrease of the peak EM field (~22 T) and pressure (~209 MPa) at 21 μs, as a result of the inward bending of the flyer away from the field shaper. The computed results show that the inward deformation of the tubular assembly starts late, at 15 μs (Figure 10c), and the net deformation remains very small until the end (Figure 10f), which demonstrates the role of the polyurethane internal support to enhance the stiffness of the target tube and to minimize the deformation of the tube assembly. With the internal polyurethane support inside the target tube, the computed results of the deformed tubular assembly depict a maximum thinning of the flyer tube wall by around 0.17 mm and little or no thinning of the target tube wall.

Figures 11 and 12 show the longitudinal cross-section of the entire welded length and the corresponding computationally obtained flyer-target deformed profile for different process conditions. The welded length, which is judged based on the intimate contact using optical microscopy, is shown by a thin red line in the experimentally observed macroscopic sections. The microscopic view of the welded interface is also probed using SEM backscattered images at six to ten random locations along the welded length for each tubular joint, as explained earlier (Figure 6).

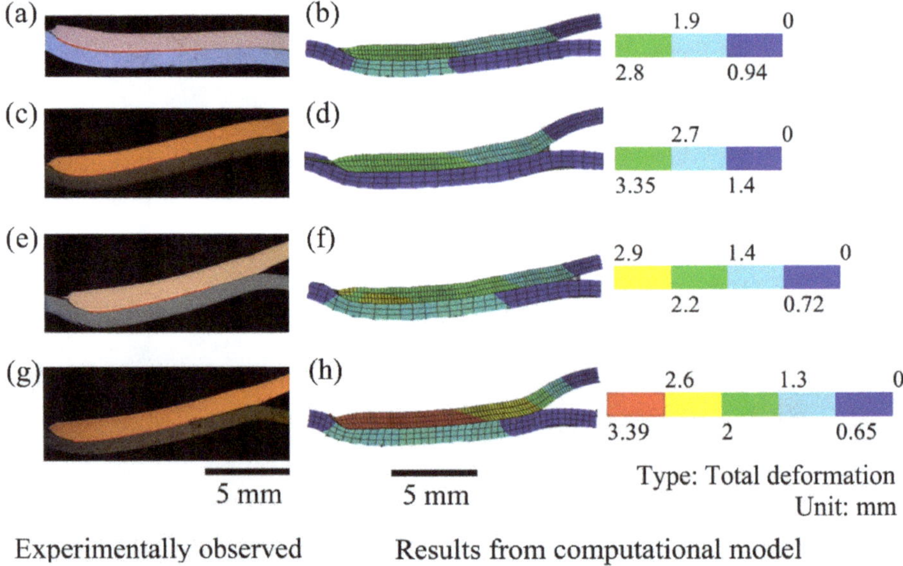

Figure 11. Experimentally observed and computationally evaluated flyer-target joint cross-sections for a discharge energy of 14 kJ (**a–d**) and 16 kJ (**e–h**), with a polyurethane internal support inside the target tube. (Initial conditions: flyer-target standoff distance = 1 mm (**a,b,e,f**) and 2 mm (**c,d,g,h**); flyer tube: diameter = 22.22 mm, wall thicknesses = 0.89 mm; target tube: wall thicknesses = 1 mm; diameter = 18.44 mm (standoff distance = 1 mm), 16.44 mm (standoff distance = 2 mm)).

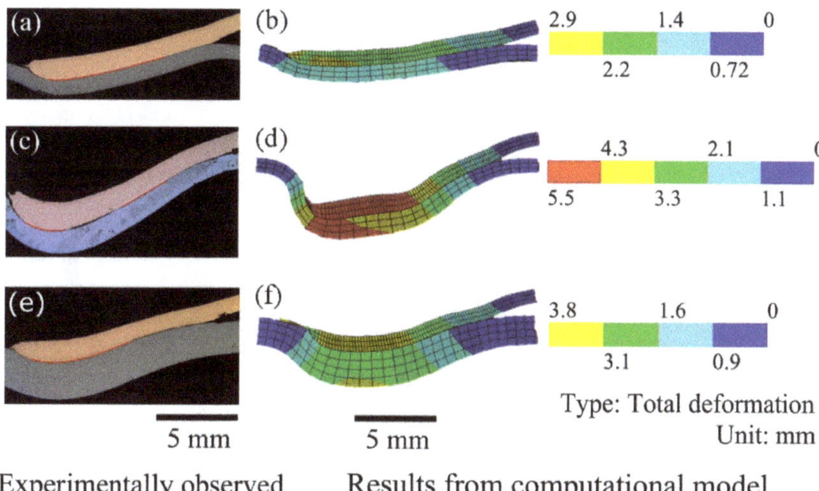

Figure 12. Experimentally observed and computationally estimated flyer-target joint cross-sections with a polyurethane internal support (**a,b**), and without any internal support (**c–f**) for the target tube. (Initial conditions: discharge energy: 16 kJ (**a,b,e,f**), 14 kJ (**c,d**); flyer: diameter = 22.22 mm, wall thicknesses = 0.89 mm; and target: diameter = 18.44 mm, wall thicknesses = 1 mm (**a–d**), 2 mm (**e,f**); standoff distance = 1 mm).

Figure 11 shows the experimentally observed and the corresponding computationally obtained joint cross-sections for a discharge energy of 14 and 16 kJ, and with the use of a polyurethane internal support placed inside the target tube. Figure 11a,b shows that the length of the experimentally observed welded interface and the corresponding calculated flyer-target contact length are approximately 8.3 and 11.5 mm, respectively, for a standoff distance of 1 mm and a discharge energy of 14 kJ. It is noteworthy that the calculated flyer-target contact length is obtained from the deformed interface of tubular assembly as a part of the mechanical analysis. The formation of the actual weld needs to consider the jetting of the impurities from the flyer-target interface, and localized plastic deformation creating mechanical interlock and atomic diffusion between the abutting surfaces of the flyer and the target, which are beyond the scope of the dynamic mechanical analysis adopted in the present work. As a result, the calculated flyer-target contact length remains always a little higher than the corresponding length of the experimentally observed welded interface.

As the standoff distance increases to 2 mm, both the length of the experimentally observed welded interface and the calculated flyer-target contact length decreases to 7.3 and 10 mm, respectively (Figure 11c,d). The decreasing tendency of the contact length with an increase of the standoff distance is also observed at the higher discharge energy of 16 kJ. For example, when the standoff distance increases from 1 to 2 mm at a discharge energy of 16 kJ, the length of the experimentally observed welded interface reduces from 7.3 to 6.3 mm (Figure 11e,g), and the corresponding calculated flyer-target contact lengths are around 11.5 and 10.5 mm, respectively (Figure 11f,h).

For an increasing standoff distance, the flyer tends to impact onto the target at a higher velocity and therefore the tubular assembly; in particular, the target tube suffers more plastic deformation, which impairs the progressive growth of the calculated flyer-target contact length. A comparison of Figure 11b,d clearly demonstrates the increasing deformation of the tubes when the standoff distance is increased from 1 to 2 mm at a discharge energy of 14 kJ. A similar effect of the standoff distance is also observed at the discharge energy of 16 kJ (Figure 11f,h). Overall, Figure 11 shows that an increase of the standoff distance can lead to excessive plastic deformation of the tubes and adversely affects the joint formation. The presence of an internal support inside the target and the wall thickness of the target tube also plays a significant role for restricting the deformation of the tubes.

Figure 12a,b shows the experimentally observed and calculated flyer-target joint cross-sections with an internal polyurethane internal support. The length of the experimentally observed welded interface and the calculated flyer-target contact length are approximately 7.3 and 11.5 mm, respectively, for a standoff distance of 1 mm and a discharge energy of 16 kJ. The wall thickness of the target tube is 1 mm. In contrast, both the length of the experimentally observed welded interface and the calculated flyer-target contact length are reduced to approximately 5.9 and 11 mm, respectively (see Figure 12c,d), when no internal support is used inside the target tube. The reduction of the weld length is attributed to the greater plastic deformation of the tubes when no internal support is used (Figure 12b,d).

For a target tube with a wall thickness of 2 mm and without any internal support, and a discharge energy of 16 kJ, the length of the experimentally observed welded interface and the calculated flyer-target contact length are approximately equal to 6.3 and 11 mm, respectively (Figure 12e,f). A comparison of Figure 12a,b,e,f shows that the internal support has helped to produce a greater welded length even with a target tube of smaller wall thickness of 1 mm (Figure 12a,b) in comparison to that with a target tube of higher wall thickness of 2 mm (Figure 12a,b). However, a comparison of Figure 12d,f shows that the tubular assembly without any internal support suffered a greater inward deformation of 5.5 mm, with a target tube of 1 mm wall thickness in comparison to an inward deformation of 3.8 mm with a target tube of 2 mm wall thickness, which can be attributed to higher stiffness posed by the thicker target tube.

The underlying phenomena of the MPW process can be examined further by following the progress of the formation of the welded interface in accordance with the flyer impact velocity and angle, and the collision velocity between the flyer and the target. As the process is extremely fast, the real-time monitoring of the welded interface, impact velocity and angle, and the collision velocity are impossible. A computational process model is therefore a practical alternative to estimate these values. Figure 13 presents the variation of the calculated flyer-target contact length, the flyer impact velocity and angle, and the collision velocity for a discharge energy of 14 kJ and two different standoff distances. Figure 13a shows that the computed impact velocity of the flyer ranges from 3.0×10^5 to 1.4×10^5 mm/s for a standoff distance of 1 mm. The corresponding experimentally measured value of the impact velocity of the flyer is around 3.1×10^5 mm/s and shown in Figure 13a. When the standoff distance is increased from 1 to 2 mm, higher values of the flyer impact velocity are obtained in the range from 4.5×10^5 to 1.5×10^5 mm/s, with the corresponding measured value equal to 4.1×10^5 mm/s, as shown in Figure 13b. Figure 13a,b also shows that the growth of the calculated flyer-target contact length tends to slow down with an increase in time as the flyer impact velocity reduces.

When the flyer continues to impact onto the target in an oblique manner, knowledge of the progressive variation of the flyer impact angle is very valuable. The collision velocity is a determining factor for the occurrence of a jet which removes the surface impurities to promote the bonding between the flyer and the target. Hence, an estimation of the variation of the collision velocity and the angle is of importance. The nature of the collision velocity and its impact on the weld formation have been studied well for explosive welding, which is similar to MPW and involves the impact between flyer and target metallic parts at a certain angle. Yuan et al. [28] reported the maximum collision velocity to be around 4.9×10^3 m/s during explosive welding of 4 mm thick AA6061 flyer sheets to magnesium alloy target plates. Bataev et al. [29] observed that the typical collision velocity of approximately 3.8×10^3 m/s produced good joints during the explosive welding of 1 mm-thick AISI1006 flyer and target sheets, while a much higher collision velocity of around 9.6×10^3 m/s resulted in welds with humps and excessive plastic deformation.

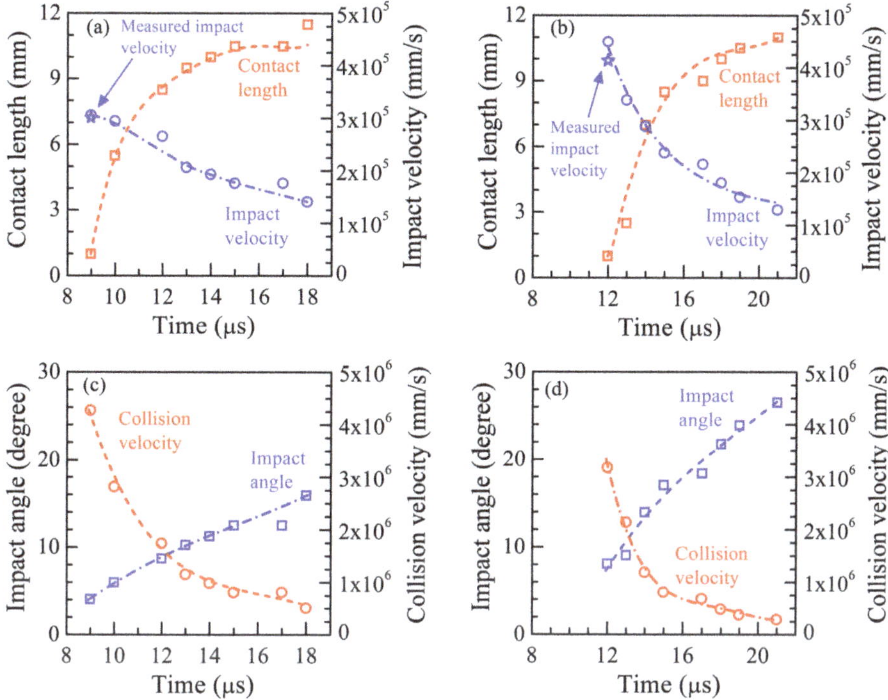

Figure 13. Computed values of the flyer-target contact length, the flyer impact velocity and angle and the collision velocity as a function of time, for an applied discharge energy of 14 kJ with a polyurethane internal support of the target tube. (Initial conditions: flyer: diameter = 22.22 mm, wall thicknesses = 0.89 mm; and target: diameter = 18.44 mm, wall thicknesses = 1 mm; standoff distance between the flyer and the target = 1 mm (**a,c**) and 2 mm (**b,d**)).

Figure 13c,d shows the computed results of the impact angle and collision velocity for two different standoff distances (1 and 2 mm) as the calculated flyer-target contact length increases with time. The flyer impact angle is calculated based on the deformed profile of the flyer at a certain time instant. The collision velocity is computed as $V_i/\sin\alpha$, where V_i is the flyer impact velocity and α is the impact angle at the end of the contact between the flyer and target [21]. For a standoff distance of 1 mm, the collision velocity and the flyer impact angle are situated in the range from 4.3×10^6 to 0.5×10^6 mm/s, and from 4 to 16°, respectively (Figure 13c). For an increase of the standoff distance to 2 mm, the flyer impact angles are higher, in the range from 8 to 27°, and consequently, the collision velocity values are reduced and are in the range from 3.3×10^6 to 0.3×10^6 mm/s (Figure 13d). This clearly shows that it is recommended to use a smaller initial standoff distance between the flyer and the target for the given tube dimensions and other process conditions.

For a higher discharge energy of 16 kJ, a standoff distance of 1 mm, and considering a polyurethane internal support inside the target tube, similar calculations were performed to obtain the typical ranges of the flyer impact velocity and angle, and the collision velocity. This resulted in 3.2×10^5 to 1.3×10^5 mm/s, 4 to 16° and, 4.5×10^6 to 0.5×10^6 mm/s, respectively. For an increase of the standoff distance to 2 mm, the calculated flyer impact velocity and angle, and the collision velocity are found to be in the range of 4.6×10^5 to 2.0×10^5 mm/s, 8 to 24° and, 3.3×10^6 to 0.4×10^6 mm/s, respectively. A comparison of these quantitative estimations with those presented in Figure 13a–d shows that the calculated ranges of the flyer impact velocity and angle, and the collision velocity remains almost the same for an increase of the discharge energy from 14 to 16 kJ for both standoff distances of 1 and 2 mm.

However, the increase of the calculated ranges of the flyer impact velocity and angle, and the collision velocity for an increased initial standoff distance remained consistent for both values of the discharge energy 14 to 16 kJ. A further set of modelling calculations for a target tube wall thickness of 1 and 2 mm and considering the same process conditions but without a polyurethane internal support inside the target tube also provided a similar range of computed values for the flyer impact velocity and angle, and collision velocity. In other words, the polyurethane internal support inside the target has shown little effect on the computed values of EM field and force, and on the resulting range of the flyer impact velocity and angle, and the collision velocity. However, the presence of the polyurethane internal support restricted the inward deformation of the tubular assembly and preserved its dimensional consistency in all cases.

5. Conclusions

A detailed investigation of MPW of Cu-DHP flyer and 11SMnPb30 steel target tubes with and without internal supports inside the target tube is performed, using a range of experimental conditions determined by the discharge energy, the standoff distance, and the wall thickness of the target tubes. The experimental observations and the computed results show that the standoff distance between the flyer and target tubes significantly influences the progressive evolution of the impact of the flyer onto the target and the resulting growth of the weld joint between the tubes. Although a target tube with a little larger wall thickness can resist the internal deformation during impact better, the presence of an internal support is a useful tool to preserve the original dimensions of the tubular assembly. Overall, the concurrent theoretical and experimental results presented in this paper provide a useful quantitative understanding of the collision behaviour between the flyer and target tubes during MPW of tubular parts, using a typical bitter plate coil. The influence of the key processing conditions on the evolution of the welded joint between the tubes with progress in time has been determined.

Based on the experimental observations and numerical modelling of MPW of Cu-DhP flyer and 11SMnPb30 steel target tubes using a multi-turn coil assembly, the following conclusions can be formulated.

- The initial standoff distance between the flyer and the target tube plays a crucial role and has a significant influence on the progressive impact of the flyer onto the target, the collision behaviour between the tubes and the evolution of the welded interface.
- The flyer tube tends to experience significant plastic deformation as it impacts onto the target. It is necessary to use an internal support for MPW of thin-walled target tubes to avoid inward plastic deformation of the tube assembly.
- For MPW of Cu-DhP flyer and 11SMnPb30 steel target tubes with a wall thickness of 1 to 2 mm, the experimentally observed weld length ranged from 6.9 to 8.5 mm. The inward distortion of the tubular assembly could be minimized significantly by using a polyurethane internal support inside the target tube.
- The computational process model is able to predict the progress of the impact of the flyer onto the target, the resulting flyer impact velocity and angle, and the collision velocity between the flyer and target during the MPW process, which are usually impossible to measure using experimental observations. Furthermore, the computed progress of the flyer-target contact length provided a measure of the actual growth of the weld length, comparable to the reality, which cannot be measured in real-time during the MPW process. The process simulation model can therefore be considered as a valuable practical tool towards the design of the MPW process.

Author Contributions: Conceptualization, A.D. and K.F.; methodology, A.D.; software, R.S. and A.D.; validation, K.F.; investigation, R.S. and K.F.; resources, K.F. and A.D.; writing—original draft preparation, R.S.; writing—review and editing, R.S. and K.F.; visualization, R.S.; supervision, A.D.; project administration, K.F.; funding acquisition, K.F. All authors have read and agreed to the published version of the manuscript.

Funding: The present studies are funded within the project "Elektromagnetisch puls lassen van gelijksoortige en ongelijksoortige materialen—Laasbaarheid en mecanische eigenschappen" for prenormative research under the grant no. CCN/NBN/PN19A02 of the FOD Economie Belgium.

Conflicts of Interest: The authors declare no conflict of interest.

Appendix A

Table A1. Process conditions and dimensions of coil and tubes assembly.

Legend	Description	Unit	Value/Range
E	Discharge energy	kJ	14, 16
a	Radial gap	mm	1.14
s	Standoff distance	mm	1, 2
L_w	Overlapping length between field shaper and flyer	mm	8
L_0	Free length of flyer tube	mm	15
d_g	Flyer tube diameter	mm	22.22
e_g	Wall thickness of flyer tube	mm	0.89
d_h	Target tube diameter	mm	16.44 to 18.44
e_h	Target tube wall thickness	mm	1, 2
θ	Field shaper taper angle	degree	30
m_b	Coil plate width	mm	12
m_c	Coil plate concentrated section width	mm	8
m_o	Plate coil diameter	mm	280
q_o	Field shaper outer diameter	mm	97.5
q_i	Field shaper inner diameter	mm	24.5
q_l	Field shaper outer width	mm	80
q_w	Field shaper inner width	mm	15

Appendix B

The dynamic mechanical behaviour of the Cu-DHP flyer, the 11SMnPb30 steel target and the S235 steel bolt (used inside the polyurethane tube for internal support) is described by the following relation,

$$\sigma_f = \left[A + B(\bar{\varepsilon}^p)^n\right]\left[1 + C\,\ln\left(\frac{\dot{\bar{\varepsilon}}^p}{\dot{\varepsilon}_0}\right)\right]\left[1 - \left(\frac{T - T_r}{T_m - T_r}\right)^m\right] \tag{A1}$$

where σ_f, T, T_r, T_m, $\bar{\varepsilon}^p$, $\dot{\bar{\varepsilon}}^p$ and $\dot{\varepsilon}_0$ refer to the flow stress, the temperature variable, the reference temperature, the melting temperature, the equivalent plastic strain, the equivalent plastic strain rate, and the reference strain rate at T_r, respectively. The values of the materials constants A, B, C, n and m are given in Table A2. The dynamic mechanical behaviour of polyurethane is described using the Cowper–Symonds plasticity model, as follows,

$$\sigma_f = \left[A + B(\bar{\varepsilon}^p)^n\right]\left[1 + \left(\frac{\dot{\bar{\varepsilon}}^p}{\dot{\varepsilon}_0}\right)^{1/q}\right] \tag{A2}$$

The values of the materials constants A, B, n, q and $\dot{\varepsilon}_0$ for polyurethane are given in Table A2.

Table A2. Material constants for different materials in Equations (A1) and (A2).

Variables	Definition	Cu-DHP	11SMnPb30	AISI1006	Polyurethane
A	Initial flow stress (MPa)	90	350.30	350	11.05
B	Hardening constant (MPa)	292	325.8	275	80
C	Strain rate sensitivity	0.025	0.04	0.36	-
N	Hardening exponent	0.31	0.90	0.022	0.7
$\dot{\varepsilon}_0$	Reference strain rate (s^1)	1.0	1.0	1.0	971
M	Thermal softening exponent	1.0	0.30	1	-
T$_m$	Melting temperature (K)	1355	1673	1811	
Q	Strain rate exponent	-	-	-	0.98

Polyurethane [30], Cu-DHP [10], 11SMnPb30 [10], AISI1006 [10].

Appendix C

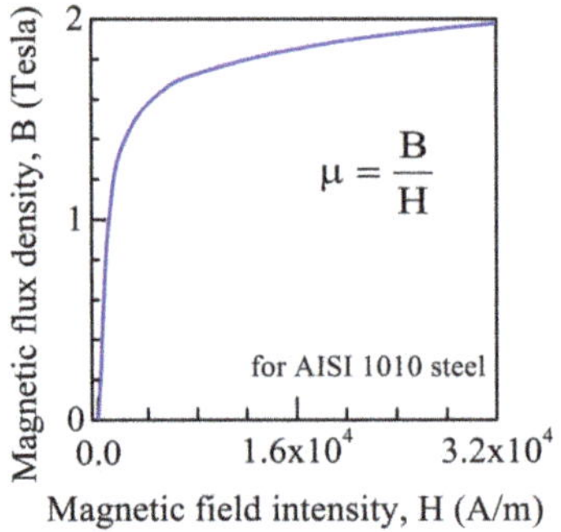

Figure A1. B-H curve used to assign magnetic permeability of 11SMnPb30 steel.

References

1. Khalil, C.; Marya, S.; Racineux, G. Magnetic pulse welding and spot welding with improved coil efficiency—Application for dissimilar welding of automotive metal alloys. *J. Manuf. Mater. Process* **2020**, *4*, 69. [CrossRef]
2. Uhlmann, E.; Prasol, L.; Zilfie, A. Potentials of pulse magnetic forming and joining. *Adv. Mat. Res.* **2014**, *907*, 349–364. [CrossRef]
3. Cuq-Lelandais, J.-P.; Avrillaud, G.; Ferreira, S.; Mazars, G.; Nottebaert, A.; Teilla, G.; Shribman, V. 3D impacts modeling of the magnetic pulse welding process and comparison to experimental data. In Proceedings of the 7th International Conference on High Speed Forming, Dortmund, Germany, 27–28 April 2016; pp. 13–22.
4. Pourabbas, M.; Abdollah-Zadeh, A.; Sarvari, M.; Movassagh-Alanagh, F.; Pouranvari, M. Roll of collision angleduring dissimilar al/cu magnetic pulse welding. *Sci. Technol. Weld. Join.* **2020**, *25*, 549–555. [CrossRef]
5. Li, J.S.; Raoelison, R.N.; Sapanathan, T.; Hou, Y.L.; Rachik, M. Interface evolution during magnetic pulse welding under extremely high strain rate collision: Mechanisms, thermomechanical kinetics and consequences. *Acta Mater.* **2020**, *195*, 404–415. [CrossRef]
6. Niessen, B.; Schumacher, E.; Lueg-Althoff, J.; Bellman, J.; Bohme, M.; Bohm, S.; Tekkaya, A.E.; Beyer, E.; Leyens, C.; Wagner, M.F.-X.; et al. Interface formation during collision welding of aluminium. *Metals* **2020**, *10*, 1202. [CrossRef]
7. Nahmany, M.; Shribman, V.; Levi, S.; Ashkenazi, D.; Stern, A. On additive manufactured AlSi10Mg to wrought AA6060-T6: Characterization of optimal- and high-energy magnetic pulse welding conditions. *Metals* **2020**, *10*, 1235. [CrossRef]
8. Cui, J.; Sun, G.; Xu, J.; Xu, Z.; Huang, X.; Li, G. A study on the critical wall thickness of the inner tube for magnetic pulse welding of tube Al-Fe parts. *J. Mater. Process. Technol.* **2016**, *227*, 138–146. [CrossRef]
9. Patra, S.; Arora, K.S.; Shome, M.; Bysakh, S. Interface characteristics and performance of magnetic pulse welded copper—Steel tubes. *J. Mater. Process. Technol.* **2017**, *245*, 278–286. [CrossRef]

10. Shotri, R.; Faes, K.; De, A. Magnetic pulse welding of copper to steel tubes-Experimental investigation and process modeling. *J. Manuf. Process.* **2020**, *58*, 249–258. [CrossRef]
11. Abrahamson, G.R. Permanent periodic surface deformations due to a travelling jet. *J. Appl. Mech. Trans. ASME* **1961**, *28*, 519–528. [CrossRef]
12. Kakizaki, S.; Watanabe, M.; Kumai, S. Simulation and experimental analysis of metal jet emission and weld interface morphology in impact welding. *Mater. Trans.* **2011**, *52*, 1003–1008. [CrossRef]
13. Stern, A.; Becher, O.; Nahmany, M.; Ashkenazi, D.; Shribman, V. Jet composition in magnetic pulse welding: Al-Al and Al-Mg couples. *Weld. J.* **2015**, *94*, 257s–264s.
14. Grubb, S.A. Coiled Tubing Lap Welds by Magnetic Pulse Welding. US 2015/0328712 AI, 19 November 2015.
15. Psyk, V.; Lieber, T.; Kurka, P.; Drossel, W.-G. Electromagnetic joining of hybrid tubes for hydroforming. *Procedia CIRP* **2014**, *23*, 1–6. [CrossRef]
16. Yu, H.; Dang, H.; Qiu, Y. Interfacial microstructure of stainless steel/aluminium alloy tube lap joints fabricated via magnetic pulsed welding. *J. Mater. Process. Technol.* **2017**, *250*, 297–303. [CrossRef]
17. Lueg-Althoff, L.; Bellman, J.; Hahn, M.; Schulze, S.; Gies, S.; Tekkaya, A.E.; Beyer, E. Joining of dissimilar thin-walled tubes by magnetic pulse welding. *J. Mater. Process. Technol.* **2020**, *279*, 116562. [CrossRef]
18. Lueg-Althoff, J.; Lorenz, A.; Gies, S.; Schulze, S.; Weddling, C.; Goebel, G.; Tekkaya, A.E.; Beyer, E. Magnetic pulse welding by electromagnetic compression: Determination of impact velocity. *Adv. Mat. Res.* **2016**, *966–967*, 489–499. [CrossRef]
19. Fan, Z.; Yu, H.; Li, C. Plastic deformation behavior of bi-metal tubes during magnetic pulse cladding: FE analysis and experiments. *J. Mater. Process. Technol.* **2016**, *229*, 230–243. [CrossRef]
20. Guigliemetti, A.; Burion, N.; Marceau, D.; Rachik, M.; Volat, C. Modelling of tubes magnetic pulse welding. *Conf. Eng. Syst. Des. Anal.* **2012**, *11*, 1–12.
21. Shotri, R.; Racineux, G.; De, A. Magnetic pulse welding of metallic tubes-experimental investigation and numerical modelling. *Sci. Technol. Weld. Join.* **2019**, *25*, 273–281. [CrossRef]
22. Cui, J.; Li, Y.; Liu, Q.; Zhang, X.; Xu, Z.; Li, G. Joining of tubular carbon-fiber-reinforced plastic/aluminium by magnetic pulse welding. *J. Mater. Process. Technol.* **2019**, *264*, 272–282. [CrossRef]
23. Faes, K.; Kwee, I.; Waele, W.D. Electromagnetic pulse welding of tubular products: Influence of process parameters and workpiece geometry on the joint characteristics and investigation of suitable support systems for the target tube. *Metals* **2019**, *9*, 514. [CrossRef]
24. Overview of Materials for Thermoset Polyurethane, Elastomer, Unreinforced. Available online: www.matweb.com (accessed on 28 July 2020).
25. Chari, M.V.K.; Salon, S.J. *Numerical Method in Electromagnetic*; Academic Press: San Diego, CA, USA, 2000; pp. 1–60. ISBN 0-12-615760-X.
26. Sadiku, M.N.O.; Kulkarni, S.V. *Principle of Electromagnetics*, 6th ed.; Oxford University Press: New Delhi, India, 2015; pp. 383–480.
27. NPTEL. Boundary Conditions for Electromagnetic Field, Electromagnetic Field. 2009. Available online: nptel.ac.in (accessed on 20 July 2018).
28. Yuan, X.; Wang, W.; Cao, X.; Zhang, T.; Xie, R.; Liu, R. Numerical study on the interfacial behavior of Mg/Al plate in explosive/impact welding. *Sci. Eng. Compos. Mater.* **2017**, *24*, 581–590. [CrossRef]
29. Bataev, I.A.; Tanaka, S.; Zhou, Q.; Lazurenko, D.V.; Jorge Junior, A.M.; Bataev, A.A.; Hokamoto, K.; Mori, A.; Chen, P. Towards better understanding of explosive welding by combination of numerical simulation and experimental study. *Mater. Des.* **2019**, *169*, 107649. [CrossRef]
30. Jamil, A.; Guan, Z.W.; Cantwell, W.J.; Zhang, X.F.; Langdon, G.S.; Wang, Q.Y. Blast response of aluminium/thermoplastic polyurethane sandwich panels—Experimental work and numerical analysis. *Int. J. Impact Eng.* **2019**, *127*, 31–40. [CrossRef]

Publisher's Note: MDPI stays neutral with regard to jurisdictional claims in published maps and institutional affiliations.

Journal of
*Manufacturing and
Materials Processing*

Article

Influence of Material Properties on Interfacial Morphology during Magnetic Pulse Welding of Al1100 to Copper Alloys and Commercially Pure Titanium

Shunyi Zhang and Brad L. Kinsey *

Department of Mechanical Engineering, University of New Hampshire, Durham, NH 03824, USA;
sz1008@wildcats.unh.edu
* Correspondence: brad.kinsey@unh.edu

Abstract: During magnetic pulsed welding (MPW), a wavy interface pattern can be observed. However, this depends on the specific material combination being joined. Some combinations, e.g., steel to aluminum, simply provide undulating waves, while others, e.g., titanium to copper, provide elegant vortices. These physical features can affect the strength of the joint produced, and thus a more comprehensive understanding of the material combination effects during MPW is required. To investigate the interfacial morphology and parent material properties dependency during MPW, tubular Al1100 and various copper alloy joints were fabricated. The influence of two material properties, i.e., yield strength and density, were studied, and the interface morphology features were visually investigated. Results showed that both material properties affected the interface morphology. Explicitly, decreasing yield strength (Cu101 and Cu110) led to a wavy interface, and decreasing density (Cu110 and CP-Ti) resulted in a wave interface with a larger wavelength. Numerical analyses were also conducted in LS-DYNA and validated the interface morphologies observed experimentally. These simulations show that the effect on shear stresses in the material is the cause of the interface morphology variations obtained. The results from this research provide a better fundamental understanding of MPW phenomena with respect to the effect of material properties and thus how to design an effective MPW application.

Keywords: magnetic pulse welding; material properties; interface morphology; numerical analysis

Citation: Zhang, S.; Kinsey, B.L. Influence of Material Properties on Interfacial Morphology during Magnetic Pulse Welding of Al1100 to Copper Alloys and Commercially Pure Titanium. *J. Manuf. Mater. Process.* **2021**, *5*, 64. https://doi.org/10.3390/jmmp5020064

Academic Editor: Steven Y. Liang

Received: 7 May 2021
Accepted: 11 June 2021
Published: 18 June 2021

Publisher's Note: MDPI stays neutral with regard to jurisdictional claims in published maps and institutional affiliations.

1. Introduction

Electromagnetic forming (EMF) process has been known and applied for several decades [1]. In the EMF process, a large amount of energy on the order of tens of kilojoules is charged to a capacitor bank and then dissipates to a designed coil in the form of sinusoid damped current trace. A magnetic field is generated around the coil, and eddy currents are induced in the nearby conductive workpiece. Thus, the repulsive Lorentz forces are generated between the coil and workpiece, which is driven away and reaches a velocity of hundred meters per second in less than 1 ms. Due to the dynamic nature of this process, the workpiece formability can be improved. The other benefits of EMF include reduced material wrinkling and springback, high repeatability, improved dimensional accuracy, etc.

One of the main applications of EMF is magnetic pulse welding (MPW), a solid-state welding technology, which has rapidly developed in recent four decades. The most outstanding feature of solid-state welding is that there is no significant heat-affected zone (HAZ) generated after welding. Hence, a stronger weld seam than the parent materials is obtained, unlike the fusion welding processes, which usually degrades the joint strength and causes residual stress with consequent cracking and corrosion issues. MPW can also be used to weld dissimilar materials regardless of their differences in the melting point, thermal expansion, and thermal conductivity.

25

Interface morphology has been seen as one of the significant expressions of welding quality in impact welding processes. Previous researchers have published many works on the influences of process parameters on interface morphologies. Vivek et al. [2] studied interface wavelengths dependency on impact velocities and impact angles during vaporized foil actuator welding (VFAW) for the titanium and copper material system. The authors investigated the impact angle effects by pre-setting angled grooves on the target metals, and used various charging energies to drive the flyer, which impacts the target, at different impact velocities. Lee et al. [3] used a similar method to investigate process parameters for another material combination, aluminum, and steel, and conducted numerical analyses to support their experimental findings. In addition, Nassiri et al. [4] reported a robust numerical method to predict the interface morphology by investigating impact welding processes between titanium and copper alloys.

For tubular MPW processes, the effect of flyer kinetics was proposed for an Al6060 and C45 material combination in [5]. Raoelison et al. [6] used the same material for both flyer and target to investigate the welding conditions and predicted the interface morphology using the Eulerian method in ABAQUS. For the material combination of Aluminum and steel, Cui et al. [7] reported the dependence of high-quality welding of MPW on the wall thickness of the targets. A theoretical model based on Tresca yield criterion was also provided to verify the experimental findings. The quality of welding was also examined by peeling tests. The fatigue resistance and weld area of Al-Fe parts were investigated in [8]. The results indicated that the fracture occurred in the transition zone of the weld seam under a cyclic loading condition due to the relatively weaker strength and ductility of the materials in this area. Lueg-Althoff et al. [9] studied the relationship between the materials wall thicknesses and impact pressure for multiple material combinations during MPW. Dependency of the tension and torsion strengths of welded parts between Al3003 and steel for various discharge voltages during MPW was investigated by Yu et al. [10]. The metallurgical joint was only obtained within a specific voltage range, and the element content of the joint was analyzed. Ben-Artzy et al. [11] attributed the wave formation at the interface to the elastic stress waves caused by the impact during MPW process. The stress waves travelled through both metals away from the interface and reflected back off of the outer surfaces.

Although researchers experimentally and numerally investigated various process parameters' effects on interface morphology and welding quality, the effects of material properties have drawn less attention. Thus, a more comprehensive understanding of the material combination effects during MPW is required. To investigate the interfacial morphology and parent material properties dependency during MPW, tubular Al1100, and two copper alloys, Cu101 and Cu110, as well as commercially pure Titanium (CP-Ti grade 2) joints were fabricated. The influence of two material properties, i.e., yield strength and density, was studied, and the interface morphology features were visually investigated. Results showed that both material properties affected the interface morphology. Numerical analyses were also conducted in LS-DYNA and validated the interface morphologies observed experimentally. These simulations show that the effect on the shear stresses in the material is the cause of the interface morphology variations we obtained.

2. Material Property Data

The material properties for the copper alloys and CP-Ti were tested on an MTS landmark servo-hydraulic test machine with a force capacity of 250kN. The specimens for the tensile tests were manufactured according to ASTM E-8 standards [12]. Figure 1 shows the dimensions of the specimens, and Figure 2 shows the tensile test setup on the MTS machine with a 3D Digital Image Correlation (DIC) system. DIC is an image analysis method that measures deformations of the specimen under load in three dimensions by using a speckle pattern and stereoscopic camera setup. The position of the speckle points on the specimen is used to calculate strain on the surface of the specimen based on the imaging parameters and orientation of the cameras [13]. For our tests, the DIC parameters used were a subset size of 19 pixels, a step size of 5 pixels, and a filter size of 5 pixels.

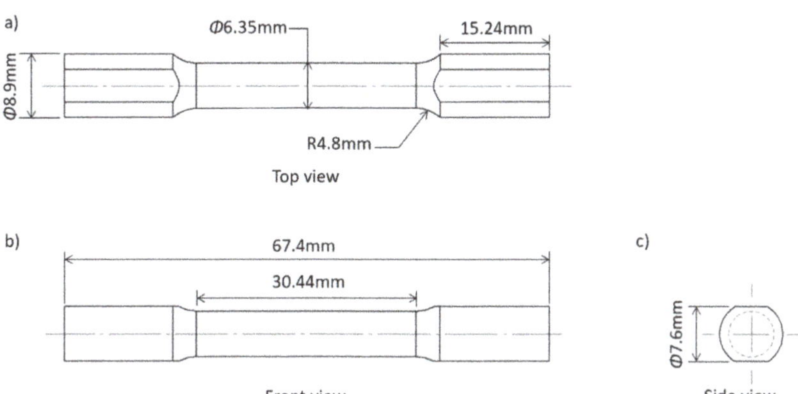

Figure 1. Dimensions of tensile test specimen (**a**) top, (**b**) front, and (**c**) side.

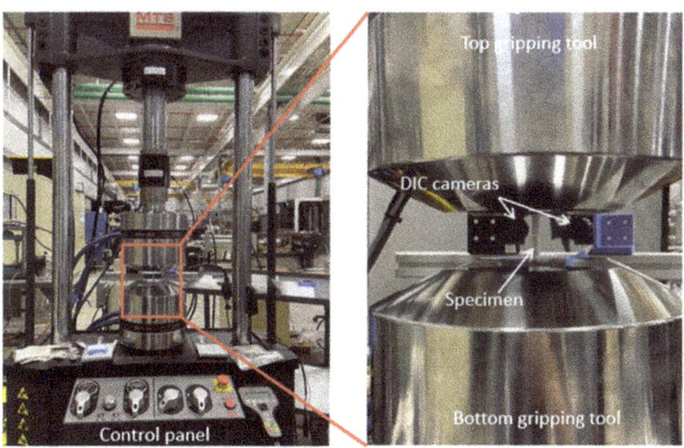

Figure 2. Tensile tests on MTS machine with DIC system.

Figure 3 shows the flow curves of Cu101, Cu110, and CP-Ti that were used in the MPW experiments. All three materials were heat-treated before being used in tensile tests and MPW. Table 1 shows the heat treatment information for the copper alloys and CP-Ti. As shown in Figure 3, the yield strengths of Cu101, Cu110, and CP-Ti, determined by the 0.2% offset method, were clearly different, and better material formability was also shown for copper alloys compared to CP-Ti that had a lower material density. The chemical composition of copper alloys and CP-Ti are given in Tables 2 and 3, respectively.

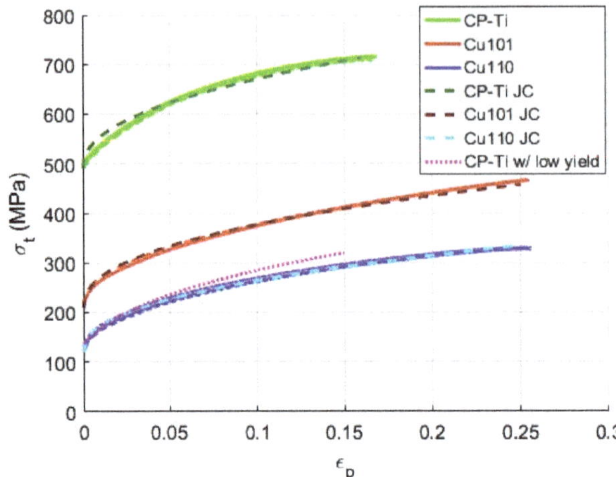

Figure 3. Tensile test results of Cu11 Cu101, and CP-Ti and Johnson-Cook models.

Table 1. Heat treatment information of each material.

Material	Temperature (F)	Duration (h)	Cooling
Cu110	650	2	Water
Cu101	650	2	Water
CP-Ti	1200	2	Air

Table 2. Chemical composition of copper alloys, % by weight.

Material	Copper	Oxygen
Cu110	99.9	0.04
Cu101	99.9	0.0005

Table 3. Chemical composition of CP-Ti, % by weight.

Material	Titanium	Oxygen	Carbon	Nitrogen	Iron	Hydrogen
CP-Ti	99.31	0.25	0.1	0.03	0.3	0.015

Three uniaxial tension specimens for each material were tested to verify the repeatability of the material properties. These results were plotted on top of each other in the same color but different line types in Figure 3. The dashed line with a different color on each material flow curve represented the fitted Johnson–Cook material model. Finally, the magenta dotted line was fictitious CP-Ti with the yield strength of Cu110, which was used in the simulations to investigate the yield strength effect for CP-Ti. The Johnson–Cook [14] material model parameters are given in Table 4.

Table 4. Johnson–Cook material model parameters, Al1100 [15], Copper alloys [16], CP-Ti [17].

Material	A (MPa)	B (MPa)	n	C	m	Tm (K)
Al1100	17	324	0.25	0.2	0.7	933
Cu110	117	389	0.43	0.025	1.09	1356
Cu101	211	448	0.43	0.025	1.09	1356
CP-Ti	491	507	0.45	0.0194	0.58	1941

2.1. Yield Strength

The first material property investigated was yield strength of target materials. Two heat-treated copper alloys, Cu101 and Cu110, were selected as the targets in this portion of the study, as they shared similar properties after heat treatment except for yield strength (see Figure 3). In addition, CP-Ti was also used as the target. To compare the interface of Al1100 & CP-Ti, a fictitious CP-Ti with low yield strength was used in numerical analyses.

2.2. Density

The effect of target material density on the interface morphology was also investigated. In this case, Cu110 was used as the target. A fictitious material with the same material properties of Cu110 except density was also used as a comparison in simulations.

3. Experimental Setup and Results

The experimental tests were conducted on a Maxwell Magneform JA7000 machine with a maximum energy capacity of 12 kJ. Other characteristics of the machine are given in Table 5. A four-turn spiral coil was used in this study, along with a field shaper to concentrate the magnetic field generated (see Figure 4). Since the inner surface of the field shaper was much smaller than the outer surface, the magnetic pressure was concentrated in the radial direction and uniformly distributed along the circumferential direction [18]. The axial length of the pressure concentration zone of the field shaper was 10 mm and the inner diameter was 25.4 mm. The primary currents were measured by a Powertek CWT 3000B Rogowski coil, and impact velocities were measured by a PDV system. The PDV laser signal was focused on the outer surface of the flyer. A Doppler-shifted light was produced by the moving surface of the flyer during MPW, then combined with the incident light signal to produce a beat frequency, which is proportional to the velocity of the moving surface [19]. A small hole was drilled into the field shaper to provide a line of sight for the PDV laser probe (see Figure 4b). The outputs of the laser detector and Rogowski coil were recorded by a LeCroy WaveSurfer WS64MXS-B oscilloscope. A workpiece holder was used to keep both the flyer and target in position and concentric to each other.

Table 5. Characteristics of the pulse generator.

Machine	Maximal Charging Energy (kJ)	Maximal Charging Voltage (kV)	Capacitance (μF)	Internal Inductance (nH)	Internal Resistance (mΩ)
Maxwell Magneform JA7000	12	8.165	360	72	4.38

To investigate the effect of a single material property on the interface morphology, process parameters were kept unchanged. The targets were impacted by the same flyer material Al1100, at the same impact velocity of 315 m/s (see Figure 5). The error bars in Figure 5 represent the highest and lowest values measured for the nine tests conducted. The gap distance was 3.4 mm, the outer diameter of the flyer was 25.4 mm, the wall thickness was 1.4 mm. The targets were solid shafts with a 15.9 mm diameter.

Figure 6 shows the interface morphologies after MPW at an energy level of 4.2 kJ. A flat interface was obtained between Al1100 and Cu101, while a regular wave occurred between Al1100 and Cu110 (see Figure 6a,b). As shown in Figure 3, the yield strengths of these two annealed copper alloys were clearly different, i.e., 220 MPa versus 135 MPa for Cu101 and Cu110, respectively. At the same time, the other material properties such as Young's modulus, density, and work hardening behavior, were comparable. For this material combination, the target with a lower yield strength resulted in a joint with a wavy interface, while the target with a higher yield strength led to a flat interface.

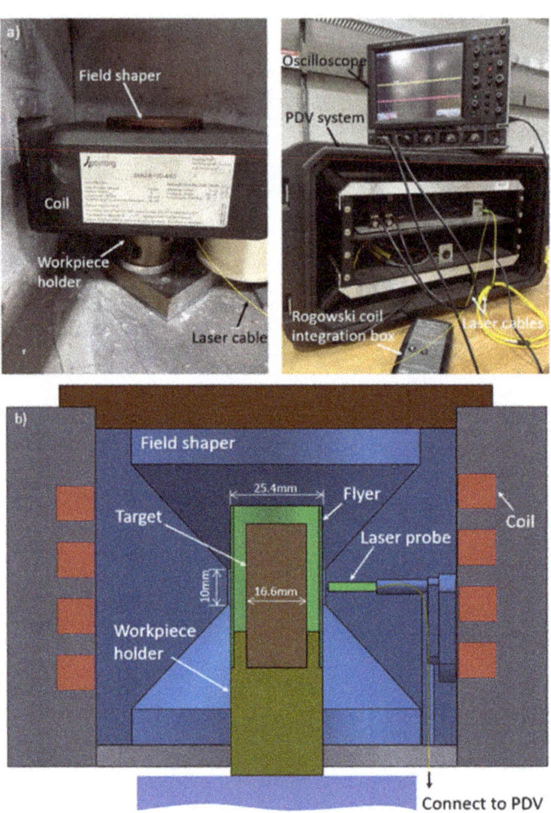

Figure 4. (**a**) Experimental setup and PDV system, (**b**) schematic of coil, field shaper, and workpieces.

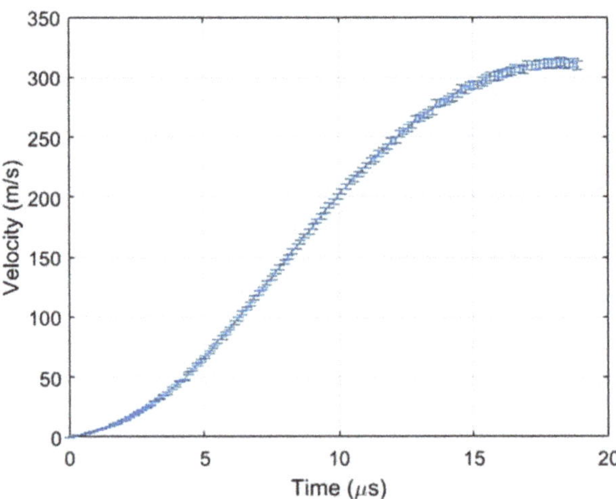

Figure 5. Velocity measured by PDV system.

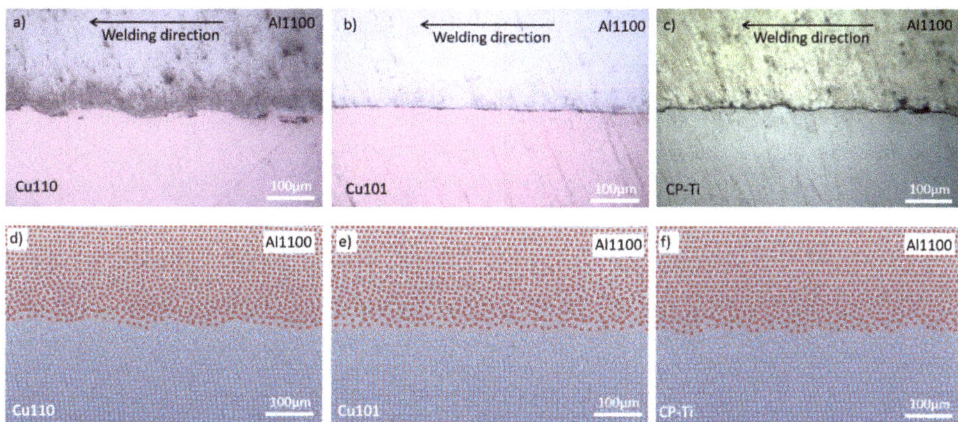

Figure 6. Interface morphology difference, experiments: (**a**) Al1100 & Cu110, (**b**) Al1100 & Cu101, (**c**) Al1100 & CP-Ti, SPH: (**d**) Al1100 & Cu110, (**e**) Al1100 & Cu101, (**f**) Al1100 & CP-Ti.

Figure 6c shows the interface between Al1100 and CP-Ti. According to the data in Figure 3, CP-Ti has the highest yield strength; however, an irregular wave interface with a small wavelength was obtained due to the differences in material ductility (see Figure 3) and density, as CP-Ti had a lower density of 4510 kg/m^3 compared to 8910 kg/m^3 for the copper alloys. Figure 6d–f are numerical results, which are in good agreement with each experimental observation, respectively. The next section will provide details related to these numerical simulations.

4. Numerical Simulations

4.1. Electromagnetic Forming

Finite element analyses of EMF and MPW processes were conducted using LS-DYNA, as their electromagnetism (EM) model is capable of capturing the effects of tube deformation on the magnetic fields in proximity to the coil. Process parameters in the numerical simulations were the same as in experimental tests, i.e., the dimensions and geometries of the coil, field shaper, and workpieces were defined as those in the experimental tests. Figure 7 shows the EMF model set up in LS-PREPOST. The material properties of the tube were described by the Johnson–Cook material model:

$$\overline{\sigma} = \left(A + B\varepsilon_p^n \right)\left(1 + Cln\frac{\dot{\varepsilon}_p}{\dot{\varepsilon}_0}\right)\left(1 - (\frac{T - T_r}{T_m - T_r})^m\right) \tag{1}$$

where $\overline{\sigma}$ is the flow stress, ε_p is the plastic strain, and $\dot{\varepsilon}_p$ the effective plastic strain rate, $\dot{\varepsilon}_{p0}$ is the reference plastic strain rate, m, n, A, B, and C are the material constants, and T_m and T_r are the melting and room temperature, respectively. The coil and field shaper were modeled as rigid bodies. The current traces used in the simulations were measured by the Rogowski coil. The model contained 70,000 elements in total. A finer mesh was defined in regions close to surfaces to capture the skin depth effects. Table 6 gives the material properties used in these simulations. Electrical conductivities were only defined for the field shaper and flyer, which were Al1100 and copper, respectively.

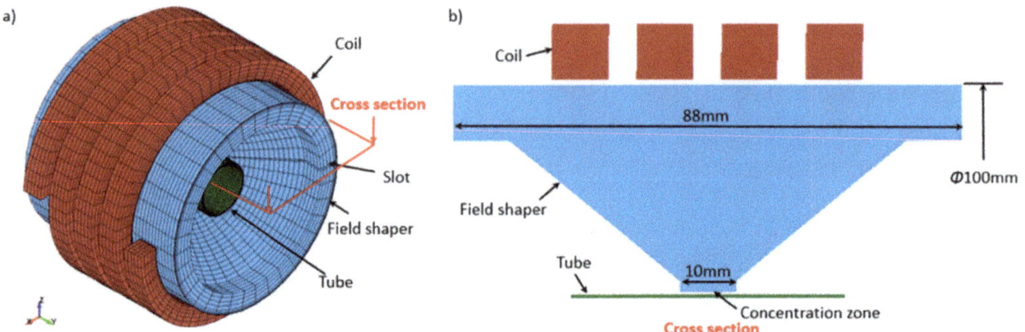

Figure 7. EMF model defined in LS-PREPOST (**a**) 3D model, (**b**) cross-section.

Table 6. Selected material properties used in the simulations.

Materials	Young's Modulus (GPa)	Density (kg/m^3)	Electrical Conductivity (S/m)
Al1100	75	2700	3.77×10^7
Copper	120	8910	5.96×10^7
CP-Ti	116	4510	-

First, EMF simulations were conducted to determine the desired gap between the flyer and target to achieve the peak velocity at impact during MPW and the corresponding charging energy level. Figure 8 shows the comparisons of experimental and numerical velocities and displacements of the flyer. The blue lines represent the peak velocities corresponding to the left Y-axis. The right Y-axis shows the displacements at peak velocities and corresponds to the red lines in the figure. To investigate the effect of yield strength on the interface, a 4.2 kJ charging energy was chosen, where the flyer travelled approximately 3.4 mm and reached a peak velocity of 315 m/s. While the velocity is still increasing in Figure 8, the displacement is leveling off at 4.2 kJ. In addition, wrinkling was occurring in the tube at higher charge energies. Therefore, higher charge energies were not investigated. The current trace for a 4.2 kJ energy discharge is shown in Figure 9, with only the first cycle included. Based on the displacement data in Figure 5, the deformation is clearly finished by 90 μs.

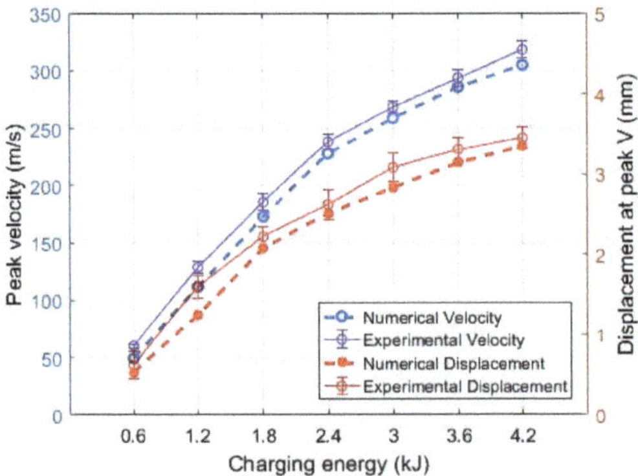

Figure 8. Peak velocities and corresponding displacements.

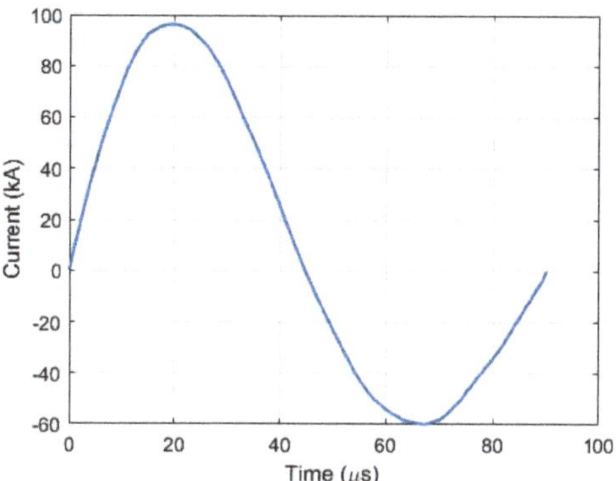

Figure 9. Current trace for 4.2 kJ charging energy.

4.2. Impact Welding

To accurately capture the wavy morphology at the interface, numerical simulations for impact welding were also conducted in LS-DYNA using the smoothed particle hydrodynamics (SPH) method, which is able to analyze the problem of large local deformations, which occur at the interface during MPW. A 2D, axisymmetric SPH analysis in LS-DYNA was used to reduce the computational time. A simplified model was set up in LS-PREPOST for this study (see Figure 10). A flat interface is created in the very center of the flyer and base material interface, where the initially parallel materials first collide during impact welding. As the process progresses, an impact angle is created away from the center of the process, and the wavy interface occurred. In the numerical model, only the portion of the process where the wavy interface occurred was simulated. Thus, the angle between the flyer and target was fixed at 14° in the model based on [20]. The length for both target and flyer was 10 mm. The thicknesses were 1.4 mm and 2 mm for the flyer and target, respectively. The impact velocity used in the model was obtained from experimental tests by PDV system, i.e., 315 m/s. The SPH particle size was defined as 5 μm to capture the interface morphology.

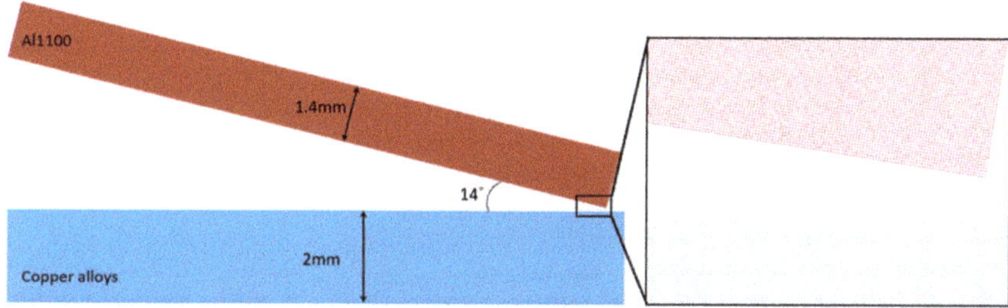

Figure 10. SPH model defined in LS-PREPOST.

Numerical results of the effect of target yield strengths are shown in Figure 6d,e. Comparable interface morphologies were observed both in experimental and numerical results. A relatively flat interface occurred when using Cu101 as the target, while the wavy interface was obtained in the case of using Cu110, which has a lower yield strength.

To investigate the density effect for Cu110 and yield strength effect for CP-Ti, two fictitious materials were created in the numerical simulations. The first material shared the material properties of Cu110, but the density was defined as 4510 kg/m^3, i.e., that of CP-Ti, and the second shared the material properties of CP-Ti except for the yield strength, which was set to be the same as Cu110 (see Figure 3). Figure 11a,b shows the interface morphology comparison of Al1100 & Cu110 and the fictitious Cu110 with the CP-Ti density. Figure 11c,d shows the interface morphology comparison of Al1100 & CP-Ti and the fictitious CP-Ti with the Cu110 yield strength. A similar trend occurred for both of these results, i.e., the target with lower density (Figure 11b) and the target with a lower yield strength (Figure 11d) led to an interface with a larger wavelength. Figure 11b,d is the interfaces for the two fictitious target materials with the same density (4510 kg/m^3) and yield strength (Cu110 yield strength), but the interfaces with different wavelengths were obtained. This was caused by the dissimilarity in material hardening behavior, various equations of state (e.g., speed of sound) for the Johnson–Cook material model, and slight differences in the Young's moduli of these two materials.

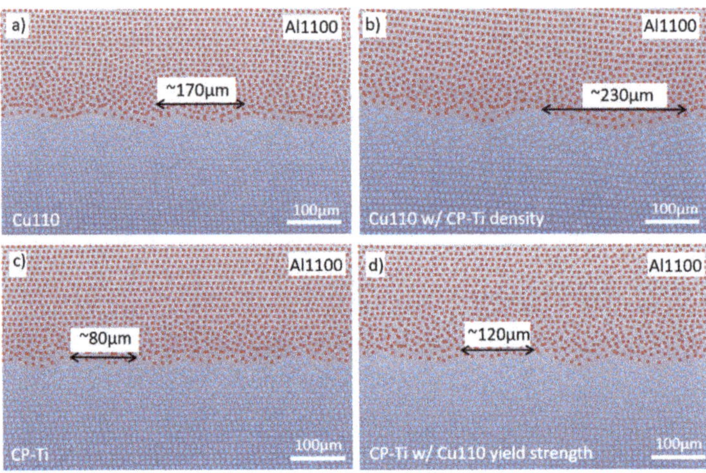

Figure 11. Interface morphologies of Al1100 & (**a**) Cu110, (**b**) fictitious Cu110 with the density of CP-Ti, (**c**) CP-Ti, and (**d**) fictitious CP-Ti with the yield strength of Cu110.

5. Discussion

Process parameters such as shear velocity, shear stress, and contact pressure were investigated in the numerical simulations to understand why lower yield strength and lower density resulted in differences in the wavy interface for these material combinations during MPW. Figure 12 shows the shear stresses at the same time increment for the material combinations of Al1100 & Cu110, Al1100 & Cu101, Al1100 & CP-Ti, Al1100 & CP-Ti with Cu110 yield strength, and Al1100 & Cu110 with CP-Ti density. Past research has shown that the shear stress is one of the reasons for the intense local deformation in proximity to the collision point, which leads to a wavy interface [21]. As shown in Figure 12, the shear stress concentrated more on the target side in proximity of the collision point when the target yield strength was lower, i.e., Figure 12a,b,d. Thus, a wavy interface was easier to create as the local deformation near the collision point occurred due to the larger shear stress difference between the region near the collision point and the other areas of the target. Despite having the largest shear stress in the target material, a wavy interface was also observed in CP-Ti, e.g., compared to Cu101, due to the other material properties that affect this behavior. Furthermore, based on the theory of Kelvin-Helmholtz instability, the interfacial wavelength is proportional to the depth of the plastically deformed layer. Specifically, the target with lower yield strength was more easily deformed at the interface under the same impact velocity. Thus, in this study, the change in the interfacial morphology was clearly associated with the target yield strength.

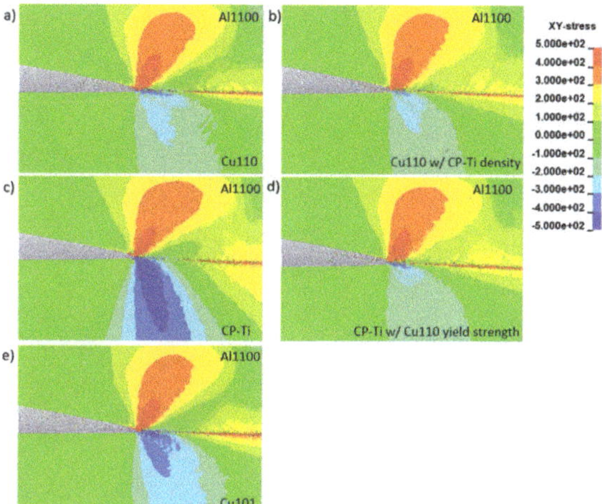

Figure 12. Shear stresses at the interfaces for Al1100 & (**a**) Cu110, (**b**) fictitious Cu110 with the density of CP-Ti, (**c**) CP-Ti, (**d**) fictitious CP-Ti with the yield strength of Cu110, and (**e**) Cu101.

Figure 13 shows the shear velocities at the interfaces. For each case, there is not a clear difference in shear velocity, which is a parameter that is mainly affected by the impact velocity. Note that the impact velocity was the same in each experiment and numerical simulation in this study. Therefore, there is no effect of shear velocity on the wavy interface observed. In Figure 14, the contact pressure is shown for the various cases investigated. The images for the copper alloys with the corresponding appropriate densities, i.e., Figure 14a for Cu110 and Figure 14e for Cu101, show higher contact pressure values due to their higher densities compared to the plots with CP-Ti densities, i.e., Figure 14b–d. The contact pressure is known to be affected by the speed of sound in the material (with CP-Ti being 15% larger than that of copper alloys, i.e., 4140 m/s versus 3570 m/s, respectively) and density [4]. Thus, the variations in the wavy interface do not correlate to the contact pressure. The fact that shear velocity and contact pressure, i.e., Figures 13 and 14 respectively, do not show an effect on the wavy interface morphology is a noteworthy finding.

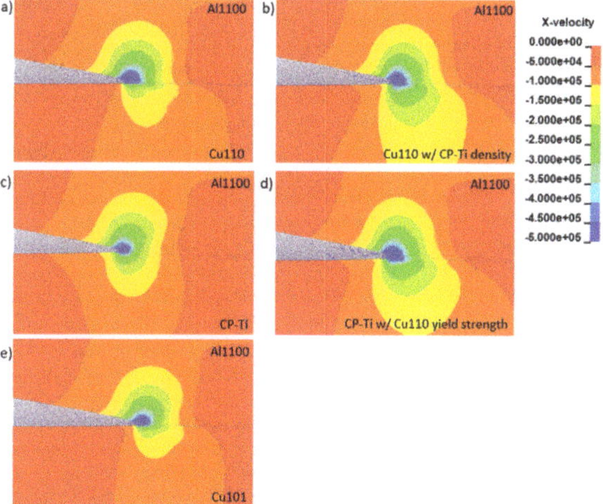

Figure 13. Shear velocities at the interfaces for Al1100 & (**a**) Cu110, (**b**) fictitious Cu110 with the density of CP-Ti, (**c**) CP-Ti, (**d**) fictitious CP-Ti with the yield strength of Cu110, and (**e**) Cu101.

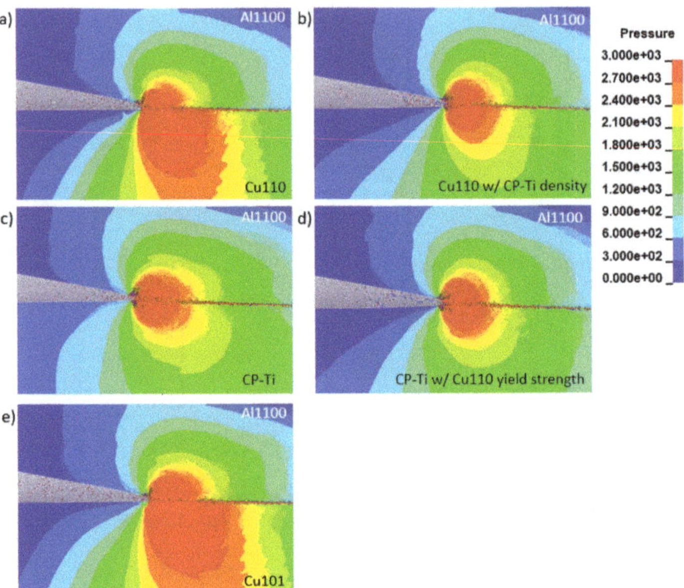

Figure 14. Contact pressures at the interfaces for Al1100 & (**a**) Cu110, (**b**) fictitious Cu110 with the density of CP-Ti, (**c**) CP-Ti, (**d**) fictitious CP-Ti with the yield strength of Cu110, and (**e**) Cu101.

6. Conclusions

In this research, results show the effect of material properties on interface morphologies during MPW. Numerical analyses were in good agreement with the experimental observations, i.e., interface morphologies, thus allowing for process parameters that cannot be physically measured to be investigated. Both experimental measurements and numerical predictions for velocity and displacement determined the gap distance and energy level to use for each case. For the material combination of Al1100 and annealed copper alloys, a lower yield strength of the target led to a joint with a wavy interface, while a flat interface was obtained for the target with a higher yield point. Using CP-Ti, i.e., a target with lower density and higher yield strength, an interface with smaller waves was obtained in the experiments and numerical simulations. Two fictitious materials were also modeled in simulations to further assess the effects of varying process parameters. Numerical results using CP-Ti material parameters, except for the yield strength of Cu110, as the target and Cu110 material parameters, except for the density of CP-Ti, were compared to that of the actual CP-Ti and Cu110 materials, respectively. Using both the fictitious materials (CP-Ti with Cu110 yield strength, and Cu110 with CP-Ti density) resulted in a wavy interface with larger wavelengths than their comparison partners. These results provide insight into the material properties that affect the wavy interface during MPW.

Author Contributions: Conceptualization, S.Z. and B.L.K.; methodology, S.Z. and B.L.K.; software, S.Z.; validation, S.Z. and B.L.K.; formal analysis, S.Z. and B.L.K.; investigation, S.Z.; resources, S.Z. and B.L.K.; data curation, S.Z.; writing—original draft preparation, S.Z.; writing—review and editing, S.Z. and B.L.K.; visualization, S.Z.; supervision, B.L.K.; project administration, B.L.K.; funding acquisition, B.L.K. Both authors have read and agreed to the published version of the manuscript.

Funding: The research was funded by the U.S. National Science Foundation, grant number CMII-1537471.

Data Availability Statement: The data presented in this study are available on request from the corresponding author.

Conflicts of Interest: The authors declare no conflict of interest.

References

1. Shribman, V. Magnetic Pulse Welding for Dissimilar and Similar Materials. In Proceedings of the 3rd International Conference on High Speed Forming, Dortmund, Germany, 11–12 March 2008; p. 10.
2. Vivek, A.; Liu, B.C.; Hansen, S.R.; Daehn, G.S. Accessing Collision Welding Process Window for Titanium/Copper Welds with Vaporizing Foil Actuators and Grooved Targets. *J. Mater. Process. Technol.* **2014**, *214*, 1583–1589. [CrossRef]
3. Lee, T.; Zhang, S.; Vivek, A.; Kinsey, B.; Daehn, G. Flyer Thickness Effect in the Impact Welding of Aluminum to Steel. *J. Manuf. Sci. Eng.* **2018**, *140*, 121002. [CrossRef]
4. Nassiri, A.; Vivek, A.; Abke, T.; Liu, B.; Lee, T.; Daehn, G. Depiction of Interfacial Morphology in Impact Welded Ti/Cu Bimetallic Systems Using Smoothed Particle Hydrodynamics. *Appl. Phys. Lett.* **2017**, *110*, 231601. [CrossRef]
5. Lueg-Althoff, J.; Bellmann, J.; Gies, S.; Schulze, S.; Tekkaya, A.E.; Beyer, E. Influence of the Flyer Kinetics on Magnetic Pulse Welding of Tubes. *J. Mater. Process. Technol.* **2018**, *262*, 189–203. [CrossRef]
6. Raoelison, R.N.; Buiron, N.; Rachik, M.; Haye, D.; Franz, G. Efficient Welding Conditions in Magnetic Pulse Welding Process. *J. Manuf. Process.* **2012**, *14*, 372–377. [CrossRef]
7. Cui, J.; Sun, G.; Xu, J.; Xu, Z.; Huang, X.; Li, G. A Study on the Critical Wall Thickness of the Inner Tube for Magnetic Pulse Welding of Tubular Al–Fe Parts. *J. Mater. Process. Technol.* **2016**, *227*, 138–146. [CrossRef]
8. Geng, H.; Sun, L.; Li, G.; Cui, J.; Huang, L.; Xu, Z. Fatigue Fracture Properties of Magnetic Pulse Welded Dissimilar Al-Fe Lap Joints. *Int. J. Fatigue* **2019**, *121*, 146–154. [CrossRef]
9. Lueg-Althoff, J.; Bellmann, J.; Hahn, M.; Schulze, S.; Gies, S.; Tekkaya, A.E.; Beyer, E. Joining Dissimilar Thin-Walled Tubes by Magnetic Pulse Welding. *J. Mater. Process. Technol.* **2020**, *279*, 116562. [CrossRef]
10. Yu, H.; Xu, Z.; Fan, Z.; Zhao, Z.; Li, C. Mechanical Property and Microstructure of Aluminum Alloy-Steel Tubes Joint by Magnetic Pulse Welding. *Mater. Sci. Eng. A* **2013**, *561*, 259–265. [CrossRef]
11. Ben-Artzy, A.; Stern, A.; Frage, N.; Shribman, V.; Sadot, O. Wave Formation Mechanism in Magnetic Pulse Welding. *Int. J. Impact Eng.* **2010**, *37*, 397–404. [CrossRef]
12. E28 Committee. *Test Methods for Tension Testing of Metallic Materials*; ASTM International: West Conshohocken, PA, USA, 2011.
13. Michael, A.M.A.; Orteu, J.-J.; Schreier, H.W. Digital Image Correlation (DIC). In *Image Correlation for Shape, Motion and Deformation Measurements: Basic Concepts, Theory and Applications*; Schreier, H., Orteu, J.-J., Sutton, M.A., Eds.; Springer: Boston, MA, USA, 2009; pp. 1–37.
14. Johnson, G.; Cook, W. A Constitutive Model and Data for Materials Subjected to Large Strains, High Strain Rates, and High Temperatures–ScienceOpen. In Proceedings of the 7th International Symposium on Ballistics, Miami, FL, USA, 19–21 April 1983; pp. 541–547.
15. Iwamoto, T.; Yokoyama, T. Effects of Radial Inertia and End Friction in Specimen Geometry in Split Hopkinson Pressure Bar Tests: A Computational Study. *Mech. Mater.* **2012**, *51*, 97–109. [CrossRef]
16. Meyers, M.A.; Andrade, U.R.; Chokshi, A.H. The Effect of Grain Size on the High-Strain, High-Strain-Rate Behavior of Copper. *Metall. Mater. Trans. A* **1995**, *26*, 2881–2893. [CrossRef]
17. Magargee, J.; Morestin, F.; Cao, J. Characterization of Flow Stress for Commercially Pure Titanium Subjected to Electrically Assisted Deformation. *J. Eng. Mater. Technol.* **2013**, *135*, 041003. [CrossRef]
18. Yu, H.; Li, C.; Zhao, Z.; Li, Z. Effect of Field Shaper on Magnetic Pressure in Electromagnetic Forming. *J. Mater. Process. Technol.* **2005**, *168*, 245–249. [CrossRef]
19. Banik, K.; Daehn, G.S.; Fenton, G.K.; Golowin, S.; Henchi, I.; Johnson, J.R.; Eplattenier, P.L.; Taber, G.; Vivek, A.; Zhang, Y. Coupling Experiment and Simulation in Electromagnetic Forming Using Photon Doppler Velocimetry. In Proceedings of the 3rd International Conference on High Speed Forming, Dortmund, Germany, 11–12 March 2008.
20. Lee, T.; Zhang, S.; Vivek, A.; Daehn, G.; Kinsey, B. Wave Formation in Impact Welding: Study of the Cu–Ti System. *CIRP Ann.* **2019**, *68*, 261–264. [CrossRef]
21. Zhang, S.; Kinsey, B. Numerical Investigation of Impact Welding by Eulerian and Smoothed Particle Hydrodynamic Methods. In Proceedings of the 13th International Conference on Numerical Methods in Industrial Forming Processes, Portsmouth, NH, USA, 23–27 June 2019.

Journal of *Manufacturing and Materials Processing*

Article

Experimental and Numerical Investigations into Magnetic Pulse Welding of Aluminum Alloy 6016 to Hardened Steel 22MnB5

Rico Drehmann [1,*], Christian Scheffler [2], Sven Winter [2], Verena Psyk [2], Verena Kräusel [2] and Thomas Lampke [1]

[1] Materials and Surface Engineering Group, Institute of Materials Science and Engineering (IWW), Chemnitz University of Technology, 09125 Chemnitz, Germany; thomas.lampke@mb.tu-chemnitz.de

[2] Fraunhofer Institute for Machine Tools and Forming Technology IWU, 09126 Chemnitz, Germany; christian.scheffler@iof.fraunhofer.de (C.S.); sven.winter@iwu.fraunhofer.de (S.W.); verena.psyk@iwu.fraunhofer.de (V.P.); verena.kraeusel@iwu.fraunhofer.de (V.K.)

* Correspondence: rico.drehmann@mb.tu-chemnitz.de

Abstract: By means of magnetic pulse welding (MPW), high-quality joints can be produced without some of the disadvantages of conventional welding, such as thermal softening, distortion, and other undesired temperature-induced effects. However, the range of materials that have successfully been joined by MPW is mainly limited to comparatively soft materials such as copper or aluminum. This paper presents an extensive experimental study leading to a process window for the successful MPW of aluminum alloy 6016 (AA6016) to hardened 22MnB5 steel sheets. This window is defined by the impact velocity and impact angle of the AA6016 flyer. These parameters, which are significantly dependent on the initial gap between flyer and target, the charging energy of the pulse power generator, and the lateral position of the flyer in relation to the inductor, were determined by a macroscopic coupled multiphysics simulation in LS-DYNA. The welded samples were mechanically characterized by lap shear tests. Furthermore, the bonding zone was analyzed by optical and scanning electron microscopy including energy-dispersive X-ray spectroscopy as well as nanoindentation. It was found that the samples exhibited a wavy interface and a transition zone consisting of Al-rich intermetallic phases. Samples with comparatively thin and therefore crack-free transition zones showed a 45% higher shear tensile strength resulting in failure in the aluminum base material.

Keywords: magnetic pulse welding (MPW); AA6016; aluminum; 22MnB5; press-hardening steel; interface characterization

Citation: Drehmann, R.; Scheffler, C.; Winter, S.; Psyk, V.; Kräusel, V.; Lampke, T. Experimental and Numerical Investigations into Magnetic Pulse Welding of Aluminum Alloy 6016 to Hardened Steel 22MnB5. *J. Manuf. Mater. Process.* **2021**, 5, 66. https://doi.org/10.3390/jmmp5030066

Academic Editor: Steven Y. Liang

Received: 22 May 2021
Accepted: 21 June 2021
Published: 24 June 2021

1. Introduction

Magnetic pulse welding (MPW), which was initially suggested by Lysenko et al. in 1970 [1], is an innovative technology for manufacturing metallic bonds of similar and dissimilar metals [2]. The setup of the process, which is based on the electromagnetic forming technique [3], consists of the pulsed power generator, the inductor (i.e., the tool), the workpieces to be joined to each other, and additional elements, ensuring that the workpieces are positioned with a defined small gap between them. The pulsed power generator provides storage and a quick release of energy. The most important machine components are the charging units, capacitor banks, high current switches, and control devices [4]. Pereira et al. [5] present an optimized machine, allowing the production of joints with lower energy than commercial machines. The inductor typically consists of a winding, which is embedded in an insulating and reinforcing housing material. Frequently, an additional fieldshaper focuses the acting loads onto the joining zone. Inductor winding and fieldshaper are typically made of a material of high electrical conductivity and acceptable mechanical strength such as CuCrZr or CuBe alloys [6]. Principally, the same machines and very similar tools can be used for MPW, for electromagnetic forming [3], for electromagnetic impact medium forming [7] or for electromagnetic acceleration of punches, e.g., for the determination of material characteristics up to very high strain rates of 10^4 s^{-1} [8].

Basically, MPW uses the Lorentz forces of induced currents to accelerate one of the joining partners to velocities at a magnitude of up to several hundreds of meters per second [9]. The impact of this joining partner (the so-called flyer) to the second, typically static, partner (the so-called target) leads to bonding in a defined zone if the collision parameters are appropriate.

Typical process variants of MPW specifically include the welding of tubes to solid [10] or hollow [11] internal parts, as well as the welding of sheets to sheets [12] and sheets to profiles [13]. Welding of tubes to hollow internal parts typically requires a support in order to reduce undesired deformations. This can be a rigid body [11] or an elastomer [14]. New developments in the field of sheet metal welding deal with the adaptation of the technology to spot welding [15].

Figure 1 shows the principle sketch of an exemplary setup for electromagnetic sheet metal welding. Here, an equivalent circuit diagram represents the pulsed power generator. The U-shaped tool coil is connected to this machine so that a damped sinusoidal current flows through the coil when the high-current switch is closed and the capacitor battery is discharged. Due to the different widths and cross section shapes of the branches of the coil, the current density differs significantly. It is high in the narrow branch (i.e., the active side of the inductor), which is positioned close to the area of the workpiece that is to be welded and significantly lower in the wider branch (i.e., the passive side of the inductor). Consequently, the magnetic field, the induced current in the workpiece, and the acting Lorentz forces are much higher in the region of the active side of the inductor when compared to the region of the passive side of the inductor. The Lorentz force ratio corresponds to the inverse ratio of the local conductor width. This strategy for adjusting the force distribution acting on the workpiece was initially suggested in [16] and applied to the electromagnetic forming of sheet metal in [17] and to an electromagnetic tube forming process in [18]. A very similar coil for MPW of sheet metal is used in [19]. The Lorentz forces acting on the workpiece initiate the deformation of the flyer sheet. In the case depicted in Figure 1, the edge of the workpiece bends downwards towards the target and impacts on the target, when it overcomes the initial distance between the two joining partners. Starting from the first contact in the area of the flyer edge, the flyer aligns to the target and the collision point moves along the surface with a collision point velocity v_c. During this process, the impact angle α and the impact velocity v_p vary with the position of the collision point [20]. The equations in Figure 1 describe the correlations between the impact angle, the impact velocity, the normal impact velocity v_n and the collision point velocity. If these collision properties correspond to a material-specific process window, a weld is formed, which frequently, but not necessarily, features a wavy character [21].

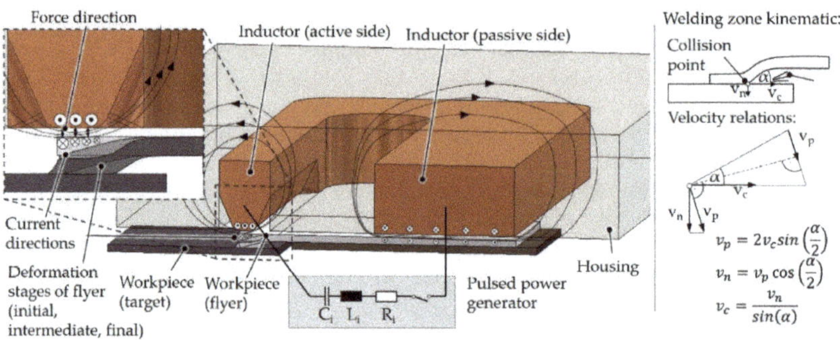

$$v_p = 2v_c \sin\left(\frac{\alpha}{2}\right)$$

$$v_n = v_p \cos\left(\frac{\alpha}{2}\right)$$

$$v_c = \frac{v_n}{\sin(\alpha)}$$

Figure 1. Electromagnetic sheet metal welding process setup and process parameter kinematic.

As a solid-state welding process, MPW avoids or at least significantly reduces the typical temperature-induced problems of the more conventional fusion welding processes, such as thermal softening or the formation of intermetallics. Therefore, it is especially promising for joining material combinations usually considered difficult to weld or non-weldable [22]. In his review on electromagnetic pulse welding, Kang presents numerous examples of material combinations that have been successfully joined by MPW [23]. In order to reach high process efficiency and avoid extreme loading of the tool and machine components, the flyer should preferably be made of a material with high electrical conductivity and moderate strength. Therefore, a lot of fundamental research has been dedicated to the MPW of aluminum and copper over recent years. These materials are highly interesting, e.g., for applications in the electrical industry including the currently highly relevant fields of development in battery technology and electro-mobility and in the fields of heating, cooling, air conditioning, and ventilation. Raoelison et al., for example, investigated the interface of aluminum/copper welds and compared it to aluminum/aluminum welds. They found that aluminum/copper welds form intermediate phases in the form of layers or pockets. Depending on the thickness of these phases, the interface becomes sensitive to microcracks and fragmentation [24]. Psyk et al. provide a combined numerical and experimental process analysis for electromagnetic pulse welding of Cu-DHP and EN AW-1050 and consolidate the results in a quantitative collision parameter based process window [20]. They proved that the forces transferable by these joints under lap shear loading can be higher than the maximum forces transferable by the weaker joining partner so that failure occurs outside and far away from the joining region. Wu and Shang have investigated the influence of the surface condition on the weld formation and quality and found that surface scratches in a tangential direction were in favor of a good weld with high strength, while oil on the surface prevented welding [25].

Especially with regard to applications in the automotive industry, another focus of the research on MPW was put on aluminum/steel joints [26]. Kimchi et al. showed that the stand-off distance between an Al tube and a steel bar is a dominant factor for achieving a sound weld and that adding receding angles to the bars can improve weldability [27].

Aizawa et al. provide a detailed study considering MPW of sheets made of multiple aluminum alloys, specifically AA1050, AA2017, AA3004, AA5182, AA5052, AA6016, and AA7075, to cold rolled carbon steel (SPCC) sheets. They considered different machine parameters and coil variants, estimated the corresponding collision velocity and characterized the resulting weld quality via micrographic investigations and lap shear tests [2]. Yu et al. investigated MPW of AA3003-O and steel 20 (0.2 wt.% C) tubes. They also showed that the tension and torsion strength values of the joint are higher than those of the aluminum tube when proper process parameters are chosen [28]. Complementing microstructural investigations indicate that the metallurgical joint composes of two interfaces, a non-uniform transition zone and basic metals with high-density dislocations and nanocrystals. The transition zone features high micro-hardness multi-direction micro-cracks and micro-apertures. Psyk et al. analyzed the influence of adjustable process parameters, specifically the capacitor charging energy, the initial gap width, the lateral position of inductor and flyer and the flyer thickness on collision conditions and the resulting weld quality in terms of transferable force, joint resistance, and weld width for different material combinations, including aluminum/copper and aluminum/stainless steel [29]. Although most of the publications dealing with MPW of steel and aluminum still consider relatively soft steels, recently some attempts have been made to transfer the technology to target materials of high strength. Wang et al. considered MPW of 3003 aluminum alloy sheets and HC340LA steel sheets (zinc-coated and non-galvanized) [30]. They found that although the zinc layer on the galvanized steel was partly removed due to the jet, the remains cause the formation of brittle and hard phases on the interface, resulting in the generation of welding defects, thereby reducing of mechanical properties of the joint. In [31], Psyk et al. exemplarily show that for press hardening steel (22MnB5) and aluminum, also, joining by MPW is basically possible. However, the specific aluminum alloy considered here (EN AW-1050) has no technological relevance with regard to structural components. In most publications, materials with a low strength and good electrical conductivity such as pure aluminum and copper were joined with MPW, whereas joining high-strength aluminum alloys to hardened steel is extremely challenging and pushes the process to its technological limits. Therefore, it is of great interest that high-strength aluminum alloys can be joined to hardened steel so that novel structural components can be provided for automotive applications.

Therefore, this paper aims for the first time at joining the technologically relevant aluminum alloy AA6016, especially used for structural components, to the press hardening steel 22MnB5 (in hardened state) by means of MPW. Sheet metal welding tests are performed and the weld quality is characterized via lap shear tests and microstructural investigations. In parallel, numerical simulations are carried out in order to quantify the conditions of the material collision during welding. Experimental and numerical results are consolidated as a process window applicable for this specific material combination. Deeper understanding of the bonding mechanisms and the microstructural effects in the joining zone is gained by comprehensive microstructural analysis in terms of light microscopic investigations, scanning electron microscopy (SEM), and energy-dispersive X-ray spectroscopy (EDX).

2. Materials and Methods

Experimental investigations and numerical simulations were carried out considering the aluminum alloy AA6016 T4 and press hardened steel 22MnB5. The chemical compositions of the joining partners as determined by EDX are shown in Table 1.

Table 1. Chemical composition (in wt.%) of flyer (AA6016) and target material (22MnB5) according to EDX measurements.

	Al	Fe	Mn	Si	Mg
AA6016	97.6	0.4	0.1	0.6	1.3
22MnB5	0.1	98.2	1.4	0.3	-

For industrial processing, 22MnB5 sheets are often protected by an AlSi anti-scaling coating, which can be assumed to have an influence on the weldability of the material combination. However, in order to reduce the complexity, this initial study on MPW of press hardening steel considers uncoated, fully martensitic 22MnB5 sheets. These were austenitized at 950 °C for 8 min in a protective atmosphere of argon and then quenched between cooled steel plates, which avoids distortion of the workpiece due to tempering. In order to ensure a defined homogeneous condition of the AA6016 raw material, which is highly susceptible to aging, the material was solution-annealed, water-quenched, and naturally aged for two weeks. The resulting condition is referred to as T4. The sheets were processed directly after ageing.

In order to describe the process parameters suitable for MPW independently of the specific setup and the properties of the used equipment (pulse power generator, tool inductor), the process window is defined via the local collision parameters, specifically the impact velocity and impact angle α as suggested, e.g., in [32]. In practice, these local process parameters can only be set and changed indirectly via adjustable process parameters such as the capacitor charging energy E, the initial gap width between flyer and target $g_{initial}$, or the relative lateral position of flyer edge and center of the active branch of the coil x_{flyer} [20]. Therefore, these parameters were systematically varied in welding tests using the setup shown in Figure 2a. Table 2 provides an overview of the most relevant process parameters characterizing these tests. The experiments were carried out using a pulsed power generator PS103–25 Blue Wave by PST Products (Alzenau, Germany). It features a maximum capacitor charging energy of 103 kJ, a maximum capacitor charging voltage of 25 kV, a stepwisely adjustable capacitance of 25.6 µF to 320 µF, a maximum discharging current of 2.2 MA in the short-circuit, and a maximum short-circuit frequency of 60 kHz. The applied tool coil was self-developed by Fraunhofer IWU. It is intended for producing a single weld seam directly under the center branch of the coil. This means that only this section is active, while the two outer branches serve for conducting the current back to the connector and the pulsed power generator, so that the current circuit is closed. The used experimental setup is shown in Figure 2c.

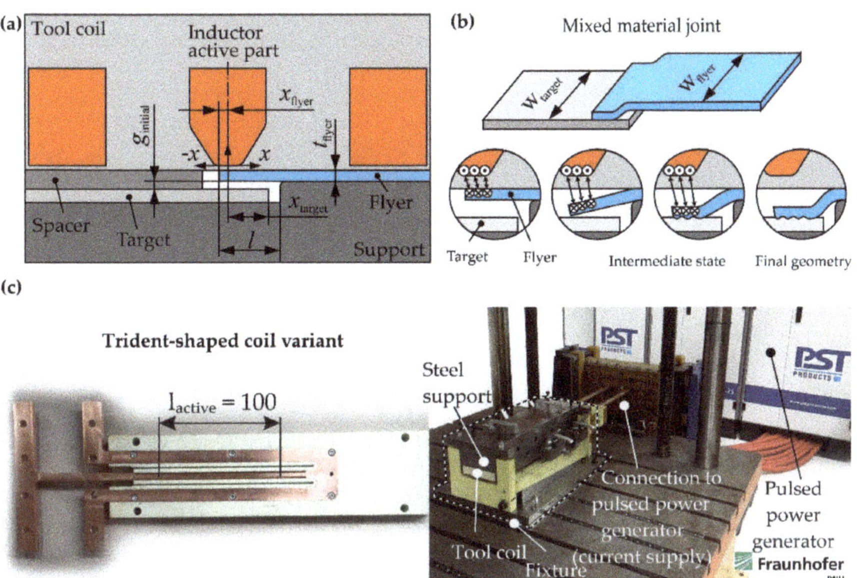

Figure 2. (**a**) Scheme of experimental setup; (**b**) Process states during the welding step; (**c**) Real MPW setup with the used trident-shaped coil and the pulsed power generator.

Table 2. Parameters of the experimental MPW process.

Generator Parameters	
Capacitor charging energy E	30–40 kJ
Capacitance C	330 μF
Tool parameter	
Active length of inductor l_{active}	100 mm
Workpiece parameters	
Flyer thickness t_{flyer}	2 mm
Target thickness t_{target}	2 mm
Width of flyer and target	100 mm
Experimental setup parameters	
Initial gap flyer to target $g_{initial}$	0.5–2.5 mm
x-position of flyer edge x_{flyer}	−3–0 mm
x-position of target edge x_{target}	fixed to 14 mm
free length l	fixed to 16 mm

Three sheets were joined per parameter set. In order to evaluate the weld quality, three lap shear tests were carried out for each joined sheet. In order to avoid failure close to the clamping area during the test, waisted specimens, similar to typical tensile specimens, were prepared from the welded sheet samples as shown in Figure 3a. Welded sheets typically feature non-welded edge zones, because here the induced current turns to form a closed current loop in the workpiece and the magnetic field lines and the Lorentz forces are re-directed correspondingly, so that the local conditions are inappropriate for welding and can cause local deformations of the flyer edges (see Figure 3a). In order to exclude this effect from the weld evaluation, three specimens per sheet sample were taken at a sufficient distance from the sample edge. The weld quality criterion (evaluation criterion) is based on the failure mode of the lap shear tests specimen, as illustrated exemplarily in Figure 3b. Weld quality is defined as high, if failure occurs in the base material of the weaker joining partner frequently far away from the joining zone (cohesive failure). The weaker partner can be either the one featuring lower material strength or significantly lower wall thickness. In the investigations considered here, it is the aluminum. In contrast, weld quality is defined as critical, if failure occurs in the joining zone by detachment of the two joining partners (adhesive failure).

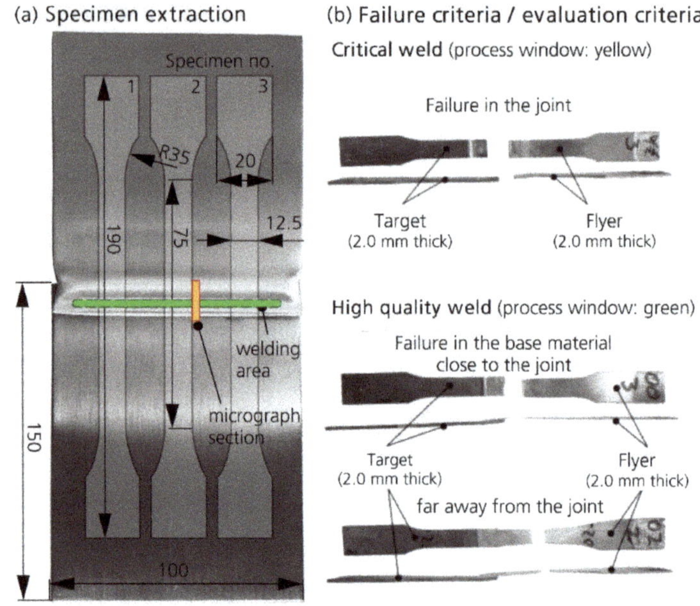

Figure 3. (**a**) Specimen extraction from welded samples; (**b**) Different failure modes of the lap shear specimens.

In addition to the global evaluation of the weld quality via lap shear tests, optical microscopic investigations were carried out on a representative section from the welded samples (see Figure 3a). Here, specifically the widths and the precise positions of the welded cross sections were identified, because it is known from literature that the contact zone of a magnetic pulse welded joint typically features welded and non-welded sections [19].

Parallel to the experimental tests, a macroscopic coupled electromagnetic and structural mechanical simulation was used in order to determine the corresponding collision parameters. The modelling was realized in LS-DYNA using the well-established FEM-BEM solver. The electromagnetic model contains the inductor, the flyer, and the target. The material is modelled as linear electromagnetic, and B-H-nonlinearity is disregarded. In the mechanical model, additional necessary supporting elements are considered. Inductor and supporting elements are modelled as rigid bodies, while elastoplastic deformation of flyer and target is possible. Here, an elastoplastic constitutive law with von Mises yield locus, isotropic hardening, and a tabulated scaling factor of the quasi-static flow curve (taken from literature: AA6016 [33] and 22MnB5 [34]), depending on the strain rate, is applied. The elastoplastic constitutive relation (Equation (1)) of the AA6016 material was assumed in the MPW process simulations as J2-plasticity (von Mises model), considering a strain rate sensitivity f_c and an approach for the material damage $f_c(\eta, \dot{\varphi}, D)$ scaling the quasi-static flow stress σ_0 in the model of the flow stress σ_y.

$$\sigma_y(\varphi, \dot{\varphi}, \eta, D) = \sigma_0(\varphi) \cdot f_c(\dot{\varphi}) \cdot f_d(\eta, \dot{\varphi}, D) \qquad (1)$$

Here, η is the stress triaxiality $\eta = \frac{-p}{\sigma_v}$, calculated by the hydrostatic pressure p and the equivalent stress σ_v, φ the true strain, $\dot{\varphi}$ the strain rate and D the damage parameter ($0 \ldots 1$, 0 = no damage, 1 = complete damage, i.e., element erosion). The damage approach based on the GISSMO model [35] and was used for the MPW simulations simplified as failure model considering an identified relatively high failure strain ε_f for the existing pressure stress state in the vicinity of the welding zone. Formula (2) describes a logarithmic approach (as known from the Johnson–Cook model) of the strain rate sensitivity, where c is the parameter identified as c = 0.01 for the AA6016 aluminum alloy and $\dot{\varphi}_0$ the reference strain rate (strain rate during quasi static stress strain curve testing, $\dot{\varphi}_0 = 0.001 \text{ s}^{-1}$).

$$f_c = \begin{cases} 1 + c \cdot ln\left(\frac{\dot{\varphi}}{\dot{\varphi}_0}\right) if \ \dot{\varphi} > 10^{-3} \text{ s}^{-1} \\ 1.0 \ if \ \dot{\varphi} \leq 10^{-3} \text{ s}^{-1} \end{cases} \qquad (2)$$

The discretization size has been optimized considering the contradicting requirements related to accuracy of the calculation result and calculation time. The result is shown in Figure 4. In order to allow a realistic modelling of the current distribution and the penetration depth of the EM field, especially the elements of the tool coil that are close to the coil surfaces facing the workpieces must feature small size and it is necessary to model several elements over the thickness of flyer and target. Here, six elements with an initial element height of 0.5 mm were used. With regard to the mechanical calculation model, fine discretization is needed in those areas of flyer and target that are deformed during the welding process. Therefore, in the workpiece areas that are close to the active branch of the coil, also the initial element width was set to 0.5 mm, while it is higher in the undeformed areas far away from the active branch of the coil.

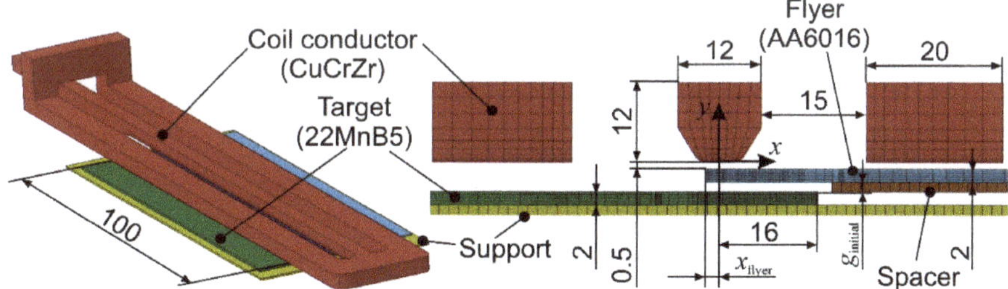

Figure 4. Discretization of the numerical model.

Obviously, the overall result of a coupled simulation significantly depends on the update time step, i.e., the duration between two subsequent recalculations of the electromagnetic system. Here, a time step of 50 ns was used. The aim of the simulation was to identify the collision parameters (impact velocity and impact angle) during the MPW. For this, however, it was not necessary to consider the influence of temperature during the simulation.

Finally, the experimental and the numerical results were consolidated as a collision parameter-based process window. For this purpose, local collision parameters are determined numerically and the position of welded and non-welded specimen sections determined via light-microscopic investigations are correlated to each other. A process window in the form of a diagram showing the impact velocity on the abscissa and the impact angle on the ordinate is presented. Collision parameters of welded specimen sections are indicated as welded (i.e., inside the process window), while collision parameters of non-welded sections and of completely non-welded specimens are indicated as not welded (i.e., outside the process window). Collision parameters of welded specimen sections are further distinguished into critically welded and high-quality welded ones according to the result of the corresponding lap shear test, specifically the failure mode. Thus, the process window was constructed by combining the results from the different process parameters, the numerical simulations, the microstructural investigation of the weld zone, and the lap-shear test.

To get a deeper understanding of the microstructural effects occurring during electromagnetic pulse welding, microstructural investigations of the aluminum/steel joints included scanning electron microscopy (SEM) in secondary electron (SE) and backscatter electron (BSE) mode that was accompanied by energy-dispersive X-ray spectroscopy (EDX). EDX investigations comprised point analyses as well as EDX mappings of the joining zone, executed with a field emission SEM Zeiss NEON 40EsB (Zeiss AG, Oberkochen, Germany). The respective cross sections were prepared with a final oxide polish (OP-S) using vibrational polishing (Buehler Vibromet 2, Mastertex, low nap, 60 min).

Furthermore, nanoindentation measurements according to ISO 14577 were conducted in order to determine the hardness of the area that is close to the interface of the aluminum/steel joint. The measurements were carried out on a UNAT nanoindentation device by ASMEC GmbH/Zwick GmbH & Co. KG (Dresden/Ulm, Germany) with a Berkovich indenter (0.394 μm tip radius). The normal force was increased from 0 to 5 mN within 10 s, followed by a hold time of 5 s and a force relief period of 4 s.

3. Results and Discussion

3.1. Determination of MPW Process Window

The result of the numerical process modelling in terms of quantified collision parameters is exemplarily shown in Figure 5. Here, a high-quality weld is achieved with a capacitor charging energy E of 40 kJ, an initial gap width $g_{initial}$ of 2.5 mm, and a lateral relative position of the coil center and the flyer edge x_{flyer} of 0 mm. All other process parameters correspond to the fixed process parameters given in Table 2. In the following, this parameter set is referred to as condition 1. Both velocity curve and angle curve are depicted as functions of the distance d from the flyer edge in Figure 5. They show a typical shape that is representative of all investigated parameter combinations. The velocity curve features a plateau of relatively constant velocity, which is followed by a steady decline of the curve. In this specific case, the plateau is rather short (approx. 2 mm) and high velocity (approx. 600 m/s) is reached because of the high capacitor charging energy in combination with a high initial gap width, which corresponds to a long acceleration distance for the flyer. With other parameter combinations considered here, the plateau reached lengths of up to 6 mm at a lower height. The subsequent steady decline of the curve is about equally steep for all regarded parameter combinations.

In contrast, an initial rising trend, which is then reversed to a decrease characterizes the angle curve. The high initial gap in combination with the short lateral relative position of the coil center and the flyer edge leads to a relatively steep rise of the curve and high absolute values as compared to other parameter combinations considered in this study. These observations are in good agreement with the general trends and influences related to the collision conditions presented in [29].

Additionally, Figure 5 directly correlates the velocity and angle curves with the result of the investigations by optical microscopy. The welded area identified in the microstructure was projected onto the diagram so that velocity-angle combinations that are beneficial for weld formation can easily be identified. In the next step, this information was transformed to collision parameter-based process windows. Figure 6 consolidated the results of all welding experiments, the corresponding numerical simulations and the evaluation of the weld quality. Collision parameter combinations corresponding to the welded zone were regarded as high-quality weld (green points) or critical weld (yellow points) depending on their failure mode (see Figure 3) and the maximum reached forces in the lap shear test (Table 3), whereas collision parameter combinations corresponding to the non-welded zone are depicted as red crosses. Green points can be considered safely within the process window, yellow points indicate the edge of the process window, and red crosses can be considered outside the process window.

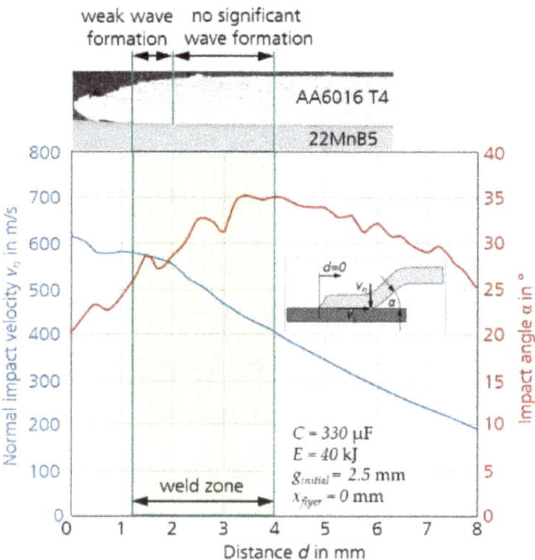

Figure 5. Comparison of the collision parameter distribution for a high-quality welding sample (condition 1).

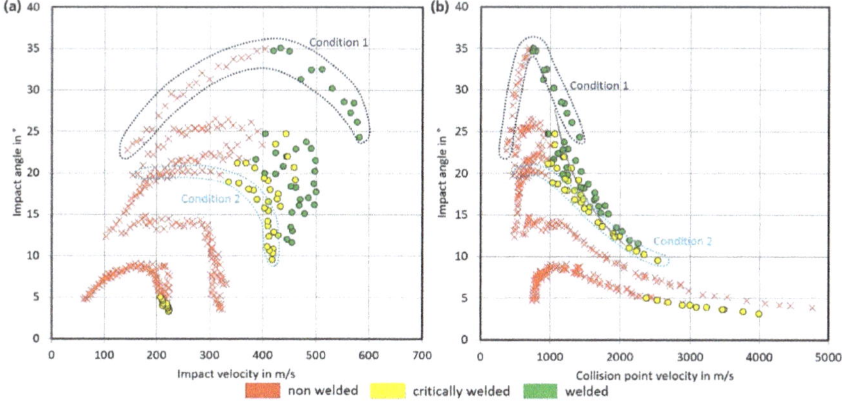

Figure 6. Identified process window for MPW of AA6016 T4 to 22MnB5 (hardened) (**a**) based on normal impact velocity v_n and impact angle α, (**b**) based on collision point velocity v_c and impact angle α. The dotted lines indicate the collision parameters of the high-quality (condition 1; dark blue line) and the critical weld (condition 2; light blue line).

Table 3. Results of the lap shear test and resulting failure mode for Condition 1 and Condition 2.

	Maximum Force F in N	Displacement at F in mm	Failure Mode
Condition 1 $E = 40$ kJ $g_{initial} = 2.5$ mm $x_{flyer} = 0$ mm	2813 ± 73	5.11 ± 0.31	**high-quality weld** (fail in base material)
Condition 2 $E = 35$ kJ $g_{initial} = 1.5$ mm $x_{flyer} = -2$ mm	1898 ± 56	1.86 ± 0.18	**critical weld** (fail in joint)

As the collision parameters were determined at discrete points with a distance of 0.25 mm between them (corresponding to the element size in this section of the numerical model), welding tests with one specific parameter set deliver a number of points in the process window. Figure 6 exemplarily highlights the group of points delivered by the welding samples processed with the parameter set referred to as condition 1. Additionally, a second group of points delivered by welding tests performed with a capacitor charging energy of 35 kJ, an initial gap of 1.5 mm between flyer and target, and a lateral relative position of the coil center and the flyer edge of -2 mm is indicated as condition 2. This parameter set was found to lead to critical welds failing by detachment of flyer and target in the joining zone und lap shear load.

In this context, the interpretation of the experimental results in the area close to the flyer edge is known to be challenging. Here, attachments of the flyer and target material to the respective other joining partner as exemplarily shown in [20] might indicate that the area was initially welded and ripped open again during the MPW process. However, as this is not fully clear, a zone of up to 1.3 mm distance from the flyer edge was disregarded when composing the process windows in order to avoid incorrect assignment either to collision parameters leading to a welded or non-welded contact.

Figure 6 shows two different variants of collision parameter-based process windows. They consider the impact angle α as a function of the normal impact velocity v_n (Figure 6a) and the collision point velocity v_c (Figure 6b), respectively. The equations for these velocities are explained in Figure 1. The latter is more common, especially in the context of explosive welding, an alternative impact welding technology, which shows some similarities to MPW although it is based on a chemical energy source instead of an electric one and the temperature regime during the process differs significantly. Both variants of the process window allow clear identification of collision parameter combinations that can be expected to lead to robust high-quality welds, i.e., failure in the base material of the weaker joining partner and those that will clearly not lead to welding. In-between these two groups there is a diffuse border area. This border area appears more distinct in the process window based on the collision point velocity, but it must be considered that the velocity scale in this diagram is roughly ten times as high compared to the scale in the diagram that is based on the impact velocity. In summary, an impact angle >15° and an impact velocity >400 m/s lead to sufficiently good joints between the aluminum alloy AA6016 and the steel 22MnB5.

3.2. Detailed Microstructural Characterization

In addition to the goal of determining a process window for MPW of AA6016 to 22MnB5, another objective of this work was the detailed investigation of the joining zone in order to gain deep knowledge about the microstructural effects in the joint zone. Specifically, differences between a high-quality weld and a critical weld were to be identified. For this purpose, the weld samples referred to as condition 1 (high weld quality) and condition 2 (critical weld quality) were regarded further. In this respect, already the examinations by optical microscopy revealed a significant difference between these two sample types. By preparing several cross-sections of the Al/steel compound at varying distances from the edge of the sheets (compare Figure 7a), the approximate width and length of the joining zone, i.e., the zone of intimate material contact between Al and steel, were determined. Figure 7b,c exemplarily show position 4 for the high-quality weld (weld parameters according to condition 1) and the weld of critical quality (weld parameters according to condition 2), respectively.

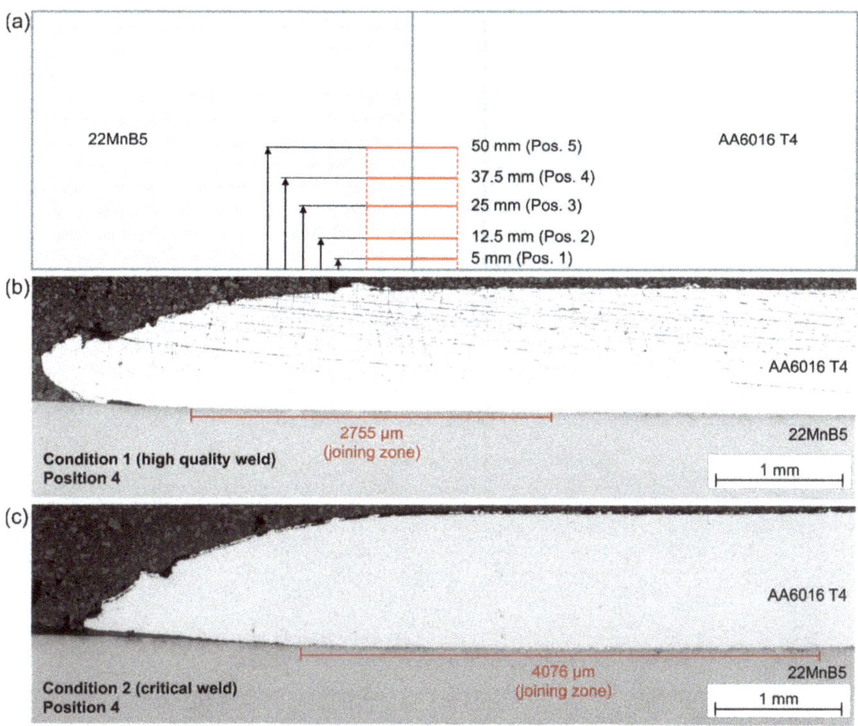

Figure 7. (**a**) Positions of cross-sections at varying distances from the edge of the Al/steel compound. (**b**) Optical microscope image of the cross-section of a high-quality weld (condition 1, position 4), (**c**) Optical microscope image of the cross-section of a critical weld (condition 2, position 4).

As confirmed by several high-quality and critical welds and to the surprise of the authors, the area of the joining zone of a high-quality weld is roughly half as large as that of a critical weld. In order to visualize this, the measured joining zone widths at positions 1 to 5 were schematically transferred to a macroscopic top view image of the Al/steel compound. The resulting assumed shape of the joining zone of both a high-quality and a critical weld is shown in Figure 8, illustrating the much smaller joining zone of the high-quality weld.

Since this observation seemed to contradict the lap shear test results and correlations observed in earlier studies on different material combinations [29], the joining zone was thoroughly investigated by SEM. This characterization revealed features that are typical for MPW joints, namely the formation of a wavy interface in the contact area of the metal sheets and the existence of an interface-near transition zone supposedly consisting of intermetallic phases. As shown in Figure 9a, the intermetallic phases (IMP) are predominantly present in the valleys of the wavy interface. Their existence is considered evidence for the temporary melting of flyer and target material in a very narrow interfacial zone due to the impact-induced conversion of the kinetic energy of the flyer into thermal energy.

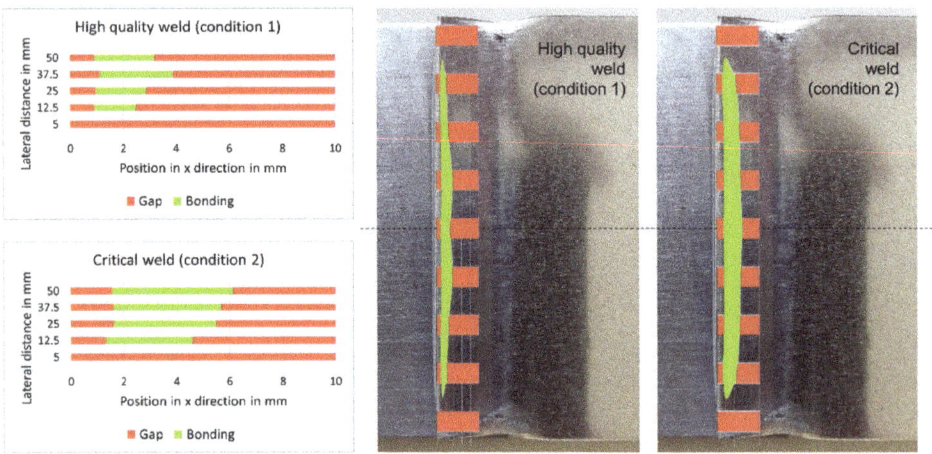

Figure 8. Measured width and assumed shape of the joining zones of a high-quality weld and a critical weld in order to visualize the significantly larger joining area (marked green) of the critical weld.

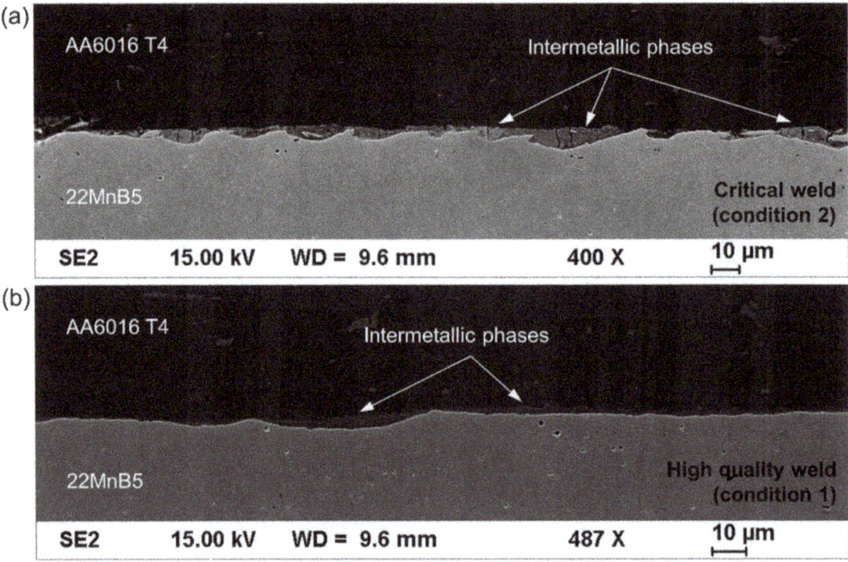

Figure 9. (**a**) SEM image (SE mode, acceleration voltage 15 kV, working distance 9.6 mm) of the joining zone of AA6016 with a fully martensitic 22MnB5 steel (critical weld). (**b**) SEM image (SE mode) of the joining zone of AA6016 with 22MnB5 steel (high-quality weld).

The comparison of the cross-section of a critical weld (parameter set during welding according to condition 2) in Figure 9a with a high-quality weld (parameter set during welding according to condition 1) shown in Figure 9b reveals a significantly less pronounced wave formation for the latter. Areas with IMP are still visible, although the transition zone is much thinner and IMP occur less frequently over the entire cross-section.

A possible explanation for the interface failure of the critical welds in the lap shear test despite the larger joining zone is provided by higher magnification SEM images. In Figure 10, the transition zone between Al and steel exhibits vortex-like structures, indicating a strong mixing of flyer and target material during the joining process. However, the formed IMP area shows numerous, mostly vertical cracks that extend over the entire transition zone and end abruptly in the adjacent Al and steel material. Since these cracks within the obviously very brittle IMP occur over the entire joining zone, this pre-damage is very likely the reason for the interfacial adhesive failure of the critical welds in the shear tensile test. In contrast, the high-quality welds, which exhibit almost no cracks in the transition zone, showed a cohesive failure in the Al sheet, thus indicating a superior interface bonding strength in comparison to the critical welds even though the joining zone is considerably smaller.

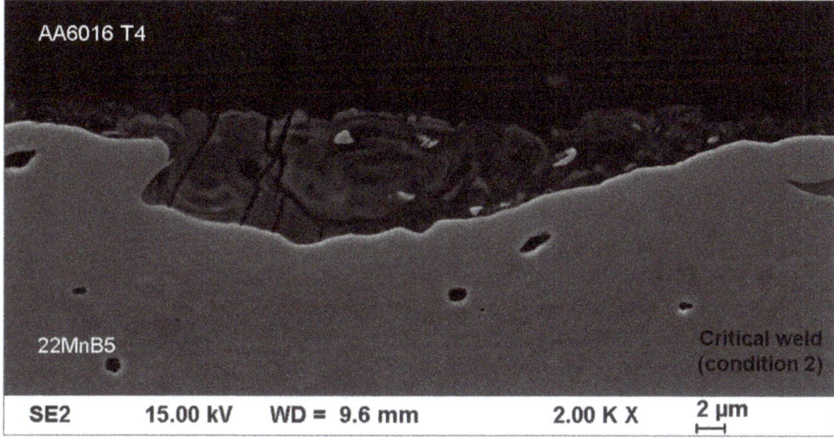

Figure 10. SEM image (SE mode) of the transition zone of a critical weld (welding parameters according to condition 2) with numerous cracks.

Further investigations were conducted in order to identify the IMP in the transition zone. An EDX mapping of an exemplarily chosen part of the interfacial area of a critical weld (Figure 11) revealed that the transition zone mainly contains Al, although the Al content seems to vary in a certain range as confirmed by the heterogeneous distribution of Fe, Mg, Mn, and Si in this zone. Additionally, oxygen can be detected, but is predominantly found in the described cracks in the transition zone.

In the SEM image shown in Figure 12a, which was taken in backscattered electrons mode, many different grayscales in the transition zone of a critical weld can be observed, indicating an inhomogeneous chemical composition. In contrast, the gray shade in the transition zone of the high-quality weld is much more homogeneous (see Figure 12b), thus suggesting a correspondingly homogenous chemical composition.

EDX point analyses at different locations in the transition zone (spots 1 to 3 in Figure 12a and 4 to 5 in Figure 12b) verify this observation. Table 4 summarizes the corresponding results. They confirm that the transition zone of the critical weld consists of Al-rich intermetallic phases with a strongly varying Al/Fe ratio and Mg, Si and Mn as accompanying elements ($\leq$1 at.% each). A comparison with the known Al-rich iron aluminide phases shows that $FeAl_2$ (66.7 at.% Al), Fe_2Al_5 (71.4 at.% Al), and Fe_4Al_{13} (76.5 at.% Al) might have been formed in the Al/steel interface during the joining process. However, due to the heterogeneous microstructure, even on a very small scale, a clear assignment of single areas to the mentioned phases is not possible by means of the characterization methods that were used in this work. Analytical methods with a very high spatial resolution such as X-ray photoelectron spectroscopy (XPS) might give more insight into the phase composition of the transition zone. Taking into account the very short time period of the joining process and the high cooling rates in the joining zone, the transition zone is expected to consist either of nanoscaled metastable phases and/or even amorphous areas.

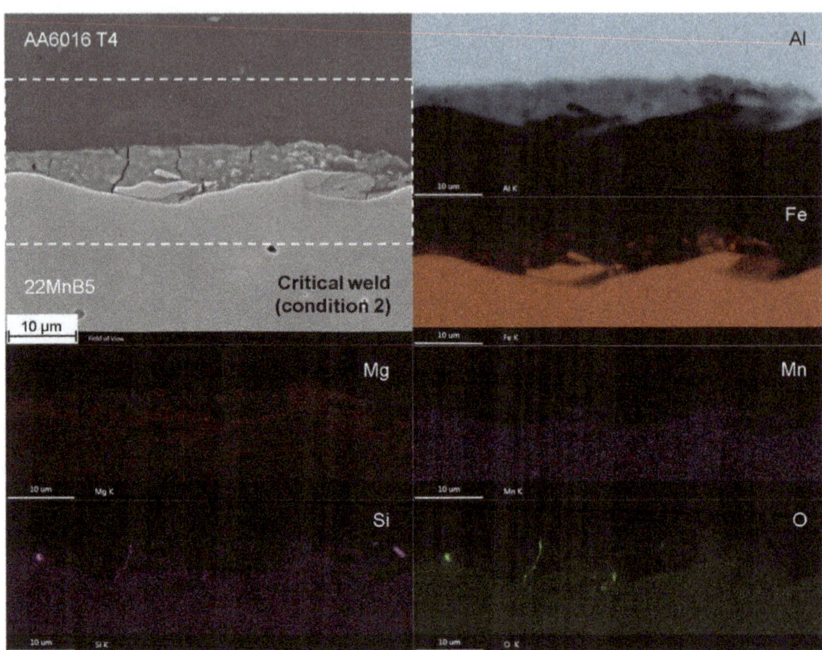

Figure 11. SEM image and EDX mapping of the Al/steel interface area of a critical weld (welding parameters according to condition 2), including the elemental distribution of Al, Fe, Mg, Mn, Si, and O.

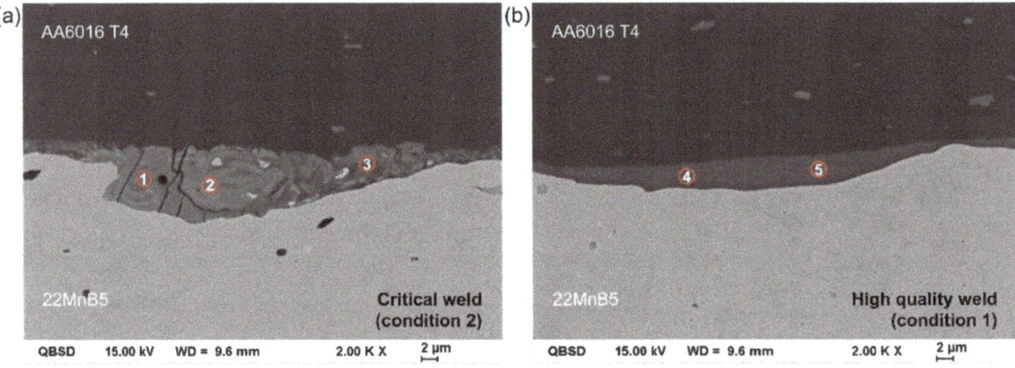

Figure 12. SEM images (BSE mode) of the joining zone of (**a**) a critical weld and (**b**) a high-quality weld with marked spots (1 to 5) for EDX point analysis.

In contrast to the critical weld, the transition zone of the high-quality weld (Figure 12b) exhibits a nearly constant chemical composition of about 88 at.% Al and 10 at.% Fe. This was also confirmed by EDX mapping (not shown here), which indicated a very homogeneous Al-rich phase in the transition zone. Regarding the mentioned composition, no stable FeAl phase with such a high Al content is known from the literature, which suggests the presence of non-equilibrium phases or a supersaturated solid solution in the transition zone of the high-quality welds, but partly also in the critical welds (see spot 3 in Figure 12a).

Table 4. Chemical composition of spots 1 to 5 from Figure 12 determined by EDX point analysis.

Element		Al		Fe		Mg		Si		Mn	
		wt.%	at.%	wt.%	at.%	wt.%	at.%	wt.%	at.%	wt.%	at.%
Critical weld	Spot 1	60.8	**75.5**	37.3	**22.4**	0.7	**1.0**	0.7	**0.8**	0.5	**0.3**
	Spot 2	52.5	**69.0**	45.6	**29.0**	0.6	**0.8**	0.6	**0.8**	0.7	**0.4**
	Spot 3	81.5	**89.2**	16.7	**8.8**	0.8	**1.0**	0.7	**0.8**	0.3	**0.2**
High-quality weld	Spot 4	79.6	**87.9**	18.2	**9.7**	1.1	**1.4**	0.7	**0.8**	0.4	**0.2**
	Spot 5	79.6	**88.0**	18.3	**9.8**	1.0	**1.3**	0.7	**0.7**	0.4	**0.2**

Figure 13 presents the results of the nanoindentation of a high-quality weld. The six hardness values on the Al side (top row) are quite similar and well within the microhardness range for severely plastically deformed AA6016 T4 described in the literature [36], except for the last one, where the indenter obviously hit a precipitation in the Al alloy. Additionally, the six hardness values on the 22MnB5 side (bottom row) are within a relatively narrow range around 480 HV, once again in good accordance with 22MnB5 hardness values known from the literature [37,38]. In contrast, the hardness values in the transition zone (middle row) exhibit a relatively large scatter. As described above, this transition zone supposedly consists of intermetallic non-equilibrium FeAl phases. Investigations by several authors have shown that especially Al-rich FeAl phases, as they were observed here, exhibit very high hardness values (>700 HV) which is usually associated with an increased brittleness [39,40]. Obviously the indents that are farther away from the interface of the transition zone with the 22MnB5 sheet, i.e., the first, fourth and fifth indent (from the left), achieve the lowest hardness values. The closer the indents get to the 22MnB5 side, the higher are the hardness values, reaching a maximum with the second indent (781 HV), which is almost exactly on the interface between the transition zone and the 22MnB5 material. As mentioned before, only high-resolution characterization methods could answer the question if this observed increase in hardness has to be attributed to a change in the phase composition within the interfacial area.

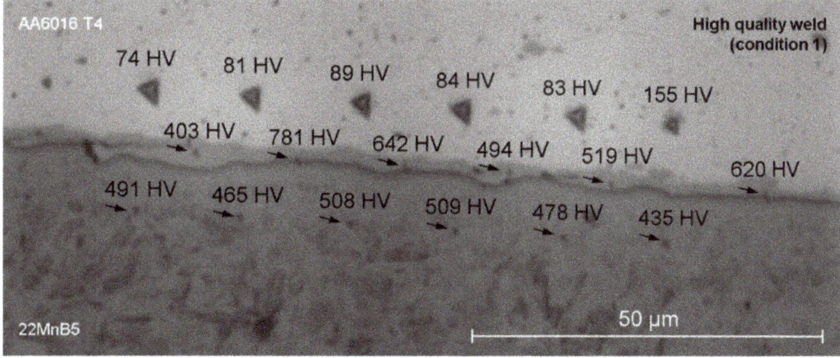

Figure 13. Nanoindentation results of the interfacial area of an Al/steel joint (high-quality weld, indentation force 5 mN).

4. Summary and Conclusions

AA6016 sheets were successfully joined for the first time to hardened 22MnB5 steel by magnetic pulse welding (MPW). By combining an extensive experimental study with a macroscopic coupled multiphysics simulation in LS-DYNA, a process window for high-quality welds was determined. Subsequent lap shear tests showed either cohesive failure in the AA6016 base material or interface failure in the joining zone. Samples that showed the latter failure mode were classified as critical welds, while those ones failing cohesively in the aluminum base material were evaluated as high-quality welds. For the chosen experimental setup, high-quality welds were formed for normal impact velocities of the AA6016 flyer starting at about 400 m/s at impact angles of around 10°. A tendency was observed that for larger impact angles, higher normal impact velocities are necessary in order to produce high-quality welds.

Microstructural characterization surprisingly revealed significantly larger joining zone areas for critical welds than for high-quality welds. However, next to a pronounced wave formation at the Al/steel interface and vortex-like structures indicating a strong mixing of AA6016 and 22MnB5, the intermetallic transition zone of the critical welds exhibited a lot of vertical cracks, thus providing a possible explanation for the inferior bonding strength in the lap shear tests. By contrast, the high-quality welds were characterized by thinner, but very homogeneous and crack-free transition zones consisting of an Al-rich intermetallic phase with about 88 at.% Al and 10 at.% Fe. Therefore, it can be concluded that for very high normal impact velocities above 400 m/s at impact angles between 10° and 35°, there is a continuous change in the way the transition zone forms. Even though the underlying mechanisms are not yet fully understood and require further research, it can be stated that process parameters that lead to a merely slightly pronounced wave formation and the development of a homogeneous transition zone with a very high Al content appear most suitable for the production of high-quality welds between AA6016 and 22MnB5. In summary, the following key results were determined and are essential for a perspective transfer of hybrid MPW joints made of high-strength aluminum alloys and hardened steel for use as structural components in automotive engineering:

- For the first time, the aluminum alloy AA6016 was successfully joined to hardened steel 22MnB5 by MPW.
- A robust process window for high-quality welds was determined by a macroscopic coupled multiphysics simulation in LS-DYNA.
- Surprisingly, the high-quality welds were characterized by thinner, but very homogeneous and crack-free transition zones consisting of an Al-rich intermetallic phase.

Author Contributions: Conceptualization, R.D., C.S., S.W. and V.P.; methodology, R.D., C.S. and V.P.; validation, R.D., C.S. and V.P.; formal analysis, R.D. and C.S.; investigation, R.D. and C.S.; resources, V.K. and T.L.; writing—original draft preparation, R.D., C.S. and S.W.; writing—review and editing, V.P. and T.L.; visualization, R.D. and C.S.; supervision, V.P. and T.L.; project administration, R.D., C.S. and V.P. All authors have read and agreed to the published version of the manuscript.

Funding: This research received no external funding.

Data Availability Statement: Not applicable.

Acknowledgments: The authors would like to express their gratitude to Elke Benedix, Steffen Clauß, Paul Seidel and Christian Loos for their support in sample preparation and microstructural characterization.

Conflicts of Interest: The authors declare no conflict of interest. The funders had no role in the design of the study; in the collection, analyses, or interpretation of data; in the writing of the manuscript, or in the decision to publish the results.

References

1. Lysenko, D.N.; Ermolaev, V.V.; Dudin, A.A. Method of Pressure Welding. U.S. Patent US3520049A, 14 July 1970.
2. Aizawa, T.; Kashani, M.; Okagawa, K. Application of magnetic pulse welding for aluminum alloys and SPCC steel sheet joints. *Weld. J.* **2007**, *86*, 119–124.
3. Psyk, V.; Risch, D.; Kinsey, B.L.; Tekkaya, A.E.; Kleiner, M. Electromagnetic forming—A review. *J. Mater. Process. Technol.* **2011**, *211*, 787–829. [CrossRef]
4. Davies, R.; Austin, E.R. *Electromagnetic Forming: Developments in High Speed Metal Forming*; Industrial Press: New York, NY, USA, 1970.
5. Pereira, D.; Oliveira, J.P.; Pardal, T.; Miranda, R.M.; Santos, T.G. Magnetic pulse welding: Machine optimisation for aluminium tubular joints production. *Sci. Technol. Weld. Join.* **2018**, *23*, 172–179. [CrossRef]
6. Psyk, V.; Linnemann, M.; Sebastiani, G. 4-Electromagnetic pulse forming. In *Elsevier Series in Mechanics of Advanced Materials*; Silberschmidt, V., Böhlke, T., McDowell, D., Zhong, C., Eds.; Elsevier: Amsterdam, The Netherlands, 2020; pp. 111–142.
7. Xu, J.; Wang, Y.; Wen, Z.; Li, Y.; Yan, L.; Cui, J. Electromagnetic impacting medium forming (EIMF): A new method forming process for magnesium alloy sheet. *Int. J. Adv. Manuf. Technol.* **2020**, *109*, 553–563. [CrossRef]
8. Psyk, V.; Scheffler, C.; Tulke, M.; Winter, S.; Guilleaume, C.; Brosius, A. Determination of Material and Failure Characteristics for High-Speed Forming via High-Speed Testing and Inverse Numerical Simulation. *J. Manuf. Mater. Process.* **2020**, *4*, 31. [CrossRef]
9. Lueg-Althoff, J.; Lorenz, A.; Gies, S.; Weddeling, C.; Goebel, G.; Tekkaya, A.E.; Beyer, E. Magnetic pulse welding by electromagnetic compression: Determination of the impact velocity. *Adv. Mater. Res.* **2014**, *966–967*, 489–499. [CrossRef]
10. Shanthala, K.; Sreenivasa, T.N.; Choudhury, H.; Dond, S.; Sharma, A. Analytical, numerical and experimental study on joining of aluminium tube to dissimilar steel rods by electro magnetic pulse force. *J. Mech. Sci. Technol.* **2018**, *32*, 1725–1732. [CrossRef]
11. Psyk, V.; Lieber, T.; Kurka, P.; Drossel, W.G. Electromagnetic joining of hybrid tubes for hydroforming. *Procedia CIRP* **2014**, *23*, 1–6. [CrossRef]
12. Kore, S.D.; Date, P.P.; Kulkarni, S.V. Electromagnetic impact welding of aluminum to stainless steel sheets. *J. Mater. Process. Technol.* **2008**, *208*, 486–493. [CrossRef]
13. Schäfer, R.; Pasquale, P. Robot automated EMPT sheet welding. In Proceedings of the 5th International Conference on High Speed Forming (ICHSF 2012), Dortmund, Germany, 24–26 April 2012; pp. 189–196.
14. Faes, K.; Shotri, R.; De, A. Probing Magnetic Pulse Welding of Thin-Walled Tubes. *J. Manuf. Mater. Process.* **2020**, *4*, 118. [CrossRef]
15. Khalil, C.; Marya, S.; Racineux, G. Magnetic Pulse Welding and Spot Welding with Improved Coil Effciency-Application for DissimilarWelding of Automotive Metal Alloys. *J. Manuf. Mater. Process.* **2020**, *4*, 69. [CrossRef]
16. Beerwald, C.; Beerwald, H. Spiralförmige Spule zur magnetischen Umformung von Blechen. Patent DE10207655A1, 16 September 2010.
17. Psyk, V.; Beerwald, C.; Henselek, A.; Homberg, W.; Brosius, A.; Kleiner, M. Integration of Electromagnetic Calibration into the Deep Drawing Process of an Industrial Demonstrator Part. *Key Eng. Mater.* **2007**, *344*, 435–442. [CrossRef]
18. Psyk, V. *Prozesskette Krümmen—Elektromagnetisch Komprimieren—Innenhochdruckumformen für Rohre und Profilförmige Bauteile*; Shaker: Dortmund, Germany, 2010.
19. Kwee, I.; Psyk, V.; Faes, K. Effect of the Welding Parameters on the Structural and Mechanical Properties of Aluminium and Copper Sheet Joints by Electromagnetic Pulse Welding. *World J. Eng. Technol.* **2016**, *4*, 538–561. [CrossRef]
20. Psyk, V.; Scheffler, C.; Linnemann, M.; Landgrebe, D. Manufacturing of hybrid aluminum copper joints by electromagnetic pulse welding—Identification of quantitative process windows. *AIP Conf. Proc.* **2017**, *1896*, 110001. [CrossRef]
21. Ben-Artzy, A.; Stern, A.; Frage, N.; Shribman, V.; Sadot, O. Wave formation mechanism in magnetic pulse welding. *Int. J. Impact Eng.* **2010**, *37*, 397–404. [CrossRef]
22. Cai, W.; Daehn, G.; Vivek, A.; Li, J.; Khan, H.; Mishra, R.; Komarasamy, M. A State-of-the-Art Review on Solid-State Metal Joining. *J. Manuf. Sci. Eng.* **2019**, *141*, 031012. [CrossRef]
23. Kang, B.-Y. Review of magnetic pulse welding. *J. Weld. Join.* **2015**, *33*, 7–13. [CrossRef]
24. Raoelison, R.N.; Sapanathan, T.; Buiron, N.; Rachik, M. Magnetic pulse welding of Al/Al and Al/Cu metal pairs: Consequences of the dissimilar combination on the interfacial behavior during the welding process. *J. Manuf. Process.* **2015**, *20*, 112–127. [CrossRef]
25. Wu, X.; Shang, J. An Investigation of Magnetic Pulse Welding of Al/Cu and Interface Characterization. *J. Manuf. Sci. Eng.* **2014**, *136*, 051002. [CrossRef]
26. Kochan, A. Magnetic pulse welding shows potential for automotive applications. *Assem. Autom.* **2000**, *20*, 129–132. [CrossRef]
27. Kimchi, M.; Shao, H.; Cheng, W.; Krishnaswamy, P. Magnetic pulse welding aluminium tubes to steel bars. *Weld. World* **2004**, *48*, 19–22. [CrossRef]
28. Yu, H.; Xu, Z.; Fan, Z.; Zhao, Z.; Li, C. Mechanical property and microstructure of aluminum alloy-steel tubes joint by magnetic pulse welding. *Mater. Sci. Eng. A* **2013**, *561*, 259–265. [CrossRef]
29. Psyk, V.; Scheffler, C.; Linnemann, M.; Landgrebe, D. Process analysis for magnetic pulse welding of similar and dissimilar material sheet metal joints. *Procedia Eng.* **2017**, *207*, 353–358. [CrossRef]
30. Wang, S.; Zhou, B.; Zhang, X.; Sun, T.; Li, G.; Cui, J. Mechanical properties and interfacial microstructures of magnetic pulse welding joints with aluminum to zinc-coated steel. *Mater. Sci. Eng. A* **2020**, *788*, 139425. [CrossRef]
31. Psyk, V.; Linnemann, M.; Scheffler, C. Experimental and numerical analysis of incremental magnetic pulse welding of dissimilar sheet metals. *Manuf. Rev.* **2019**, *6*, 7. [CrossRef]

32. Cuq-Lelandais, J.-P.; Avrillaud, G.; Ferreira, S.; Mazars, G.; Nottebaert, A.; Teilla, G.; Shribman, V. 3D Impacts Modeling of the Magnetic Pulse Welding Process and Comparison to Experimental Data. In Proceedings of the 7th International Conference on High Speed Forming, Dortmund, Germany, 27–28 April 2016; pp. 13–22.
33. Vogel, M.; Lechner, M. Manufacturing of process adapted tailored blanks by flexible rolling process using aluminum alloy AA6016. *Procedia Manuf.* **2018**, *15*, 1224–1231. [CrossRef]
34. Karbasian, H.; Tekkaya, A.E. A review on hot stamping. *J. Mater. Process. Technol.* **2010**, *210*, 2103–2118. [CrossRef]
35. Basaran, M. *Stress State Dependent Damage Modeling with a Focus on the Lode Angle Influence*; Shaker: Aachen, Germany, 2011.
36. Topic, I.; Höppel, H.W.; Göken, M. Friction stir welding of accumulative roll-bonded commercial-purity aluminium AA1050 and aluminium alloy AA6016. *Mater. Sci. Eng. A* **2009**, *503*, 163–166. [CrossRef]
37. Bok, H.-H.; Kim, S.N.; Suh, D.W.; Barlat, F.; Lee, M.-G. Non-isothermal kinetics model to predict accurate phase transformation and hardness of 22MnB5 boron steel. *Mater. Sci. Eng. A* **2015**, *626*, 67–73. [CrossRef]
38. Aziz, N.; Aqida, S.N. Optimization of quenching process in hot press forming of 22MnB5 steel for high strength properties. *IOP Conf. Series Mater. Sci. Eng.* **2013**, *50*, 012064. [CrossRef]
39. Matysik, P.; Jóźwiak, S.; Czujko, T. Characterization of low-symmetry structures from phase equilibrium of Fe-Al system—Microstructures and mechanical properties. *Materials* **2015**, *8*, 914–931. [CrossRef] [PubMed]
40. Potesser, M.; Schoeberl, T.; Antrekowitsch, H.; Bruckner, J. The characterization of the intermetallic Fe-Al layer of steel-aluminum weldings. In Proceedings of the TMS Annual Meeting & Exhibition, San Antonio, TX, USA, 12–16 March 2006; pp. 167–176.

Journal of
*Manufacturing and
Materials Processing*

Article

Magnetic Pulse Welding and Spot Welding with Improved Coil Efficiency—Application for Dissimilar Welding of Automotive Metal Alloys

Chady Khalil, Surendar Marya and Guillaume Racineux *

Research Institute in Civil and Mechanical Engineering (GeM, UMR 6183 CNRS), Ecole Centrale de Nantes, 1 rue de la Noë, F-44321 Nantes, France; chadykhalil@gmail.com (C.K.); surendar.marya@ec-nantes.fr (S.M.)
* Correspondence: guillaume.racineux@ec-nantes.fr

Received: 4 June 2020; Accepted: 6 July 2020; Published: 8 July 2020

Abstract: Lightweight structures in the automotive and transportation industry are increasingly researched. Multiple materials with tailored properties are integrated into structures via a large spectrum of joining techniques. Welding is a viable solution in mass scale production in an automotive sector still dominated by steels, although hybrid structures involving other materials like aluminum are becoming increasingly important. The welding of dissimilar metals is difficult if not impossible, due to their differential thermo mechanical properties along with the formation of intermetallic compounds, particularly when fusion welding is envisioned. Solid-state welding, as with magnetic pulse welding, is of particular interest due to its short processing cycles. However, electromagnetic pulse welding is constrained by the selection of processing parameters, particularly the coil design and its life cycle. This paper investigates two inductor designs, a linear (I) and O shape, for the joining of sheet metals involving aluminum and steels. The O shape inductor is found to be more efficient both with magnetic pulse (MPW) and magnetic pulse spot welding (MPSW) and offers a better life cycle. Both simulation and experimental mechanical tests are presented to support the effect of inductor design on the process performance.

Keywords: magnetic pulse welding; spot welds; linear coils; shear lap test; automotive alloys; numerical analysis; LS-DYNA

---

1. Introduction

The emissions targets set by various climate summits, which started in the 1970s, as well as the increase in oil prices since then, has made the automotive OEMs focus their efforts on optimizing vehicle emissions and keeping fuel consumption at a minimum to meet both the environmental and end customers' economic concerns. These efforts were translated by extensively optimizing the combustion engines, introducing the hybrid concept, and introducing, during the 1990s, the first commercial electric vehicles. The common helping factor for all the versions proposed was reducing the weight of the structural and non-structural components by using a combination of materials.

The metal percentage of the materials used in the production of a typical passenger vehicle, even by the year 2025, is expected to maintain a big part of the share: 60% steel (combination of low-carbon, mild-, and high-strength steels), 18% for aluminum, and 5% magnesium [1]. Consequently, the use of these materials implies the need to deal with the challenges related to dissimilar metal joining by overcoming the constraints stemming from unmatching properties (mechanical, thermal, and chemical) and by ensuring manufacturing conditions (limit the damage on joining partners, reliable and cost-efficient processes, and recyclability).

The traditional joining technologies which are well established in the industry are the mechanical and thermal techniques. The former includes processes such as screwing, riveting, punch riveting and clinching.

The main disadvantages of these processes are the additional steps required in the production cycle, stress concentrations at the points of fastening, and weight increase inherent to the fasteners themselves [2,3]. For fusion welding technologies that implicitly involve a heat source with or without a filler addition, the main concern is microstructural changes, residual stresses and defects during the solidification of the molten zone between the adjoining components. For instance, weld fusion zones are not always exempt from porosity, solidification cracks and oriented structures with grain orientations that induce unpredictable variability in the welded components [4,5]. When a filler is additionally used to mitigate weld metal chemistry, the weight increase is inherent, as in mechanical joints. In automotive applications of fusion welding, resistance spot welding has been in practice for a very long time and, for the last two decades, remote laser spot welding is increasingly preferred due to the processing speed [6,7]. Further lightweight automotive body structures imply hybrid materials as diverse as aluminum and steels and need fast joining technologies to keep pace with the large-scale automotive sector [8]. The fusion welding of such dissimilar materials is more complex due to the formation of intermetallic compounds in the weld zone and the presence of differential thermal effects stemming from different physical properties, like the thermal expansion coefficient and melting points. This leads to the unacceptable weakening of the welded components. In the case of aluminum to steel thermal joining application, two important limitations need to be mentioned: the formation of brittle, aluminum, rich-intermetallic compounds (Fe_2Al_5, $FeAl_3$) and the negligible solid solubility of Fe in Al [9]. For a viable dissimilar joint, one possible way of doing this is to reduce, as much as possible, the size of the intermetallic phases. For this to happen, amongst other alternatives to minimize the formation of brittle compounds, some prospective studies involve laser roll bonding through brazing [10] and mixing, as in friction spot welding [11], but with limited success in large-scale applications.

To overcome the mechanical and thermal limitations, the studies of high-velocity impact processes which create a direct joining between metals at solid-state and at microscopic level presented an opportunity. These studies started first in the 1950s with the explosive welding (EXW) [12], and continued with magnetic pulse welding (MPW), laser impact welding (LIW) [13], and vaporizing foil actuator welding (VFAW) [14]. The cold and rapid natures of these processes—some microseconds—limit the risk of HAZ and intermetallic compounds' formation, respectively [15].

The principle of the high-velocity impact processes is based on accelerating one metal, called a flyer, at very high velocities, towards another fixed metal, called parent, where the local progressive collision creates the bond between the two materials. The difference lies in the accelerator types: in EXW, for example, the detonators are used, while in MPW, electromagnetic driving forces are used [2–5]. For safety and ergonomic reasons of massive industrial applications, the MPW process presents itself as a more appropriate candidate, as it uses the electromagnetic fields as a source.

The driving force generation is based on the Laplace force principle: the presence of a current-carrying conductive metal in a time-varying magnetic field generates Laplace forces on this metal. Therefore, and to achieve large amounts of Laplace forces, the MPW processes use:

- A high pulse generator capable of generating an intense, time-varying current that will be the source of the time-varying magnetic field;
- A coil capable of producing the magnetic field due to the discharged current delivered by the generator;
- A conductive metal in the vicinity of the coil in which the magnetic field penetrates and induces currents, leading to the generation of large amounts of Laplace forces, causing its acceleration.

MPW for a tubular geometry was the most used and developed in the MPW field [16,17]. Starting in the year 2000, more focus was given to the sheet metal applications where the operational positioning is presented in Figure 1a,b and Figure 2A. Manogaran et al. [18] developed an additional process concept for metal sheets: Magnetic Pulse Spot Welding (MPSW). In this process, a prior local stamping is created on the intended spot-welding location in the flyer metal, which is called the hump (Figure 1c,d); this hump will guarantee the required standoff distance, and hence the flyer could be directly placed on the parent metal without any concern for the overlapping distance between the two materials to comply with the industrial ease for automated applications. This hump will then be accelerated by the magnetic pulse to create a spot weld after collision (Figure 2B).

After this brief description of the MPW/MPSW, the process's operational parameters can be concluded: the electrical coil's geometry, the configuration's geometrical parameters and the discharge energy of the generator. These operational parameters condition the collision mechanism, where the two impact physical parameters are the impact velocity and the impact angle, which are mandatory to ensure two clean surfaces ready for welding and the formation of the jet [2–6,19].

It is worth mentioning that the impact would invariably generate heat by the transformation from mechanical to thermal energy, thereby limiting the highest impact velocity that would not engender melting at the welded interface. For dissimilar joints, as mentioned earlier, melting has to be curtailed in order to eliminate the possible formation of intermetallic compounds. It is worth noting that, due to short impact times, heat generation remains nearly adiabatic and localized in the interfacial zone. In the case of Al to Cu impact joining, Marya et al. [20] have reported localized melting in tube to tube welding, which has been substantiated by others [15,19]. Briefly, the velocity range for successful welding has a limited window and consequently an equal choice of parameters like standoff distance, discharge energy and configuration of sheets/parts.

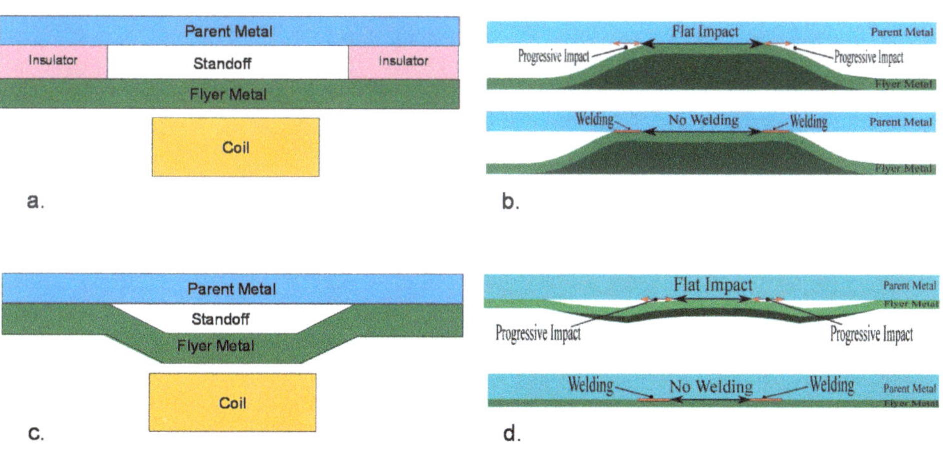

Figure 1. (**a,b**) Magnetic Pulse Welding (MPW); (**c,d**) Magnetic Pulse Spot Welding (MPSW).

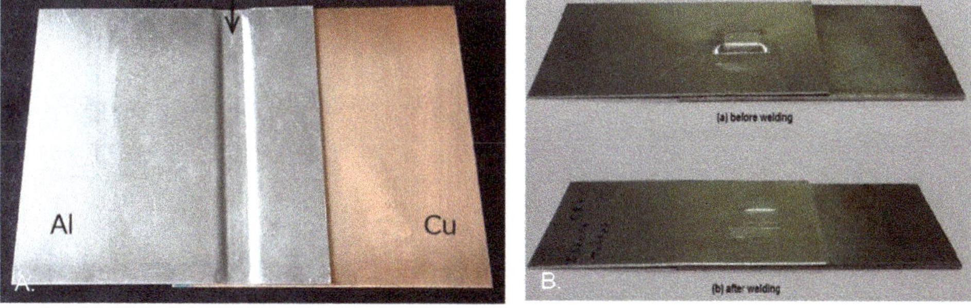

Figure 2. (**A**) Al/Cu MPW [21]; (**B**) MPSW [18].

Objectives of Study

The MPW/MPSW for sheet metal applications has been successfully applied and achieved for similar and dissimilar metals welding [1,4,7–10,22]: aluminum to aluminum, aluminum to steel, aluminum to magnesium. In the literature, a lot of alloys were tested, and the applications, in general, concerned thin sheet metal thicknesses, omitting the real thicker sheets that can be applied in real applications for the automotive industry. On the other hand, the most used electrical coil during these studies were linear with rectangular cross-section coils, which present a good efficiency to simplicity ratio for aluminum 1xxx and thin sheet applications [1,11].

However, in automotive applications, the 5000- and 6000- series aluminum alloys' use is dominant (body panels, seat structures, reinforcement members, etc.). On the steel side, the low-carbon drawing steels are used for various components and, during the last three decades, the developments of the advanced high-strength steels (AHSS) led to the increase in the use of the dual phase steels (DP) [23].

In this context, the current study aimed to develop an electrical coil geometry allowing a better efficiency and wider application scope of the MPW/MPSW for the automotive metal alloys. The new geometry coils, as will be seen in the results, led to a wide range of successful welds for different combinations of similar and dissimilar alloys. Several combinations were also tested under quasi-static, dynamic and fatigue loads to see the welds strengths level.

First, the equipment used, including the two coils (linear and the newly designed), the experimental setup (Figure 3) and an LS-Dyna numerical model which will be used in the analysis of the improved efficiency of the new coil design, will be presented. After that, we will present the comparison of the two coils and proceed to the experimental results.

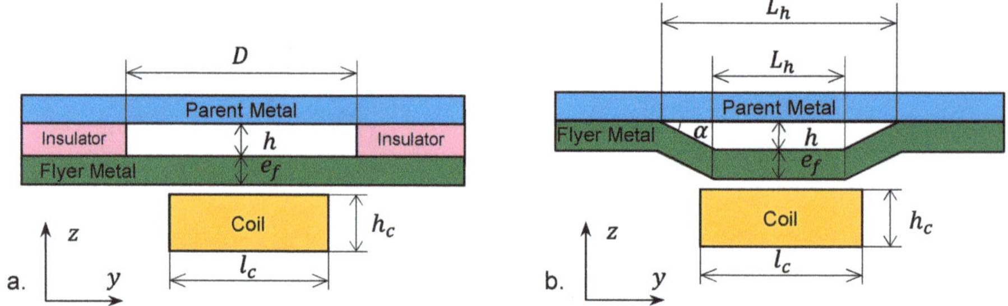

Figure 3. Experimental setup: (**a**) MPW; (**b**) MPSW.

2. Materials and Methods

2.1. Materials

The aluminum and steel alloys chemical compositions used during this study are presented in Tables 1–3. Their mechanical properties are given in Tables 4 and 5.

Table 1. Aluminum alloys, chemical compositions (%at.).

	Si max	Fe max	Cu max	Mn max	Mg max	Cr max	Zn max	Ti max	Other max
1050	0.25	0.40	0.05	0.05	0.05	-	0.07	0.05	0.03
5182	0.20	0.35	0.15	0.50	4.0–5.00	0.10	0.25	0.10	0.15
5754	0.40	0.40	0.10	0.50	2.6–3.6	0.30	0.20	0.15	-
6013	0.78	0.28	0.97	0.40	1.02	0.06	0.10	0.05	0.15
6016	0.5–1.5	0.50	0.20	0.20	0.25–0.70	0.10	0.20	0.15	0.15

Table 2. DC04 steel chemical compositions (%at.).

	C max	Mn max	Si max	P max	S max	Al max
DC04	0.08	0.40	0.10	0.025	0.025	0.020

Table 3. DP steels chemical compositions (%at.).

	C max	Mn max	Si max	P max	S max	Al max	Ti + Nb max	V max	Cr max	Mo max	B max	N max	Ni max	Nb max
DP450	0.10	0.16	0.4	0.04	0.015	0.015–0.08	0.05	0.01	0.8	0.3	0.005	0.008	-	-
DP1000	0.139	1.50	0.21	0.009	0.002	0.046	-	0.01	0.02	-	0.0002	0.003	0.03	0.015

Table 4. Steel's mechanical properties.

	YS (MPa)	UTS (MPa)	A % ISO 20 × 80
DC04 ($e \leq 1.47$ mm)	160–200	280–340	37
DP450	290–340	460–560	27
DP1000	787	1059	8.5

Table 5. Aluminum alloys' mechanical properties.

	YS (MPa)	UTS (MPa)	A % ISO 20 × 80
1050-H14	85	105	4
5754-H111 ($e \leq 1.5$ mm)	90–130	200–240	21
5182 ($e \leq 1.5$ mm)	120–160	260–310	23
6013-T4	174	310	26
6016-T4	110–150	220–270	23

2.2. Equipment and Experimental Procedure

2.2.1. Pulse Generator

The pulse generator used is a 50 kJ, developed at ECN, with the following characteristics:

$$C_{Gen} = 408 \ \mu F, \ L_{Gen} = 0.1 \ \mu H; \ R_{Gen} = 3 \ m\Omega; \ V_{max} = 15 \ kV; \ I_{max} = 500 \ kA; \ f_{short} = 25 \ kHz.$$

The highest limit for the discharge energy is hence fixed at 16 kJ, so that the discharge current does not exceed 80% of the maximum allowable current for the generator:

$$I_{operation_{max}} = 0.8 \times I_{max} = 400 \ kA$$

2.2.2. Coils

The two coils used in this study are a linear rectangular cross-section coil—dimensions are presented in Figure 4—and an O-shaped rectangular cross-section coil—dimensions are presented in Figure 5. The active areas of the two coils are the same: 20 × 8 mm.

2.2.3. Discharge Energy and Discharge Current Relation

The discharge energy E depends on the voltage that is charging the capacitors and it can be expressed by

$$E_{discharge} = \frac{1}{2} C_{Gen} V_0^2 \tag{1}$$

where C_{Gen} is the generator capacitance and V_0 is the charging voltage.

The resulting discharge current in the coil, which is highly damped sinusoidal due to the RLC equivalence of the circuit, is

$$I(t) = I_0 e^{-\frac{t}{\tau}} \sin(\omega t) \tag{2}$$

where

$$I_0 = V_0 \sqrt{\frac{C_{Gen}}{L}} \tag{3}$$

$$\tau = \frac{2L}{R} \tag{4}$$

$$\omega = \frac{1}{\sqrt{LC_{Gen}}} = \frac{f}{2\pi} \tag{5}$$

with L and R as the equivalent inductance and resistance of the system, respectively

$$L = L_{Gen} + L_{Coil} \tag{6}$$

$$R = R_{Gen} + R_{Coil} \tag{7}$$

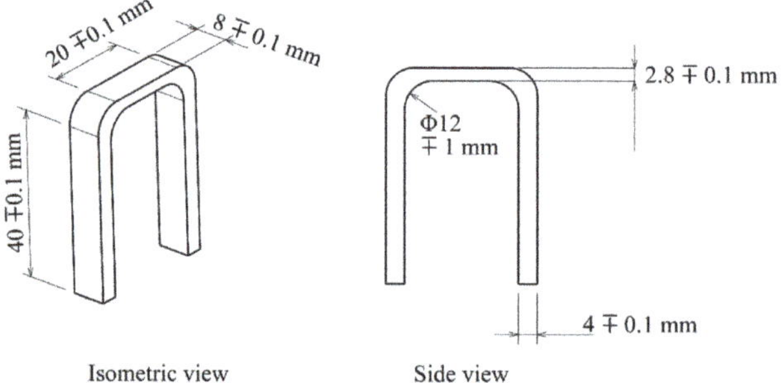

Figure 4. Linear rectangular cross-section coil.

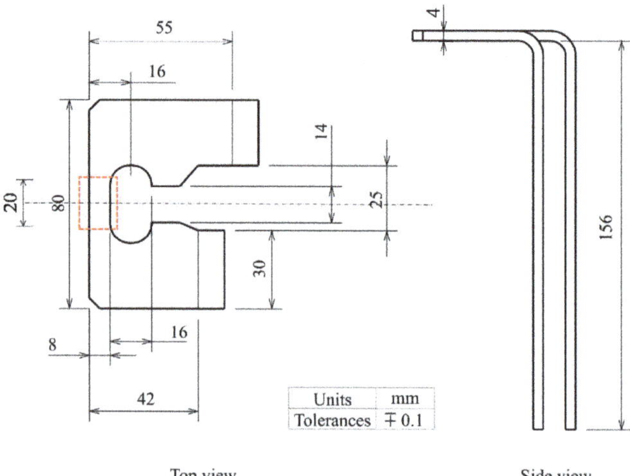

Figure 5. O-shape coil dimensions.

2.2.4. Experimental Procedure and Design

In the case of the MPW (Figure 1a,b), the configuration involves the use of insulators to create the needed airgap between the two metals (Figure 3). The insulators are made of PE and PVC and they are designed to have the same thickness of the required standoff distance between metals. They are fixed on the flyer metal using adhesive tapes. In Figures 6 and 7, the positioning of the flyer metal regarding the coil is represented for both linear and O-shaped coils, respectively.

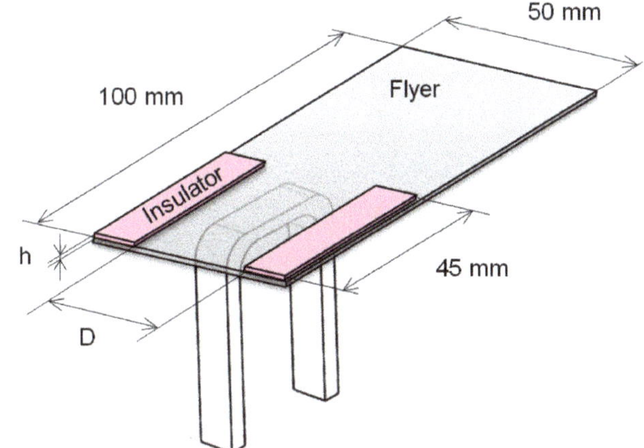

Figure 6. Magnetic Pulse Welding configuration with linear coil.

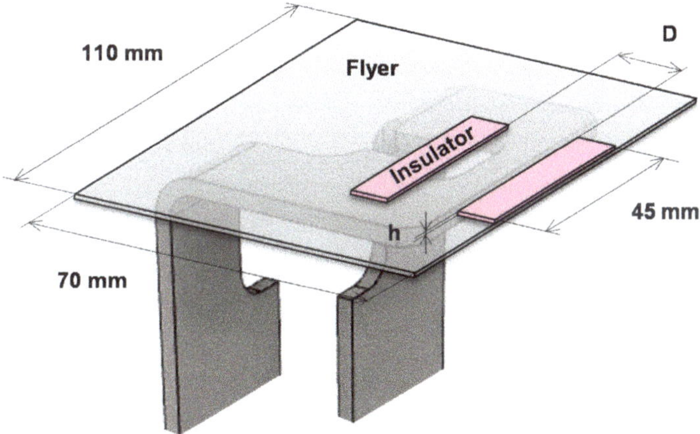

Figure 7. Magnetic Pulse Welding configuration with O-shape geometry coil.

After the cleaning step, the flyer metal is positioned facing the coil where a Kapton insulation sheet with a 0.1 mm thickness is used to separate it from the coil. The positioning is controlled by a laser so that the part of the flyer metal to be deformed is centered regarding the coil's active area. The parent metal is then positioned on the insulators, over it a massive steel die, and finally the whole system is clamped using a special system designed to avoid any displacement during the welding process. The standoff distance is controlled using a feeler gauge before every test.

In case of MPSW (Figure 1c,d and Figure 3b), the first step of the process is to create the hump in the flyer metal. The general geometry of the hump chosen is a rectangular one. This geometry was chosen based on the previous study done by Manogaran et al. [19], during which different geometries were tested, and it was proven that the rectangular one is the more efficient. Figure 8 presents this general geometry with different dimensions. The hump is stamped in the flyer metal using a hydraulic press die.

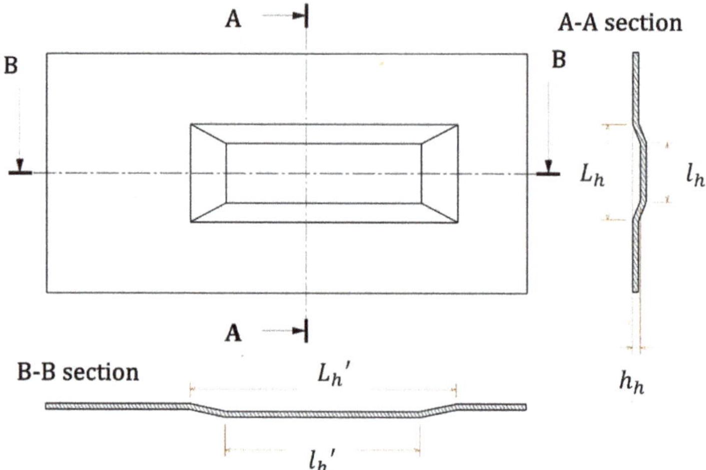

Figure 8. General geometry of the hump for MPSW.

After cleaning the metal surfaces from oil using acetone solution, the flyer metal is then positioned with a laser so that it is facing and centered on the coil's active area. The parent metal is then positioned above the flyer metal with the massive die on, and finally the system is clamped like the MPW case.

2.2.5. Welds Strength Evaluation

For the welding mechanical strength evaluation, the welded specimens were tested as schematically represented in Figure 9 in lap-shear conditions:

- Quasi-static at 10^{-2} mm/s (with an Instron 5584 mechanical tensile machine, Norwood, MA, USA);
- Dynamic at 614 mm/s (with an MTS 819 hydraulic high-speed tensile machine, Eden Prairie, MN, USA);
- Fatigue under unidirectional conditions ($R = 0$, frequency of 20 Hz, $F_{max} = 0.6 \times F_{maxquasi-static}$ with an MTS Electropulse E10000 machine, Eden Prairie, MN, USA).

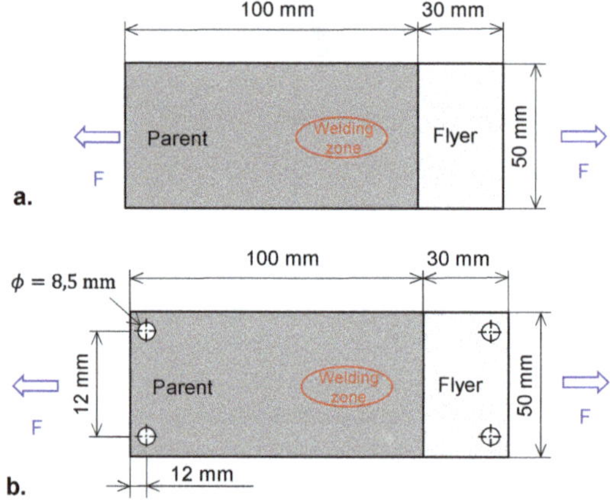

Figure 9. (**a**) quasi static and fatigue lap shear tests specimen, (**b**) dynamic lap-shear test specimen.

2.2.6. Numerical Simulation

The model used is a coupled mechanical/thermal/electromagnetic using LS-DYNA 3D FEM-BEM code (Finite Element Method – Boundary Element Method).

The simulations during this study were performed using the High-Performance Computing (HPC) resources of the Centrale Nantes Supercomputing Center on the cluster Liger by the High-Performance Computing Institute (ICI).

2.3. Materials Properties and Boundary Conditions

The coils material used in this study are the OFHC copper and the ASTM A36 steel. The flyer metals used in the numerical models are the aluminum alloys 5754 and 5182. The parent metal is the DC04 deep drawing law carbon steel.

A Johnson–Cook model was used for the mechanical solver and the parameters for all materials are taken from the literature [24–27] and they are presented in Table 6. The thermal and electrical parameters of different materials are presented in Tables 7 and 8 and they are also taken from the literature [24–27]. The initial room temperature T_{RT} was set at 293 K.

Table 6. Johnson–Cook model parameters.

Coil	Alloy	A (MPa)	B (MPa)	n	C	m
Linear Coil	OFHC	90	292	0.31	0.025	1.09
O-Shape	OFHC	90	292	0.31	0.025	1.09
Coil	ASTM A36	286.1	500.1	0.2282	0.022	0.917
Flyers	5754	67.456	471.242	0.424	0.003	2.519
	5182	106.737	569.120	0.485	−0.001	3.261
Parent	DC04	162	598	0.6	2.623	0.009

Table 7. Thermal properties.

Coil	Alloy	Thermal Conductivity (W/mk)	Specific Heat Capacity (J/kgK)
Linear Coil	OFHC	386	383
O-Shape Coil	OFHC	386	383
	ASTM A36	50	450
Flyers	5754	130	897
	5182	123	902
Parent	DC04	52	470

Table 8. Electrical Properties—International Annealed Copper Standard.

Coil	Alloy	%IACS (International Annealed Copper Standard)	S/m
Linear Coil	OFHC	100	5.8001×10^7
O-Shape Coil	OFHC	100	5.8001×10^7
	ASTM A36	12	6.9600×10^6
Flyers	5754	33	1.9140×10^7
	5182	28	1.6240×10^7
Parent	DC04	13	7.5400×10^6

The variation in the electrical conductivity, with temperature for the copper and aluminum, is based on the Meadon model, which is a simplified Burgess model [28], and it gives the conductivity as a function of temperature and density at solid phase [29]

$$\rho = \left(C_1 + C_2 T^{C_3}\right) f_c\left(\frac{V}{V_0}\right) \tag{8}$$

where ρ is the electrical resistivity, T is the temperature, V is the specific volume, V_0 is the reference specific volume (zero pressure, solid phase) and

$$f_C\left(\frac{V}{V_0}\right) = \left(\frac{V}{V_0}\right)^{2\gamma-1} \tag{9}$$

where γ is the Gruneisen value given by

$$\gamma = \gamma_0 - \left(\gamma_0 - \frac{1}{2}\right)\left(1 - \frac{V}{V_0}\right) \tag{10}$$

with γ_0 the reference Gruneisen value. In Table 9, the set of parameters for aluminum and copper from Burgess paper are given [20,21].

Table 9. Meadon-Burges parameters.

Parameter	Cu	Al
V_0 (cm^3/g)	0.112	0.370
γ_0	2	2.13
C_1	-4.12×10^{-5}	-5.35×10^{-5}
C_2	0.113	0.223
C_3	1.145	1.210

For the steel case, a simple linear model was used

$$\rho = \rho_0[1 + \alpha_0(T - T_{RT})] \tag{11}$$

where ρ is the electrical resistivity at the actual temperature T, ρ_0 is the electrical resistivity at the reference temperature T_{RT} and $\alpha_0 = 6 \times 10^{-3}\ \text{K}^{-1}$ is the temperature coefficient of resistivity for the taken T_{RT} reference temperature.

The source currents in the coil are due to the imposition of boundary conditions, where it is possible to have an imposed current or an imposed voltage [29]. In our case, the voltage $V(t)$ is imposed through an equation depending on the charging voltage V_0, the resistance R_{Gen}, the inductance L_{Gen} and capacity C_{Gen} of the generator, as well as the mesh resistance and inductance (Figure 10). Dirichlet and Newman boundary conditions are applied, and no further constraint is applied on the BEM [30].

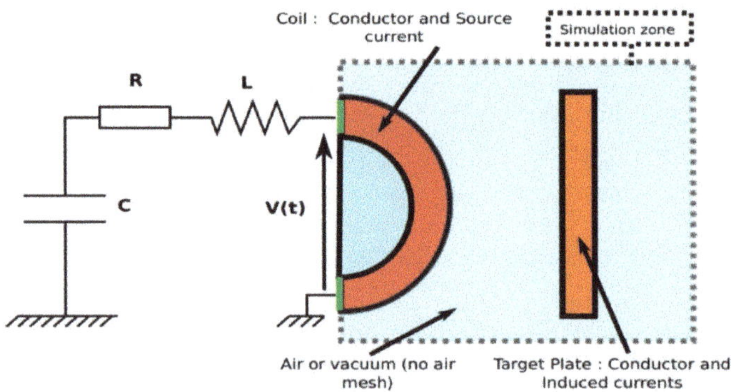

Figure 10. Problem with (R, L, C) imposed voltage on the coil [28].

The inductor nodes' movement was constrained in all directions (translation and rotation), since in the MPW applications, the coils are fixed in such a way as to prevent their movement. It is the same for the parent metal nodes. The flyer sheet nodes' movement in MPW overlapping case (Figure 11) was blocked outside the distance that represents the distance between the insulators. In the case of the MPSW (Figure 12), and since the hump is the only part that will be accelerated from the flyer, all the other nodes were also constrained.

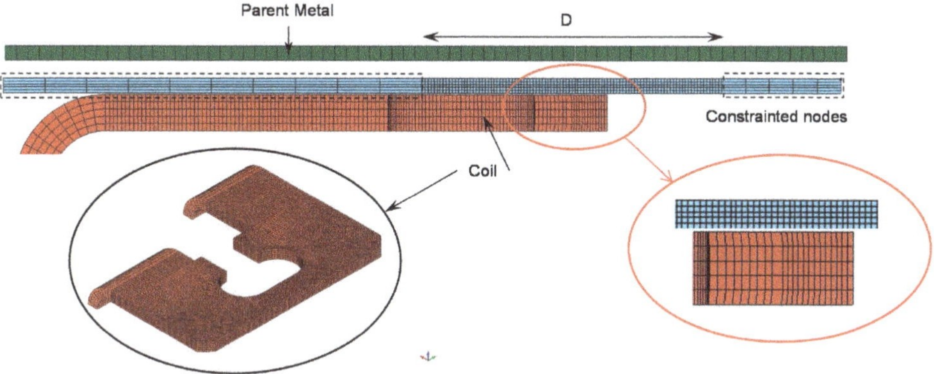

Figure 11. Model in the MPW overlapping configuration with O shape coil.

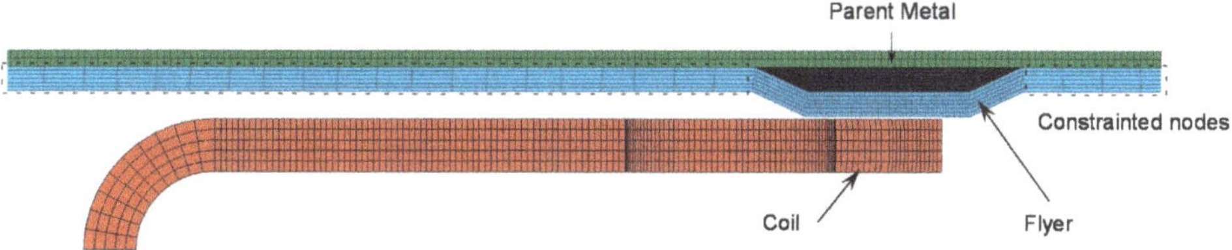

Figure 12. Model cross section view in the MPSW configuration with O shape coil.

2.4. Mesh and Time Step

The electromagnetic solver uses eight nodes' brick elements only, and hence the mesh was controlled to not have any other type of elements in the model. Since we are in a 3D full-coupled simulation, and to avoid very long simulation times, the length of one element edge l_e in the non-active zones was equal to the skin depth δ, and in the active zones it was smaller to have accurate numerical results:

$$l_e = \frac{\delta}{3} \tag{12}$$

The time-step Δt for the electromagnetic solver is computed as the minimal elemental diffusion time step over the elements [30]

$$\Delta t = \frac{\mu_0 \sigma}{2} l_e^2 \tag{13}$$

where μ_0 is the magnetic permeability of the material and σ its electrical conductivity.

3. Results and Discussion

3.1. Numerical Comparison of Linear and O-Shape Coils Efficiencies

The presence of a current-carrying conductive metal in a time-varying magnetic field generates Laplace forces on this metal. The Laplace force acting on an elementary volume dV is expressed in (14). In the MPW case, the flyer metal is the current-carrying conductive where the mentioned current is induced due to the time-varying magnetic field generated by the discharged current in the coil. Consequently, and in accordance with the following Equation (14),

$$\vec{dF} = \vec{J} \times \vec{B}\, dV \tag{14}$$

The forces generated are directly proportional to the induced current density in the flyer metal and the magnetic induction generated by the discharge current, which are also related, since the latter is the source of the former.

Going back to the Laplace force equation, the O-shape coil design presented in Figure 13 is based on the idea of improving efficiency by having higher discharge currents and induced currents at the same levels of energies based on:

- Reducing the inductance of the coil to increase the peak currents (3);
- Putting more charge carriers on the flyer metal in movement to increase the induced currents magnitude and increasing, at the same time, the current density in the area where the material should be accelerated.

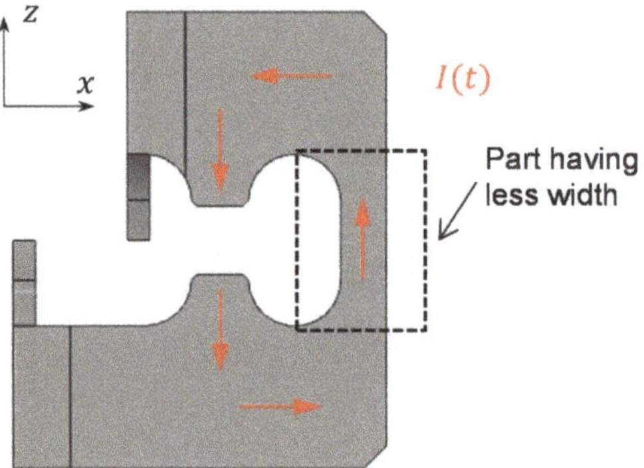

Figure 13. O-Shape rectangular cross-section.

The measured inductance of the linear coil presented a value equal to 81% of the generator's inductance L_{Gen}, giving a system total inductance of

$$L_1 = 1.81 L_{Gen} \tag{15}$$

The measured inductance of the O-shape coil presented a value equal to 49% of the generator's inductance L_{Gen}, giving a system total inductance of

$$L_2 = 1.49 L_{Gen} \tag{16}$$

Comparing the inductances, the system with the O-shape had a 10% higher peak current and frequency.

$$L_2 = 0.82 L_1 \tag{17}$$

$$\frac{I_{02}}{I_{01}} = \frac{\omega_2}{\omega_1} = 1.1 \tag{18}$$

To compare the efficiency of the O-shape vs. the linear coil, the developed numerical model was used to compare the flyer velocities in the MPW configuration case for the thick aluminum 5754-H111 ($e_f = 1.2$ mm). Since the discharge currents are higher in the O-shape coil case, the mechanical strength of the coil should be higher, and hence two approaches were considered. In the first, the material of the coil is kept as OFHC copper, but the thickness was increased to 4 mm (i.e., 1.4 times the thickness in the linear coil case). The second approach involves the use of a 4-mm thickness coil made of steel ASTM

A36, which has better mechanical properties, and seeing whether the loss in electrical conductivity will have a high influence on its global efficiency.

Figure 14 presents a comparison between impact velocities for the cases of use of an OFHC copper O-shape and a linear OFHC copper coil. The velocities at $E = 10$ kJ in the O-shape coil case are much higher than these at $E = 16$ kJ in the linear case. The velocity in the O-shape case is, on average, 1.75 times the velocity in the linear coil shape.

When using a steel O-shape coil, the velocities are 1.55 times the velocities in the linear coil case. The velocities' curves at different standoff distances in both cases are presented in Figure 15.

The data presented in Figures 14 and 15 show clearly that the O-shaped coil is far better than the linear coil and, even when steel is used for the coil, it still has higher efficiency than the linear coil.

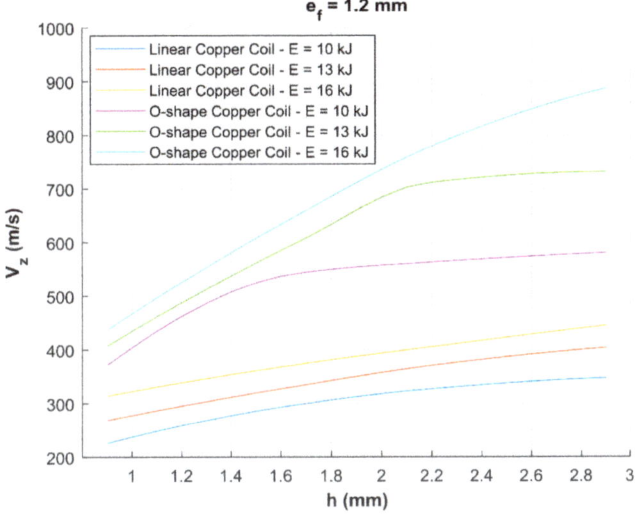

Figure 14. Impact velocity comparison between the OFHC O-shape and linear coils ($e_f = 1.2$ mm).

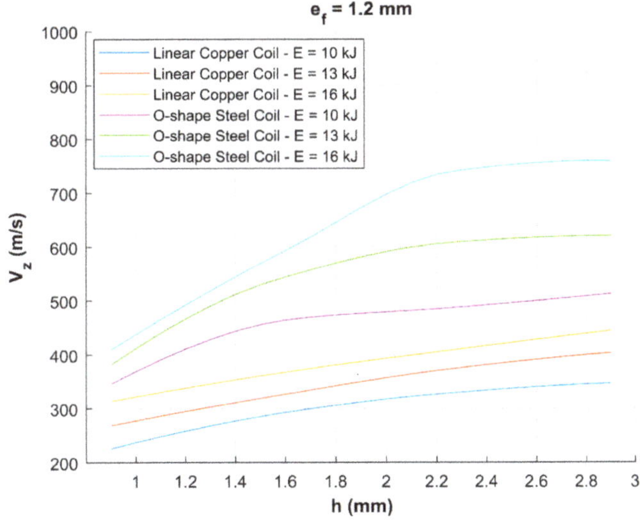

Figure 15. Impact velocities comparison between the steel O-shape coil and linear copper coil ($e_f = 1.2$ mm).

3.2. Experimental Validation

The first step in the experimental analysis was to explore the weldability between various metal sheets with various mechanical properties and thicknesses, using both coils in the same range of discharge energies (4 kJ $\leq E_{discharge} \leq$16 kJ). The configuration used is the MPSW and the hump dimensions (Figure 8) are: $l_h = 8$ mm; $L_h = 12$ mm; $l'_h = 10$ mm, $L'_h = 18$ mm, and $h_h = 1.6$ mm.

Table 10 summarizes the results for different combinations. The limit of the linear coil can be observed when flyer metal thickness exceeds 1.2 mm, while the O-shape coil succeeded the welds up to 2 mm.

Table 10. Weldability of different metals alloys using copper linear coil (L) and O shape coil (✗: not welded/✓: welded/-: not tested).

Parent Flyer	5754		5182		6016		DP450		DP1000		DC04	
	L	O	L	O	L	O	L	O	L	O	L	O
5754 (0.5 mm)	✓	✓	✓	✓	✓	✓	✓	✓	✓	✓	✓	✓
5182 (1 mm)	✓	✓	-	-	-	-	✓	✓	-	-	✓	✓
5182 (1.2 mm)	✓	✓	-	-	-	-	✓	✓	-	-	✓	✓
5182 (1.4 mm)	✗	✓	-	-	-	-	✗	✓	-	-	✗	✓
5182 (2 mm)	✗	✓	-	-	-	-	✗	✓	-	-	✗	✓
6013 (1.4 mm)	-	-	-	-	-	-	-	-	✗	✓	✗	✓
6016 (1 mm)	✓	✓	-	-	-	-	-	-	-	-	✓	✓

3.3. Mechanical Testing for the Welding Joints

Now that the weldability of a variety of combinations is proven using the O-shape coil for thick metal sheets, the strength of welds will be tested. As the focus is the dissimilar welding for automotive applications, five combinations are tested in quasi-static, dynamic, and fatigue lap-shear configuration:

- Aluminum 5182 ($e_f = 1.2$ mm) with steel DC04 ($e_p = 0.8$ mm) in MPSW configuration;
- Aluminum 6016 ($e_f = 1$ mm) with steel DC04 ($e_p = 0.8$ mm) in MPSW configuration;
- Aluminum 5182 ($e_f = 1.2$ mm) with steel DP450 ($e_p = 1.17$ mm) in MPSW configuration;
- Aluminum 6013-T4 ($e_f = 1.4$ mm) with steel DP1000 ($e_p = 1$ mm) in MPSW configuration;
- Aluminum 6013-T4 ($e_f = 1.4$ mm) with steel DP1000 ($e_p = 1$ mm) in MPW configuration.

For the five combinations, the same steel O-shape coil and the same discharge energy ($E_{discharge} = 16$ kJ) are used.

For the MPSW configurations, the same hump dimensions are used: $l_h = 12$ mm; $L_h = 20$ mm; $l'_h = 40$ mm, $L'_h = 55$ mm, and $h_h = 1.6$ mm.

For the MPW configuration, the distance between insulators is $D = 18$ mm and the standoff distance is $h = 1.4$ mm.

In each case, three specimens are tested, and the results are presented in Table 11 for quasi-static lap-shear tests, in Table 12 for dynamic lap-shear tests and in Table 13 for fatigue tests.

For 6016/XES dynamic tests, the failure occurred at the aluminum fixation holes at 9000 N without any failure in the welding for all the tested specimens.

The 5182/DC04 weld shows, in the quasi-static lap shear tests, maximum loads between 6250 and 6750 N, with a displacement between 1.62 and 1.7 mm. The failure was in the weld and a deformation of the steel in the welding area was observed. The maximum dynamic loads that the weld attained were between 7900 and 8300 N, with a displacement between 1.38 and 1.53 mm, and the failure occurred in the welding also. Finally, the number of cycles reached during the fatigue tests was between 25,000 and 28,000 cycles, where a tearing is observed in both aluminum and steel plates during the test (Figure 16).

5182/DP450 weld shows in the quasi-static lap shear tests maximum loads between 9200 and 10,000 N and the displacements are between 1.32 and 1.88 mm. The failure occurred in the welds in all the 3 tests. The maximum dynamic loads were between 6100 and 8500 N and the displacements were between 1.36 and 2.9 mm. The fatigue tests showed that the number of cycles is between 21,000 cycles and 24,000 cycles.

Table 11. Quasi-static lap-shear tests

Materials (Flyer/Target)	Welding Configuration	Fmax—Quasi-Static Lap-Shear Test							
		Average	Average	Specimen 1		Specimen 2		Specimen 3	
		F_{max} (N)	ΔL_{max} (mm)	F_{max} (N)	ΔL_{max} (mm)	F_{max} (N)	ΔL_{max} (mm)	F_{max} (N)	ΔL_{max} (mm)
5182/DC04	MPSW	**6500**	**1.63**	6250	1.56	6500	1.7	6750	1.62
5182/DP450	MPSW	**9400**	**1.64**	10,000	1.88	9000	1.32	9200	1.73
6016/DC04	MPSW	**8433**	**2.15**	8500	2.15	8700	2.3	8100	2
6013/DP1000	MPSW	**7800**	**0.98**	8400	1.1	7600	1.06	7400	0.77
6013/DP1000	MPW	**8767**	**0.98**	9500	1.1	9200	0.95	7600	0.9

Table 12. Dynamic lap-shear tests.

Materials (Flyer/Target)	Welding Configuration	Fmax—Dynamic Lap-Shear Test (0.614 m/s)							
		Average	Average	Specimen 1		Specimen 2		Specimen 3	
		F_{max} (N)	ΔL_{max} (mm)	F_{max} (N)	ΔL_{max} (mm)	F_{max} (N)	ΔL_{max} (mm)	F_{max} (N)	ΔL_{max} (mm)
5182/DC04	MPSW	**8067**	**1.46**	8300	1.48	7900	1.53	8000	1.38
5182/DP450	MPSW	**7033**	**1.88**	8500	2.9	6500	1.37	6100	1.36
6016/DC04	MPSW	**>9000**		>9000		>9000		>9000	
6013/DP1000	MPSW	**7667**	**3.85**	9300	2.75	7400	3.6	6300	5.2
6013/DP1000	MPW	**14,233**	**1.7**	14,600	2.32	14,400	1.3	13,700	0.78

Table 13. Fatigue tests results (R = 0; f = 20 Hz).

Materials (Flyer/Target)	Welding Configuration	F_{max} (kN)	Number of Cycles			
			Average	Specimen 1	Specimen 2	Specimen 3
5182/DC04	MPSW	4	**26,333**	26,000	28,000	25,000
5182/DP450	MPSW	5	**22,333**	21,000	24,000	22,000
6016/DC04	MPSW	4.4	**43,000**	44,000	42,000	43,000
6013/DP1000	MPSW	4.4	**31,000**	32,000	30,000	31,000
6013/DP1000	MPW	5	**64,333**	61,000	69,000	63,000

6016/DC04 weld shows in the quasi-static lap shear tests maximum loads between 8100 N and 8700 N and the displacements are between 2 mm and 2.3 mm. The fatigue tests showed number of cycles between 42,000 and 44,000 and tearing was observed in both aluminum and steel sheets (Figure 17). Concerning the dynamic test in this case and for all tested specimens the failure occurred at the aluminum fixation holes at 9000 N without any failure in the welding.

6013/DP1000 welds were tested in both MPW and MPSW configurations.

- In MPW case: the welds showed quasi-static maximum loads between 7600 and 9500 N with displacements between 0.9 and 1.1 mm. The dynamic loads are between 13,700 and 14,600 N and the corresponding displacements are between 0.78 and 2.32 mm. The fatigue tests showed the number of cycles oscillating between 61,000 and 69,000;
- In the MPSW case: the welds showed quasi-static maximum loads in this case were between 7400 and 8400 N and the displacements were between 0.77 and 1.1mm. The dynamic loads attained were between 6300 and 9300 N and the displacements were between 2.75 and 5.2 mm. The fatigue tests showed a number of cycles between 30,000 and 32,000.

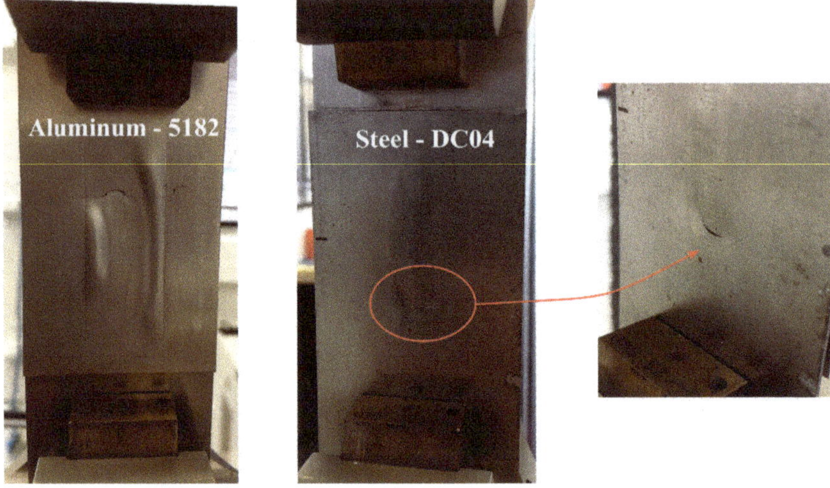

Figure 16. Aluminum 5182 and steel DC04 tearing during fatigue tests.

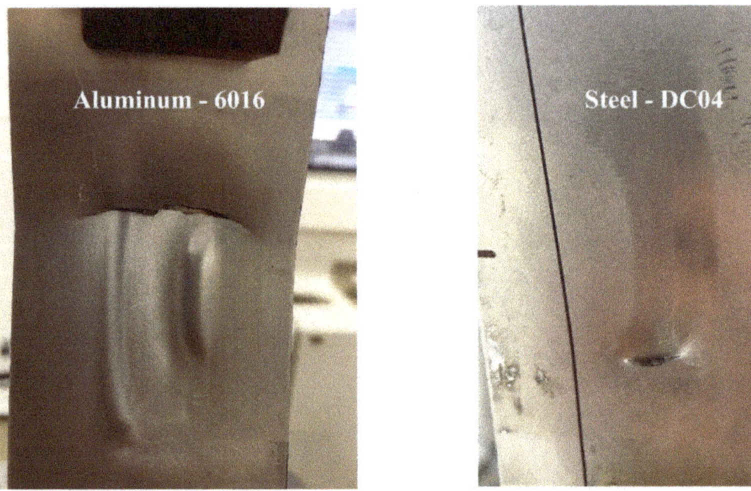

Figure 17. Aluminum 6016 and steel DC04 tearing during fatigue tests.

3.4. Inductor Life Cycle

For heavy duty applications prevalent in automotive industry, the inductor must be capable of sustaining repeated shocks during impulse welding. As mentioned earlier in Section 3, the acceleration was imparted to the flyer metal to produce impact velocities, particularly in the case of hard materials and/or thicker sections, which requires either increasing the discharge current and/or reducing the coil cross-section. Subsequently, the coil will then itself experiment higher thermo-mechanical stresses because of the action–interaction properties of the magnetic forces and the significant temperature increase due to Joule effects. The result will be a faster failure of the coil with increasing discharge energies (Figure 18a) for a linear coil. As previously mentioned, O-shape coils provide higher impact velocities compared to straight coil, but what about the life cycle? Table 14 presents the matrix of tests with different inductors and provides information on the cumulative sustainable and average energy/shot. The data analysis gives some trends with discharge energy and coil shape:

- For a straight conductor, the life cycle is drastically reduced from 50 to 10 shots when the discharge energy is increased from 13 to 16 kJ;
- For O-shape conductors, 50 shots with a discharge energy of 18 kJ are observed. This is the highest limit, but with decreasing discharge energy, a much higher number of weld shots could be seen, but this was not evaluated in real terms.

An additional factor is the coil mounting system that has to maintain the inductor in place by counteracting the magnetic forces experimented by coil due to the action–reaction nature during repeated shocks. The magnetic forces occur not only between flyer and coil, but also within the internal parts of the coil itself. This is explicit in Table 14, which shows the effect of a strong coil hold (polyurethane, Figure 18b or by fiberglass/epoxy resin Figure 1c). As a general observation, polyurethane mount leads to coil distortion after a small number of shots (about five shots), while fiber glass/resin hold composite shows only joule heating (50 shots).

Table 14. Matrix of tests with different inductors (in blue O-shape coil with polyurethane support, in yellow linear coil and in a green O-shape coil with fiberglass/epoxy resin composite support).

Inductor #	Shots Per Energy													Total Shots	Total Cumulative Energy (kJ)	Average Energy Usage
	5	6	7	8	9	10	11	12	13	14	15	16	17	18			
9	0	0	0	0	0	0	0	0	0	0	0	3	0	0	3	48	16
8	0	0	0	0	0	3	0	0	3	0	0	1	0	0	7	85	12
7	0	0	0	0	0	3	0	0	3	0	0	3	0	0	9	117	13
3	0	0	0	0	0	3	0	0	2	0	0	6	0	0	11	152	14
4	0	0	0	0	0	4	0	0	3	0	0	4	0	0	11	143	13
6	0	0	0	0	0	3	0	0	5	1	0	3	0	0	12	157	13
5	0	1	3	3	5	2	1	0	1	0	0	3	0	0	19	188	10
14	0	0	0	0	0	3	0	0	5	0	0	25	0	0	33	495	15
10	0	0	0	1	0	6	0	0	10	3	0	14	0	0	34	464	14
13	0	0	0	0	0	7	0	8	4	7	0	16	0	1	43	590	14
12	0	0	0	0	0	15	0	0	16	0	0	16	0	0	47	614	13
2	5	11	10	8	6	6	0	0	7	0	0	0	0	0	53	430	8

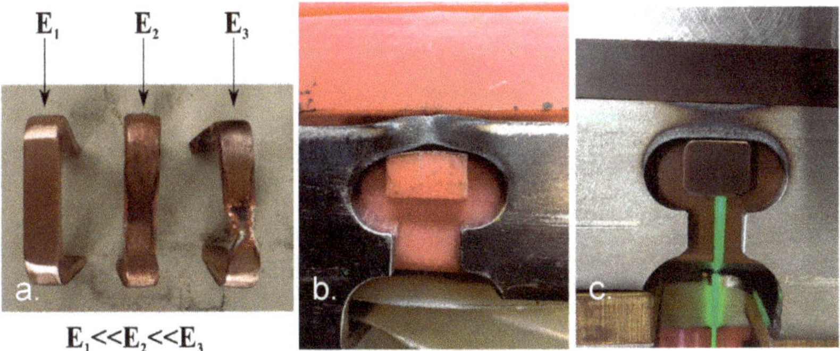

Figure 18. Effect of different levels of thermo-mechanical stresses for: (**a**) linear coil, (**b**) O-shape coil with PU support, (**c**) O-shape coil with fiberglass/epoxy resin composite support.

4. Conclusions

Electromagnetic pulse welding between similar (Al/Al) and dissimilar metals (Al/Steel) with straight (I-shape) and O-shape coils has been investigated using two different layouts, i.e., straight (MPW) or humped (MPSW) for the lap joints. The conclusions presented herein are relevant to maximum discharge energies up to 16 kJ using copper and steel coils, and a flyer–parent standoff between 1 and 3 mm. The numerical simulation of straight and O-shape coils using coupled mechanical/thermal/electromagnetic using LS-DYNA 3D Finite Element Method–Boundary Element Method code (FEM-BEM) is employed as an asset to estimate the effect of coil shape on the welding of different alloyed Al–Steel couples. The conclusions of the numerical model are then validated by experimental production and characterization of lap welds in Al/Al, Al/steel couples of interests in automotive industry. The main conclusions can be summarized as follows:

1. Simulation suggests that O-shape coils have higher process efficiency when measured on impact velocity criterion compared to straight (I-shape) coils for identical process parameters (discharge Energy, flyer–parent gap). Irrespective of the coil material, O-shape coils produce higher impact velocities which are propitious to effective welding. The maximum impact velocities of 550 and 900 m/s are, respectively, estimated for straight and O-shape copper coils. Impact velocity increases with standoff linearly for straight coils, contrary to O-shape coils where a velocity plateau seems to shift towards higher standoffs with increasing discharge energy. For instance, the plateau onset point moves from 1.6 to 2.1 mm when discharge energy goes up from 10 to 13 kJ for copper O-shape coil. At 16 kJ, the plateau onset is not observed for copper O-shape coil within the standoff of 2.8 mm;

2. Compared to steel, copper coil is estimated to generate superior impact velocities (900 m/s for copper O-shape coil vs. 710 m/s for steel O-shape coil at 16 kJ). Copper has a higher electrical conductivity—lower joule losses—but lower mechanical resistance than steel. Joule losses in steel coils end up with coil heating and would require cooling to maintain its stiffness in heavy duty applications as in automotive plants. The disadvantage with copper as coil material stem from the fact that coil gets heavy recoil pressure from the outgoing flyer sheet, which significantly reduces the life cycle of coil. Copper coils' lifetime is reduced to a very few welds as the discharge energy increases;

3. The life cycle of the O-shape coil is significantly higher than that of the straight coil and is subject to how the coil is mounted in its holding system. Though reasonable from a theoretical point of view, this requires further study for a quantitative measure;

4. Higher impact velocities with O-shape coil, derived from numerical calculations, are indirectly validated by the possibility of welding additional couple sheets that were not straight coil at the tested discharge energies. For example, 1.4-mm thick 5182 was welded with 5754, DP450, and DC04; 2-mm thick 5182 with 5754, DP450, DC04, and, finally, 6013 with DP1000. As a general rule, thicker and materials with higher mechanical characteristics require superior discharge energies capable of providing higher impact velocities. When the generator has limited disposable energy, O-shape coil is a preferential choice for electromagnetic pulse welding;

5. Quasi-static, dynamic shear tests along with fatigue test results on different sheet combinations attest a higher resistance of the spot welds as fracture occurred outside the welds, mostly around the corners of the welded spots where applied stresses are concentrated.

The investigations reported here are focused on the application of electromagnetic pulses for the sheet metal welding in automotive applications in view of lightweight hybrid tailored structures, like body parts. The results do indicate a viable process approach and feasibility, as reported here.

Author Contributions: Conceptualization, C.K., S.M., G.R.; Methodology, C.K., G.R.; Software, C.K., G.R.; Validation, C.K., S.M., G.R.; Formal Analysis, C.K., S.M., G.R.; Investigation, C.K., G.R.; Resources, C.K., G.R.; Data Curation, C.K., S.M., G.R.; Writing—Original Draft Preparation, C.K., S.M., G.R.; Writing—Review and Editing, C.K., S.M., G.R.; Visualization, C.K., S.M., G.R.; Supervision, G.R.; Project Administration, G.R. All authors have read and agreed to the published version of the manuscript.

Funding: This research received no external funding.

Acknowledgments: The authors wish to acknowledge Faurecia Group for the financial support to conduct this study which was a part of an innovation program that took place from March 2014 to November 2017.

Conflicts of Interest: The authors declare no conflict of interest.

References

1. Messler, R.W. *Joining of Materials and Structures: From Pragmatic Process to Enabling Technology*; Butterworth-Heinemann: Burlington, MA, USA, 2004.
2. Grote, K.-H.; Antonsson, E.K. *Springer Handbook of Mechanical Engineering*; Springer Science & Business Media: New York, NY, USA, 2009; Volume 10.
3. Gould, J.E. Joining aluminum sheet in the automotive industry—A 30 year history. *Weld. J.* **2012**, *91*, 23–34.
4. Solomon, H.D. Fundamentals of Weld Solidification. In *Welding, Brazing and Soldering*; ASM International: Novelty, OH, USA, 1993; pp. 45–54. ISBN 978-1-62708-173-3.
5. Cieslak, M.J. Cracking Phenomena Associated With Welding. In *Welding, Brazing and Soldering*; ASM International: Novelty, OH, USA, 1993; pp. 88–96. ISBN 978-1-62708-173-3.
6. Wagner, M.; Jahn, A.; Brenner, B. Innovative joining technologies for multi-material lightweight car body structures. In Proceedings of the International Automotive Body Congress (IABC), Dearborn, MI, USA, 29–30 October 2014; pp. 29–30.
7. Kulkarni, N.; Mishra, R.S.; Yuan, W. *Friction Stir Welding of Dissimilar Alloys and Materials*; Butterworth-Heinemann: Oxford, UK, 2015.
8. Matthew Gilloon, C.G. Remote laser welding boosts production of new Ford Mustang. *Ind. Laser Solut.* **2017**, *32*, 28–31.
9. Agudo, L.; Eyidi, D.; Schmaranzer, C.H.; Arenholz, E.; Jank, N.; Bruckner, J.; Pyzalla, A.R. Intermetallic Fe x Al y-phases in a steel/Al-alloy fusion weld. *J. Mater. Sci.* **2007**, *42*, 4205–4214. [CrossRef]
10. Rathod, M.; Kutsuna, M. Use of pulsed CO_2 laser in laser roll bonding of A5052 aluminium alloy and low carbon steel. *Weld. World* **2006**, *50*, 28–36. [CrossRef]
11. Van der Rest, C.; Jacques, P.J.; Simar, A. On the joining of steel and aluminium by means of a new friction melt bonding process. *Scr. Mater.* **2014**, *77*, 25–28. [CrossRef]
12. Blazynski, T.Z. *Explosive Welding, Forming and Compaction*; Springer Science & Business Media: New Yok, NY, USA, 2012.
13. Daehn, G.S.; Lippold, J.; Liu, D.; Taber, G.; Wang, H. Laser impact welding. In Proceedings of the 5th International Conference on High Speed Forming, Dortmund, Germany, 24–26 April 2012.
14. Vivek, A.; Hansen, S.R.; Liu, B.C.; Daehn, G.S. Vaporizing foil actuator: A tool for collision welding. *J. Mater. Process. Technol.* **2013**, *213*, 2304–2311. [CrossRef]
15. Manoharan, P.; Manogaran, A.P.; Priem, D.; Marya, S.; Racineux, G. State of the art of electromagnetic energy for welding and powder compaction. *Weld. World* **2013**, *57*, 867–878. [CrossRef]
16. Bellmann, J.; Lueg-Althoff, J.; Schulze, S.; Gies, S.; Beyer, E.; Tekkaya, A.E. Measurement and analysis technologies for magnetic pulse welding: Established methods and new strategies. *Adv. Manuf.* **2016**, *4*, 322–339. [CrossRef]
17. Shotri, R.; Racineux, G.; De, A. Magnetic pulse welding of metallic tubes—Experimental investigation and numerical modelling. *Sci. Technol. Weld. Join.* **2020**, *25*, 273–281. [CrossRef]
18. Manogaran, A.P.; Manoharan, P.; Priem, D.; Marya, S.; Racineux, G. Magnetic pulse spot welding of bimetals. *J. Mater. Process. Technol.* **2014**, *214*, 1236–1244. [CrossRef]
19. Manogaran, A.P. Développement du Procédé de Soudage par Point par Implusion Magnétique: Assemblage Hétérogène Al/Fe. Doctoral Dissertation, Ecole centrale de Nantes, Nantes, France, 2013.
20. Marya, M.; Marya, S. Interfacial microstructures and temperatures in aluminium—Copper electromagnetic pulse welds. *Sci. Technol. Weld. Join.* **2004**, *9*, 541–547. [CrossRef]
21. Aizawa, T. Magnetic Pulse Welding of Al/Cu Sheets Using 8-Turn Flat Coil. *J. Light Met. Weld.* **2020**, *58*, 97s–101s.
22. Avettand-Fènoël, M.-N.; Khalil, C.; Taillard, R.; Racineux, G. Effect of steel galvanization on the microstructure and mechanical performances of planar magnetic pulse welds of aluminum and steel. *Metall. Mater. Trans. A* **2018**, *49*, 2721–2738. [CrossRef]
23. Mallick, P.K. *Materials, Design and Manufacturing for Lightweight Vehicles*; Elsevier: Amsterdam, The Netherlands, 2010.

24. Johnson, G.R.; Cook, W.H. A constitutive model and data for metals subjected to large strains, high strain rates and high temperatures. In Proceedings of the 7th International Symposium on Ballistics, The Hague, The Netherlands, 19–21 April 1983; Volume 21, pp. 541–547.

25. Seidt, J.D.; Gilat, A.; Klein, J.A.; Leach, J.R. High strain rate, high temperature constitutive and failure models for EOD impact scenarios. In Proceedings of the SEM Annual Conference & Exposition on Experimental and Applied Mechanics, Springfield, MA, USA, 3–6 June 2007; p. 15.

26. Smerd, R.; Winkler, S.; Salisbury, C.; Worswick, M.; Lloyd, D.; Finn, M. High strain rate tensile testing of automotive aluminum alloy sheet. *Int. J. Impact Eng.* **2005**, *32*, 541–560. [CrossRef]

27. Verleysen, P.; Peirs, J.; Degrieck, J. Experimental study of dynamic fracture in Ti6A14V. In Proceedings of the 4th International Conference on Impact Loading of Lightweight Structures (ICILLS 2014), Cape Town, South Africa, 12–16 January 2014.

28. Burgess, T.J. Electrical resistivity model of metals. In Proceedings of the 4th International Conference on Megagauss Magnetic-Field Generation and Related Topics, Santa Fe, NM, USA, 14–17 July 1986.

29. Hallquist, J.O. *LS-DYNA Theory Manual*; Livermore Soft-Ware Technology Corporation: Livermore, CA, USA, 2006.

30. Mamutov, A.V.; Golovashchenko, S.F.; Mamutov, V.S.; Bonnen, J.J. Modeling of electrohydraulic forming of sheet metal parts. *J. Mater. Process. Technol.* **2015**, *219*, 84–100. [CrossRef]

Journal of
*Manufacturing and
Materials Processing*

Article

The Energy Balance in Aluminum–Copper High-Speed Collision Welding

Peter Groche * and Benedikt Niessen

Institute for Production Engineering and Forming Machines (PtU), The Technical University (TU) of Darmstadt, 64287 Darmstadt, Germany; niessen@ptu.tu-darmstadt.de
* Correspondence: groche@ptu.tu-darmstadt.de; Tel.: +49-6151-16-23143

Abstract: Collision welding is a joining technology that is based on the high-speed collision and the resulting plastic deformation of at least one joining partner. The ability to form a high-strength substance-to-substance bond between joining partners of dissimilar metals allows us to design a new generation of joints. However, the occurrence of process-specific phenomena during the high-speed collision, such as a so-called jet or wave formation in the interface, complicates the prediction of bond formation and the resulting bond properties. In this paper, the collision welding of aluminum and copper was investigated at the lower limits of the process. The experiments were performed on a model test rig and observed by high-speed imaging to determine the welding window, which was compared to the ones of similar material parings from former investigation. This allowed to deepen the understanding of the decisive mechanisms at the welding window boundaries. Furthermore, an optical and a scanning electron microscope with energy dispersive X-ray analysis were used to analyze the weld interface. The results showed the important and to date neglected role of the jet and/or the cloud of particles to extract energy from the collision zone, allowing bond formation without melting and intermetallic phases.

Keywords: collision welding; impact welding; welding window; aluminum and copper; high-speed imaging; jet; cloud of particles; energy balance; energy extraction; melting

Citation: Groche, P.; Niessen, B. The Energy Balance in Aluminum–Copper High-Speed Collision Welding. *J. Manuf. Mater. Process.* **2021**, *5*, 52. https://doi.org/10.3390/jmmp5020062

Academic Editor: Steven Y. Liang

Received: 21 May 2021
Accepted: 11 June 2021
Published: 15 June 2021

Publisher's Note: MDPI stays neutral with regard to jurisdictional claims in published maps and institutional affiliations.

1. Introduction

Mitigating climate change with all its consequences requires in comprehensive strategy. This includes a massive reduction in CO_2 emission by the manufacturing industry and its products. One factor to reach this goal is consistent lightweight design. In this context, new joining techniques play an important role by enabling the high-strength joints of multi-material assemblies.

A promising, environmentally friendly joining technology is electromagnetic pulse welding (EMPW), which belongs to the group of collision welding processes, such as explosive welding or foil actuator vaporization [1]. All of these welding techniques are based on the collision of two joining partners, often referred to as moving flyer and stationary target, at high relative velocity (called impact velocity v_{imp}) and a certain collision angle β [2]. During the oblique collision, a collision front moves along the colliding surfaces, which is often described by the collision point velocity v_c. Due to the high impact velocities around $200\,\mathrm{m\,s^{-1}}$ to $600\,\mathrm{m\,s^{-1}}$, extreme conditions occur in the collision front resulting in strain rates of up to $10^6\,\mathrm{s^{-1}}$ and compressive stresses of several GPa [3]. This leads to the formation of a metal stream by the near-surface layers, called jet, which is pushed ahead the collision front. Furthermore, brittle layers of oxides and other surface contaminations are spalled from the strongly deformed surfaces and form a dispersed cloud of particles (CoP) ahead of the collision front [4]. Both phenomena can interact with each other, with the colliding surfaces and also with the ambient gas in the welding gap by entrapment and the stored thermal and chemical energy, as discussed in the literature [5–9]. This interaction determines the possible joining mechanism (solid-state, solid–liquid coexisting state or liquid-state bonding) and depends not only on the selected material and geometry of the joining partners, but on the collision kinetics, which are often described by welding windows [9,10]. As EMPW is a highly transient process, the collision kinetics change continuously. Therefore, it is challenging to predict the weld formation, especially when dissimilar metals have to be joined.

A prominent dissimilar material combination for EMPW is aluminum (Al) and copper (Cu), as it is of high importance for e-mobility. Therefore, several studies have been conducted on this topic in recent years. It was focused on how different process parameters influence the formation of the weld interface with interdiffusion layers, intermetallic layers and resulting defects [11–15] and how these parameters can be optimized to obtain high-quality welds in terms of mechanical and electrical properties [16–19].

However, none of this research addressed bond formation close to the lower limit of the welding process, which could provide a deeper understanding of the mechanisms involved due to the higher sensitivity to the process parameters. Therefore, in this paper, a welding window for aluminum and copper bonds is determined based on previous studies for the same material combination of aluminum [20] and copper in [21] on a model test rig for comparable low-impact energies. The following questions are addressed:

1. At which process parameters can Al–Cu-joints be formed and how do the welding window boundaries differ compared to those of similar material combinations?
2. How does the weld interface develop as a function of the process parameters?

2. Materials and Methods

2.1. Series of Experiments

The experiments of the presented study were performed on a model test rig, see Section 2.2. The samples used were made by laser cutting from sheets of aluminum (EN AW-1050A Hx4, thickness: 2 mm) and copper (Cu EPT, thickness: 1 mm), see material data in Table 1.

Table 1. Used material data of aluminum EN-AW1050A and copper Cu-ETP.

Material	EN-AW1050A	Cu-ETP
Yield strength R_e	105 MPa	202 MPa
Ultimate tensile strength R_m	116 MPa	263 MPa
Density [22] ρ	2.7 g cm^{-3}	8.9 g cm^{-3}
Speed of sound c_{sonic} [7]	6250 m s^{-1}	4660 m s^{-1}
Secific heat capacity c [7]	899 J kg^{-1} K^{-1}	386 J kg^{-1} K^{-1}

For the main study with Al as the pre-bent flyer and Cu as the target, the impact velocity was set in a range of $v_{imp} = 205$ m s^{-1} to 262 m s^{-1} to cover the experimental region of similar material combinations. At each impact velocity, the collision angle was varied (max. spread: $\beta = 2°$ to 20°) to determine the weldable region. In a complementary series, flyer and target material were changed for selected impact velocities ($v_{imp} = 214, 234, 254$ and 262 m s^{-1}). Furthermore, another complementary study of similar combinations of Al and Cu at three impact velocities ($v_{imp} = 214, 234$ and 262 m s^{-1}) was performed and compared with the welding window determination in [20,21] to verify whether the modification of the model test rig, described below, influences the bond formation process.

2.2. Model Test Rig and Process Observation

The model test rig, developed at the PtU, Darmstadt, Germany, allowed the collision welding process to be investigated with constant and precisely adjustable process parameters using a purely mechanical concept. Figure 1a shows its main components: two rotors turning synchronously, each with mounted joining partner samples (colliding area:

length × width: 12 mm × 12 mm). One of the samples was pre-bent to set the collision angle β, as can be seen in Figure 1b. Each sample was accelerated to half the desired impact velocity v_{imp} and collided exactly in the center between the two turning points of the rotors. Since the final impact velocity could not be reached within one revolution, the rotors had to accelerate with an offset. In a former setup, the offset was realized by a phase offset of 45°, which was compensated to initiate the collision [20,23]. After the collision, the samples are torn off at a predetermined breaking point because the rotors could not stop immediately. Due to the high dynamics during the compensation, the rotors and other turning components were geometrically optimized to reduce the mass inertia. Nevertheless, the concept was limited at an impact velocity of 262 m s^{-1}. Furthermore, the rotors showed a shortened lifetime due to the occasional collision with the welded or non-welded samples after the actual welding process [20].

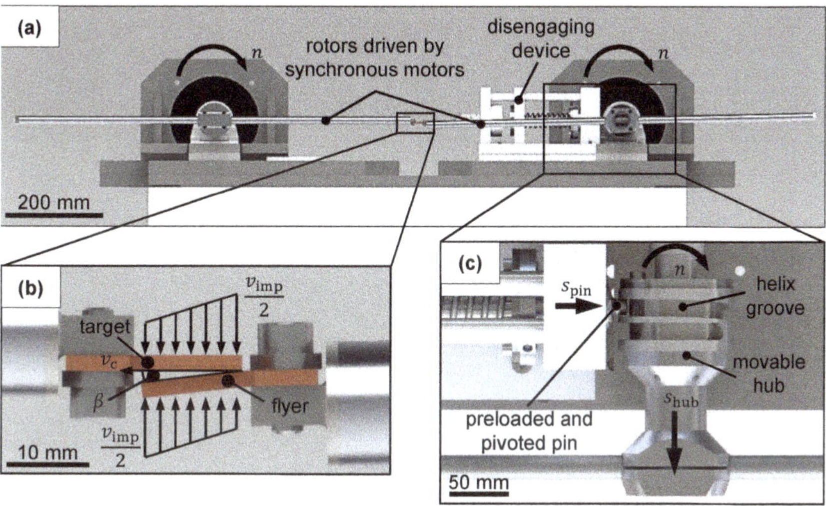

Figure 1. The model test rig (**a**) consisted of two synchronous motor-driven rotors and a disengaging device. A sample (**b**) was attached to one end of each rotor, one of which is pre-bent to determine the collision angle. In the disengaging device a pivoted pin (**c**) is preloaded by an electric holding magnet. When the magnet is switched off, the pin engages the helix groove (s_{pin}) and the entire rotor hub (s_{hub}) is extended within one revolution.

Therefore, a new mechanism for initiating the collision process was developed. The necessary offset of the samples was provided by a movable rotor hub which allowed an axial shift of 15 mm in the direction of the drive shaft during rotor acceleration, as can be seen in Figure 1c. At the desired turning speed, a preloaded, pivoted pin is released and engaged in a helix groove on the hub. Due to the defined motion by the cam gear, the hub's axial offset could be compensated up to impact velocities of 523 m s^{-1}. Moreover, the rotors could be made of solid aluminum alloy to increase the service life.

The model test rig concept provided good process observability. Therefore, a high-speed process observation was implemented by an image intensifier camera *hsfc pro* (by *PCO*, Kelheim, Germany) with a macro lens (*Milvus Makro 100 mm f2* by *Zeiss*, Oberkochen, Germany) and a *CAVILUX Smart* lightning laser (power: 400 W and wavelength: 640 nm, by *Cavitar*, Tampere, Finland). To determine the collision angle β, a *MATLAB* script (version: 2018b, *MathWorks*, Natick, Massachusetts, MA, USA) with edge detection was applied to the high-speed images. More detailed information about the observation and image processing methods can be found in [9,20,21].

2.3. Analysis of the Weld Interface

The influence of the process parameters on the weld interface was determined in two different ways. First, the length of the welded interface was measured in the welding direction in the middle of the welded sample with an optical microscope (*Smartzoom 5* by *Zeiss*, Oberkochen, Germany) and related to the length of the overlapped samples for a variation of impact velocity and collision angle.

Furthermore, the weld interface characteristics at three different sets of process parameters were analyzed using cross-sections. Due to the small thickness of the weld interface zone, the cross-sectional plane was tilted by 80° along the welding direction vector to increase the occurring interlayer. Hence, a perspective distortion had to be considered in the following. The sections were examined with the optical microscope and with a scanning electron microscope (SEM) with energy-dispersive X-ray spectroscopy (EDX) (*Mira* by *Tescan*, Brno, Czech Republic).

3. Results

3.1. Welding Windows

In this section, the welding windows for the different series of experiments are presented. The examined welding windows of Al–Al and Cu–Cu are shown in Figure 2, whereas the examined welding windows of Al as flyer and Cu as target (a) and of Cu as flyer and Al as target (b) are shown in Figure 3. An experiment was defined as bonded, when the two joining partners could not be manually separated.

In order to analyze the results, the following boundaries known from the literature were added to the welding windows. The minimum impact velocity to form a bond is described as surpassing the dynamic elastic limit in the near-surface layers, due to the local pressure from the impact [24]. In this region, metal behaves like a fluid and can form a jet [25,26]. An approximation of the minimal impact velocity normal to the flyer surface can be calculated by an empirical expression of the ultimate tensile strength R_m and the material density ρ [27]:

$$v_\mathrm{p,krit} = \sqrt{\frac{R_\mathrm{m}}{\rho}}. \tag{1}$$

As the controlled impact velocity at the test rig (see Figure 1b) is normal to the target surface, the equation was supplemented by the trigonometric relationship:

$$v_\mathrm{imp,krit} = \frac{v_\mathrm{p,krit}}{\cos\beta}. \tag{2}$$

On the other hand, no jet can be formed, when the collision point velocity exceeds the speed of sound of the material [25,26]. The resulting $\beta_\mathrm{sonic} = f(v_\mathrm{imp})$ was done by Equation (A1) and plotted as the dash dotted line in the welding windows. The dashed line represents the determined upper collision angle boundaries β_max from experiments at a former setup with the same materials [20,21], as can be seen in the welding windows in the Appendix B. The corresponding empirical equations of the boundaries are listed in Appendix A. Both material combinations deviated at higher impact velocities to larger collision angles and showed no clear transition for copper. A closer look at the bonded area at the joints of these experiments revealed that only small regions were welded, which could also be considered as (almost) not welded.

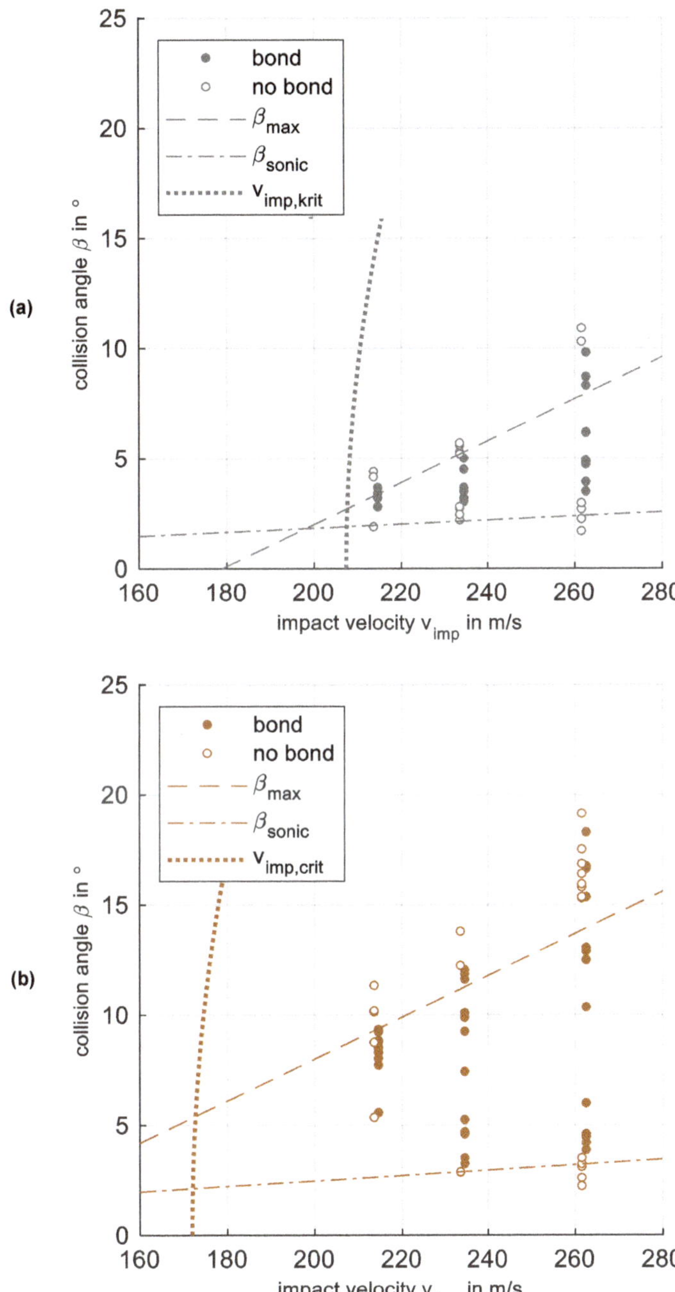

Figure 2. The resulting welding windows for similar combinations of Al–Al (**a**) and Cu–Cu (**b**). To improve the visibility of the data points, welded experiments are plotted with $+0.5\,\mathrm{m\,s^{-1}}$, non-welded with $-0.5\,\mathrm{m\,s^{-1}}$. $v_{\mathrm{imp,krit}}$ is the minimal impact velocity calculated by Equation (2). The subsonic boundary angle β_{sonic} represents the surpassing of the speed of sound in the material by the collision point velocity. The upper collision angle boundary β_{max} was obtained from the previous welding window investigation [21].

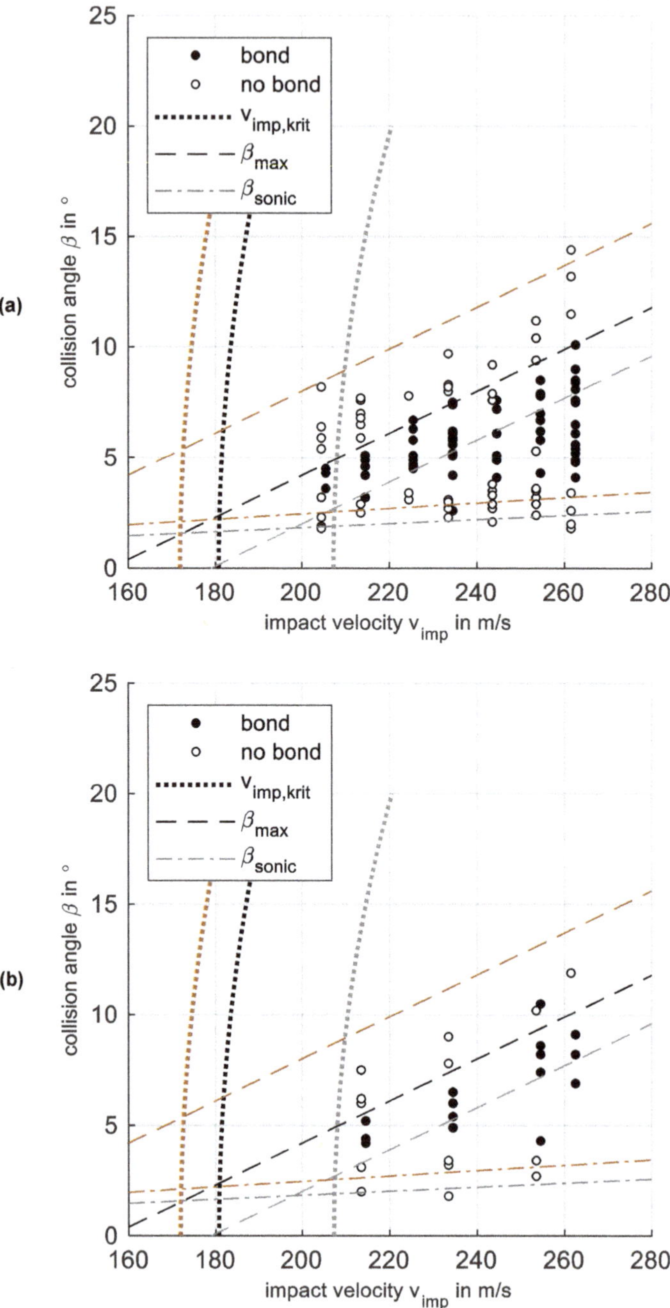

Figure 3. The welding windows of Al as flyer and Cu as target (**a**) and of Cu as flyer and Al as target (**b**). To improve the visibility of the data points, welded experiments are plotted with $+0.5\,\mathrm{m\,s^{-1}}$, non-welded with $-0.5\,\mathrm{m\,s^{-1}}$. The determined boundaries of the welding windows of Al–Al (gray) and Cu–Cu (orange) were added, as can be seen in Figure 2. The upper collision angle boundary β_{max} of Al–Cu (black) was linear approximated as those of the similar material combinations. The minimal impact velocity $v_{\mathrm{imp,krit}}$ of Al–Cu (black) was calculated with Equation (2) and the corresponding mean values of ultimate tensile strength and density of both materials, as can be seen in Section 4.

Regarding the mixed material combination, the minimal collision angles for experiments with a bond were mainly larger than the calculated supersonic limits of both similar material combinations, see Figure 3a. The upper boundary could be linearly approximated with a similar gradient to the former determined values of Al–Al and Cu–Cu, while its ordinate intercept was located between the ones of the similar combinations, as can be seen in Appendix A. Due to the mixed material combination, no minimal impact velocity could be calculated by Equation (2). The complementary series with a changed flyer (Cu) and target (Al) did not show a significant difference considering the determined boundaries, as can be seen in Figure 3b.

3.2. Formation of the Weld Interface

Figure 4 shows the size distribution of the bond formation across the welding window. The variation of the impact velocity (a) exhibited that large areal welded joints could be obtained even for low-impact velocities. Considering the results for different collision angles (b), a plateau of at least 80% welded interface was formed almost over the whole weldable region at an impact velocity of $v_{imp} = 262\,\mathrm{m\,s^{-1}}$. Furthermore, a sharp transition with only small formed welds towards the upper and lower boundary was visible.

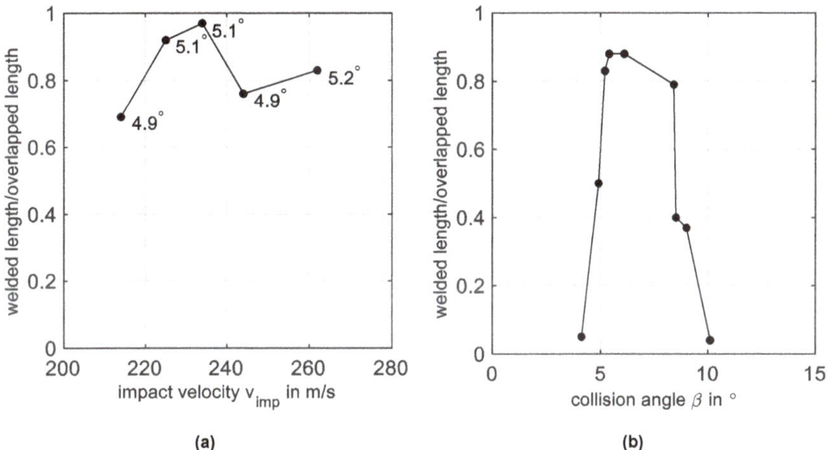

Figure 4. The ratio of welded to overlapped length of the samples for different impact velocities at a constant collision angle of approximately $\beta = 5.0°$ (**a**) and for different collision angles at a constant impact velocity of $v_{imp} = 262\,\mathrm{m\,s^{-1}}$ (**b**).

Regarding the microstructure of the weld, the micro sections revealed that the weld interface strongly differs depending on combination of the process parameters, even if one of them was kept constant, as can be seen in Figure 5. The weld interface of (a) ($\beta = 4.9°$ at $v_{imp} = 214\,\mathrm{m\,s^{-1}}$) was quite steady over the whole length with small and short waves. No transition layer was visible, even by this tilted preparation of the cross-section, as can be seen in Section 2.3. This was confirmed by EDX-analysis, which showed a distinct transition of element distribution at a representative location of the weld interface, as can be seen in Figure 6a.

In contrast, at the same angle but at a higher impact velocity ($262\,\mathrm{m\,s^{-1}}$), the interface exhibited two welded regions separated by a porous and rough non-welded region, as can be seen in Figure 5b. The first welded part showed a thick interlayer with different islands of element distribution, as can be seen in Figure 6b. Such layers were also found by other research groups for this material combination and were identified as intermetallic phases, formed by local melting [28,29]. In the following, the interlayer of the second welded region thinned out more and more until it was barely visible and faded out at the end of the samples.

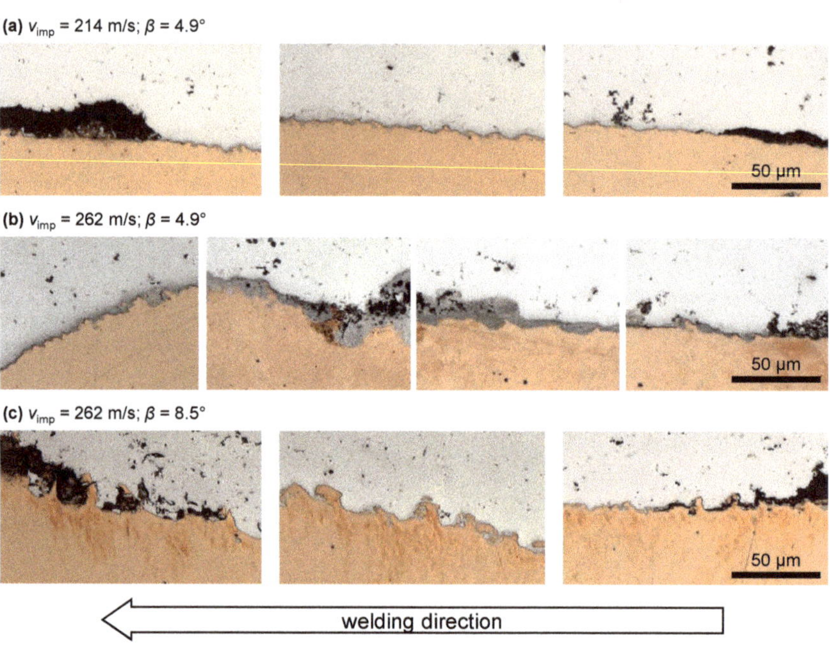

Figure 5. Cross-sections of the weld interface at three different parameter sets. The optical microscopy images show the initiation and the end of the welds and for (**a**,**c**) the middle of the weld interface. The sample in (**b**) was largely not welded but exhibited two welded areas in the second and fourth quarters of the section, the beginnings and ends of which are shown.

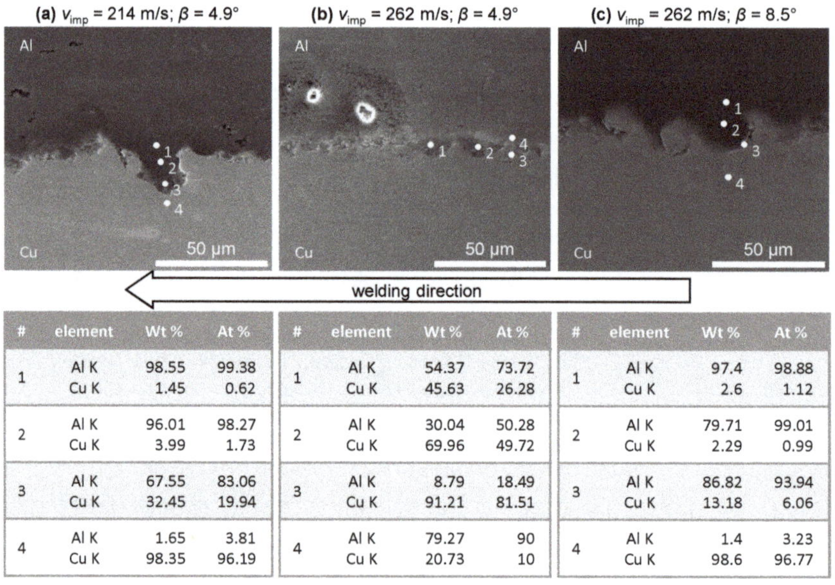

#	element	Wt %	At %	#	element	Wt %	At %	#	element	Wt %	At %
1	Al K	98.55	99.38	1	Al K	54.37	73.72	1	Al K	97.4	98.88
	Cu K	1.45	0.62		Cu K	45.63	26.28		Cu K	2.6	1.12
2	Al K	96.01	98.27	2	Al K	30.04	50.28	2	Al K	79.71	99.01
	Cu K	3.99	1.73		Cu K	69.96	49.72		Cu K	2.29	0.99
3	Al K	67.55	83.06	3	Al K	8.79	18.49	3	Al K	86.82	93.94
	Cu K	32.45	19.94		Cu K	91.21	81.51		Cu K	13.18	6.06
4	Al K	1.65	3.81	4	Al K	79.27	90	4	Al K	1.4	3.23
	Cu K	98.35	96.19		Cu K	20.73	10		Cu K	98.6	96.77

Figure 6. SEM images with EDX analysis at prominent locations of the cross-sections shown in Figure 5.

When the collision angle was increased to 8.5°, another unsteady type of weld interface occurred, as can be seen in Figure 5c. While at the beginning, the interface was similar to (a), it developed large waves with typical vortices and detachments. These waves continued even after the end of the weld, and particles of the mating material adhered to the surfaces as if it had first been welded and then separated. However, along the whole weld interface, no thick interlayer was visible or detected by the EDX.

No residue resulting from jet formation was visible at any of the weld interfaces, possibly due to the limited magnification of the optical microscopy. Moreover, no inclusion of the CoP could be detected by EDX search for oxides, although some of the pores in the interfaces might suggest this, as shown in [12].

4. Discussion

Before discussing the boundaries of the process windows below, it is worth taking a closer look at the size distribution of the weld interface and the definition of bond formation. As shown in Figure 4, the size of the weld interface varied across the welding window and dropped from a plateau to almost zero—close to the collision angle boundaries. This is in good agreement with the former investigation for the similar material combinations [9,30]. Joints in this region may not be relevant for industrial application. However, the definition of a joint in Section 3.1 was chosen to consider the range of process parameters which physically enables bond formation and to investigate the influencing phenomena. In this context, the comparison of the welding windows of Al–Cu joints with the similar material combinations of Al–Al and Cu–Cu reveals a related trend with respect to the lower and upper collision angle boundaries, as can be seen in Figures 2 and 3.

In previous research, the upper collision angle boundary of the welding window has been related to the collision kinetics and the mechanical properties of the materials, the relationship of which influences the formation of a jet in terms of a dense metal stream [9,21]. The jet formation depends on the local surpassing of the dynamic elastic limit of the material by high pressures at the collision zone [24]. The local pressure conditions can be related to the momentum and the collision angle. The momentum depends on the total impact velocity and the inertia of the colliding joining partners. In this context, it is plausible that the upper boundary of the dissimilar combination is located between the two of the similar combinations, which is also visible in the comparison of the ratios of the momentum and the upper boundary angles for Al–Al/Cu–Cu and Al–Cu/Cu–Cu, as can be seen in Table 2.

Table 2. Ratios of the momentum and upper boundary angle $\beta_{\max}$ of Al–Al and Al–Cu related to Cu–Cu. The calculation of the momentum is listed at Appendix C. The upper boundary angle was calculated by the equations in Appendix A.

	Ratio of Momentum	Ratio of $\beta_{\max}$
Al–Al/Cu–Cu	0.60	0.57
Al–Cu/Cu–Cu	0.80	0.73

In this context, a different behavior of jet formation could be expected in the experiments with changed flyer and target due to the reversed inertia in the asymmetric collision setup and assuming that the flyer undergoes most of the plastic deformation. However, no significant deviation of the upper collision boundary was exhibited. An explanation for this could be the results by Kakizaki et al. because they showed by numerical simulation that for dissimilar material combinations with large density difference, such as Al and Cu, the jet was not formed from the flyer material, but was mainly composed of the lower density material [31].

In previous studies, the minimum impact velocity to initiate bond formation for Al–Al was experimentally determined as $205\,\mathrm{m\,s^{-1}}$, which is close to the intersection of the upper boundary line (Equation (A2)) and the supersonic boundary curve (Equation (A1)), see

Figure A1a and [20,21]. Furthermore, the empirical approximation for the minimum impact velocity by Wittman (Equation (1)) is close to this value, calculated as approximately $207\,\mathrm{m\,s^{-1}}$. The minimal impact velocities for Cu–Cu and Al–Cu have not yet been determined experimentally. Both can be estimated by the experimental results as described above. However, for Cu–Cu, the values of the intersections of the upper collision boundary line with the supersonic boundary curve ($132\,\mathrm{m\,s^{-1}}$) and the empirical approximation for the minimum impact velocity by Wittman ($172\,\mathrm{m\,s^{-1}}$) diverge. Thus, experimental verification is needed. For the dissimilar material combination Al–Cu, the intersection of the upper collision angle boundary and the minimal weldable collision angle boundary, which is higher than the supersonic limit of both materials and will be discussed below, occurs at about $185\,\mathrm{m\,s^{-1}}$. Wittman's approach is not defined for a dissimilar material combination, but using the corresponding mean values of ultimate tensile strength and density of both materials delivers a suitable measure with $181\,\mathrm{m\,s^{-1}}$. However, this extended approximation has to be validated for other material combinations.

As for the lower collision angle boundary of the dissimilar material combination, the supersonic boundary was not responsible for the failed bond formation. However, previous studies revealed that the CoP, which is pushed ahead in the joining gap by the collision front, may prevent bond formation at small collision angles. In fact, the high roughness of the colliding surfaces can cause the CoP to be trapped by the collision front, inhibiting the particles from contacting the virgin base material [21,30,32]. However, since the EDX analysis did not detect any oxides, this is not the reason in the present study. However, the formation of large intermetallic phases at the weld interface in Figure 5b and the porous structure indicate excessive melting, which partially prevented bond formation. Recently, a correlation between the extent of the emitted light of the CoP and the temperature in the joining gap was shown [8]. Comparing the conditions in the welding gap close to the lower collision angle boundary, the CoP at the Cu–Cu and the Al–Cu configuration emitted a bright glow, which is not visible for Al–Al at the high-speed images, see Figure 7. The total kinetic energy of the experiments with Cu is higher due to the higher accelerated mass and it can be assumed that higher temperatures were reached in the collision zone for both Cu–Cu and Al–Cu. As the melting temperature of Al is distinctly lower than for Cu, the melting of Al can be expected in the latter configuration. Due to the mechanical concept of the model test rig, an estimation of the temperatures at the weld interface is possible. The following simplifications are assumed:

1. The energy input to the welding process E_{kin} is the sum of the kinetic energies of the joining partners, which is partially stored as potential energy in the material during plastic deformation (about 10–15%) [7]. The remaining amount is transformed to heat, which is conservatively estimated with 80% in the following calculations.
2. This heat is concentrated in the near-surface layers under adiabatic conditions, where the majority of the plastic strain occurs [33].

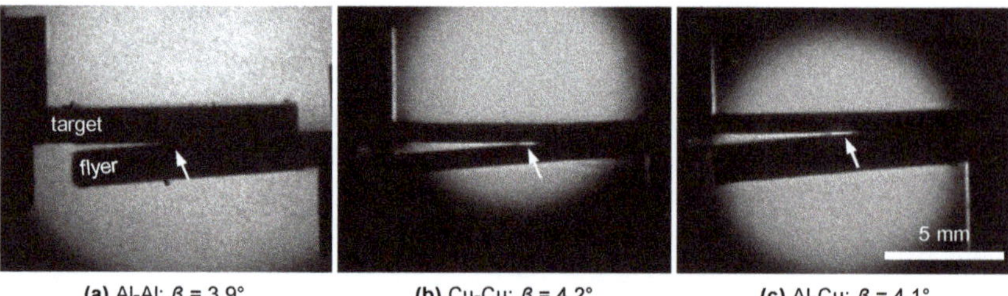

Figure 7. High-speed imaging series of the collision with different material parings at a similar progression point of the collision: Al–Al (**a**); Cu–Cu (**b**); Al–Cu (**c**). All experiments were driven at $v_{\mathrm{imp}} = 262\,\mathrm{m\,s^{-1}}$ and closely above the lower collision angle boundary of Cu–Cu and Al–Cu. The exposure time was 20 ns for all experiments. The arrows indicate the CoP.

Based on these assumptions and adapting the equation of heat transfer, the following equation allows to analytically calculate the thickness t_m of a melted layer:

$$t_m = \frac{0.8 \cdot E_{kin}}{l \cdot w \cdot (T_{m,Cu} + 273K) \cdot (c_{Al} \cdot \rho_{Al} + c_{Cu} \cdot \rho_{Cu})}, \tag{3}$$

where l is the length and w is the width of the samples, $T_{m,Cu}$ is the melting temperature of Cu, c_{Al} and c_{Cu} is the specific heat capacities and ρ_{Al} and ρ_{Cu} the densities of Al and Cu. Inserting the corresponding values predicted a melted layer of 9.3 μm for an impact velocity of 214 m s^{-1} and a melted layer of 13.6 μm continuous thickness over the whole weld interface for 262 m s^{-1}.

However, even in the cross-section in Figure 5b, no continuous melted layer of these thicknesses was visible. These findings are in good agreement with the literature [11–16,33]. Li et al. described the formation of intermediate zones, especially waves and vortices, by local melting due to local plastic deformation causing high shear instabilities and an adjacent inter-diffusion layer of 70 nm thickness, formed below the melting point of aluminum combined with ultrahigh heating and cooling rates of about 10^{13} K s^{-1} [14,15]. However, these cited values are from simulations and experiments with process parameters ($v_{imp} = 600$ m s^{-1}, $\beta = 20°$) much higher than those used in the present paper. In another study of this group, the development of the microstructure at the weld interface during EMPW was investigated. For this purpose, the weld interface was divided into different zones by its appearance and in dependence on the collision angle and impact velocity [33]. One region was formed without any trace of melting, although the determined minimal impact velocity was higher than 300 m s^{-1}. Consequently, a considerable amount of energy must be extracted from the weld interface.

Lysak et al. deduced an energy balance for explosive welding, where an energy loss W_{jet} is defined by the kinetic and thermal energy of the jet [34]:

$$W_{jet} = W_{jet,k} + W_{jet,t} = m_{jet} \left(\frac{v_{jet}^2}{2} + c \cdot \Delta T \right), \tag{4}$$

where m_{jet} is the mass of the jet, v_{jet} is the velocity of the jet, ΔT is the temperature of the jet and c is the specific heat capacity of the jet material. However, their calculated contribution to the stored energy in the jet was negligible compared to the large amount of energy induced by the detonation. In contrast, the energy input at the model test rig and also at the EMPW is much lower, which is why the energy loss of the jet could have a greater impact and explain the lower temperature in the collision zone. Since no residuals of the jet in the form of a solid stream could be detected in the cross-section, the CoP, that accumulates during the collision process, see Figure 8, can largely remove heat from the collision front.

Hence, the temperature of the combined mass stream consisting of the jet and/or the CoP can be estimated by rearranging Equation (4) and making the subsequent further assumptions:

1. The mass stream must extract at least enough heat from the collision zone so that the melting temperature of Al ($T_{m,Al}$) is not reached at the interface. The volume of the adiabatic region $V_{interface}$ is estimated as 12 mm · 12 mm · 10 μm, where it is assumed to be symmetric around the interface for both joining partners. The thickness of adiabatic heating is chosen as 10 μm, which is in accordance with the findings of [33]. This assumption results in a heat input to the interface of:

$$Q_{t,interface} <= (c_{Al}\rho_{Al} + c_{Cu}\rho_{Cu}) \cdot V_{interface} \cdot (T_{m,Al} - 293\,K). \tag{5}$$

Considering the micro-sections in Figure 5, the actual thickness of the adiabatic region might be smaller, which is why even more energy would have to be extracted by the jet and/or the CoP.

2. As mentioned before, it is assumed that 80% of the kinetic energy is transformed into heat. Together with the previous estimation, the total energy of the CoP and/or the jet is:

$$W_{\text{CoP/jet}} = 80\% \cdot W_{\text{kin}} - Q_{\text{t,interface}}. \tag{6}$$

3. In previous studies, no loss of mass during the welding of Cu–Cu could be detected by a scale with a resolution of 0.001 g, [21]. This determines the maximum total mass of the jet and the CoP ($m_{\text{CoP/jet}}$) for the calculation.
4. The velocity of the CoP is presumed to be the velocity of the jet and can be calculated by its progression in the high-speed images in Figure 8 and the frame delay, as can be seen in Table 3.
5. Most of the mass stream mainly consists of material with lower density (in this case Al) which is predicted by various simulations in the literature [24,31]. Therefore, the specific heat capacity of Al is used for the calculation of the temperature of the CoP and/or jet.

(a) v_{imp} = 214 m/s; β = 4.9°

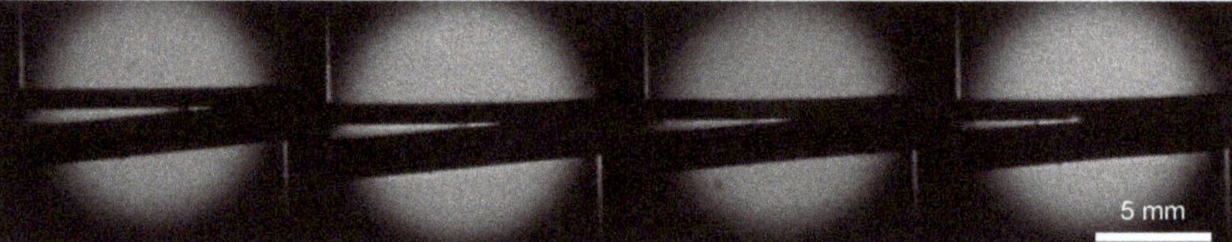

(b) v_{imp} = 262 m/s; β = 4.9°

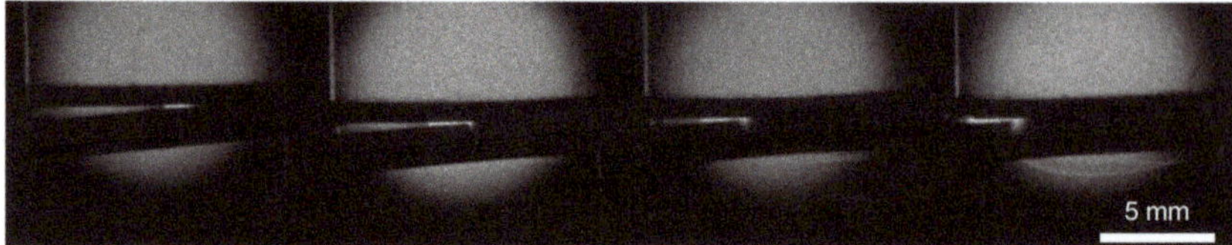

(c) v_{imp} = 262 m/s; β = 8.5°

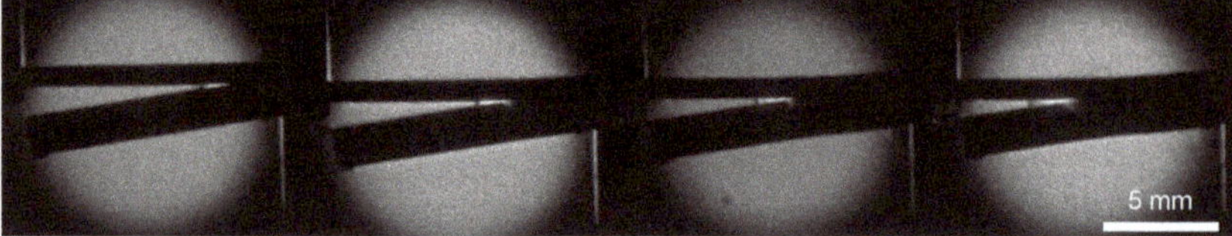

Figure 8. High-speed imaging of three different process parameter sets. The exposure time was 20 ns for all experiments.

All these estimations lead from Equation (4) to the following equation for the temperature of CoP and/or the jet:

$$T_{\text{CoP/jet}} = \frac{W_{\text{CoP/jet}}}{c_{\text{Al}} \cdot m_{\text{CoP/jet}}} - \frac{v_{\text{CoP/jet}}^2}{2 \cdot c_{\text{Al}}} + 293\,\text{K}. \tag{7}$$

Inserting the determined and assumed values delivers the results for the two configurations without melting in the weld interface in Table 4. The temperature for configuration $v_{\text{imp}} = 262\,\text{m s}^{-1}$ and $\beta = 4.9°$ was not calculated as melting largely occurred along the interface, which does not allow us to estimate the maximal temperature at the weld interface.

Table 3. The velocity of the CoP v_{CoP} at different impact velocities v_{imp} and collision angles β. It was calculated by the averaged progression p_{CoP} of pikes of the CoP between two sequenced high-speed images with a frame delay t_{delay} in Figure 8.

v_{imp} in m s^{-1}	β in °	p_{CoP} in mm	t_{delay} in ns	v_{CoP} in m s^{-1}
214	4.9	1.5	520	2836
262	4.9	2.4	600	3990
262	8.5	1.9	600	3165

Table 4. The temperature of the CoP and/or jet $T_{\text{CoP/jet}}$ at different impact velocities v_{imp} and collision angles β calculated by Equation (7).

v_{imp} in m s^{-1}	β in °	$T_{\text{CoP/jet}}$ in K
214	4.9	836
262	8.5	4810

It has to be stated that the calculation above neglects the interaction of the CoP and/or the jet with the ambient medium in the closing gap which may also be heated by the high compression rate. Furthermore, some parts of the CoP and/or the jet may interfere with the surfaces in front of the collision zone which may return heat in the collision zone. However, both interactions are also dependent on the process parameters' collision angle and impact velocity. Regarding the assumptions and simplifications above, the actual temperatures might be even higher. Nevertheless, the determined value for $262\,\text{m s}^{-1}$ corresponds to the regime studied earlier for the CoP at an EMPW configuration of Al and steel, which provided a temperature of 5600 K at a maximal impact velocity of $245\,\text{m s}^{-1}$ [8]. As the temperature was measured by the light emission of the process glare, it can be considered as an average of the temperature development during the unsteady process. In this context, the approach above seems to be a promising extension for the energy balance in collision welding.

5. Conclusions

The following conclusions concerning the welding window boundaries of Al–Cu can be drawn by the findings of the present paper:

- The boundaries of the welding window for the dissimilar material combination largely correspond to those of the welding windows of Al–Al and Cu–Cu.
- The upper boundary of the collision angle of Al–Cu is between the boundaries of the similar material configurations with a ratio that can be related to the total momentum of the collision. However, no dependence of the boundary on the configuration of switched flyer and target material (Cu–Al) was found. A different behavior is expected for a collision weld setup with a moving flyer and stationary target due to different collision kinetics.
- The critical impact velocity of Al–Cu for the initiation of bond formation shows a dependence on the intersection of the upper and lower collision angle boundaries, like the similar material configurations. The modified empirical approximation by Wittman can also be applied for Al–Cu, whose universal applicability has to be validated for other dissimilar material combinations.
- The lower collision angle boundary for Al–Cu is higher than the supersonic limits of both materials and is subject to a different mechanism. The presented findings indicated that excessive melting of Al prevented bond formation at small collision angles due to the higher energy input compared to the configuration of Al–Al.
- Therefore, the energy extraction by the CoP and/or the jet from the collision zone is crucial for prevailing mechanisms in low-energy collision welding. If too much energy remains as heat in the collision zone, melting will occur and influence the weld

interface formation or even prevent permanent bonding. In the future, the adjustment of the parameters will be further investigated in order to define the transition region for melting in the weld interface in dependence on the process and material parameters. This will allow for the more precise prediction of high-strength joints between dissimilar materials without the negative influence of excessive brittle intermetallic phases.

Author Contributions: Conceptualization: P.G. and B.N.; methodology, design of experiments, investigation, data analysis and data curation: B.N.; visualization and writing—original draft preparation, B.N.; writing—review and editing: P.G.; resources, funding acquisition, project administration and supervision: P.G. All authors have read and agreed to the published version of the manuscript.

Funding: The results have been generated within the framework of the research project "Investigation of the formation mechanisms of the bonding zone in collision welding" (GR 1818/49-3) which was part of priority program 1640 "joining by plastic deformation" gratefully funded by the German research foundation (DFG).

Data Availability Statement: Data sharing not applicable.

Acknowledgments: The experiments for the determination of the welding window of Al and Cu were conducted by Kumaran Sutharsan during their Bachelor Thesis under the supervision by Benedikt Niessen. The approach of the heat dissipation in the jet and/or the CoP was initiated during fruitful discussions with Eugen Schumacher from the Department for Cutting and Joining Manufacturing Processes at the University of Kassel. The authors greatly appreciate the effort of Claudia Wasmund and Petra Neuhäusel, Department of Physical Metallurgy at TU Darmstadt, for the sample preparation and SEM and EDX analysis. Furthermore, we would like to acknowledge the help of Stephan Ditscher of Baumüller who supported the programming of the electronic control system of the test rig. Last but not least, we thank Daniel Martin for the review of the calculations and the proofreading of the manuscript and Aaron Zotz and Faraz Mehmood for their support during experiments and analysis.

Conflicts of Interest: The authors declare no conflict of interest.

Abbreviations

The following abbreviations are used in this manuscript:

Al	aluminum EN-AW1050A
CoP	cloud of particles
Cu	copper Cu-ETP
EDX	energy-dispersive X-ray spectroscopy
EMPW	electromagnetic pulse welding
PtU	Institute for Production Engineering and Forming Machines
SEM	scanning electron microscope

Appendix A. Material Data and Equations for the Calculation of the Boundaries of the Determined Welding Windows

Appendix A.1. Lower Collision Angle Boundary by the Supersonic Limit

The supersonic limit is reached, when the collision point velocity v_c reaches the speed of sound of the material c_{sonic}. The corresponding boundary angle β_{sonic} is calculated by the trigonometric relationship from Figure 1b.

$$\tan \beta = \frac{v_{imp}}{v_c} \Leftrightarrow \beta_{sonic} = \arctan \frac{v_{imp}}{c_{sonic}} \tag{A1}$$

Appendix A.2. Upper Boundary Collision Angle

Aluminum–Aluminum

$$\beta_{\text{ub,AlAl}}(v_{\text{imp}}) = 0.095\,^{\circ}\text{s}\,\text{m}^{-1} \cdot v_{\text{imp}} - 17^{\circ} \tag{A2}$$

Copper–Copper

$$\beta_{\text{ub,CuCu}}(v_{\text{imp}}) = 0.095\,^{\circ}\text{s}\,\text{m}^{-1} \cdot v_{\text{imp}} - 11^{\circ} \tag{A3}$$

Aluminum–Copper

$$a_{\text{ub,AlCu}}(v_{\text{imp}}) = 0.095\,^{\circ}\text{s}\,\text{m}^{-1} \cdot v_{\text{imp}} - 14.8^{\circ} \tag{A4}$$

Appendix B. Complementary Welding Windows

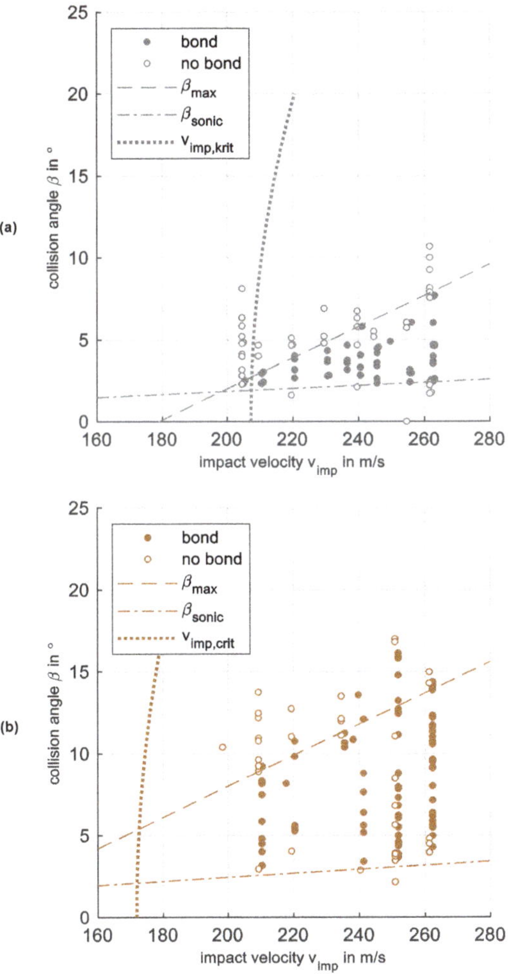

Figure A1. The welding windows for similar combinations of Al–Al (**a**) from [20] and Cu–Cu (**b**) from [21]. To improve the visibility of the data points, welded experiments are plotted with $+0.5\,\text{m}\,\text{s}^{-1}$, non-welded with $-0.5\,\text{m}\,\text{s}^{-1}$. $v_{\text{imp,krit}}$ is the minimal impact velocity calculated by Equation 2. The subsonic boundary angle β_{sonic} represents the surpassing of the speed of sound in the material by the collision point velocity. The upper collision angle boundaries β_{max} were linear approximated.

Appendix C. Calculation of the Momentum

$$p_{\text{AlAl}} = 2 \cdot m_{\text{Al}} \cdot v_{\text{imp}} \tag{A5}$$

$$p_{\text{CuCu}} = 2 \cdot m_{\text{Cu}} \cdot v_{\text{imp}} \tag{A6}$$

$$p_{\text{AlCu}} = m_{\text{Al}} \cdot v_{\text{imp}} + m_{\text{Cu}} \cdot v_{\text{imp}} \tag{A7}$$

References

1. Kapil, A.; Sharma, A. Magnetic pulse welding: An efficient and environmentally friendly multi-material joining technique. *J. Clean. Prod.* **2015**, *100*, 35–58. [CrossRef]
2. Carpenter, S.H.; Wittman, R.H. Explosion Welding. *Annu. Rev. Mater. Sci.* **1975**, *5*, 177–199. [CrossRef]
3. Stern, A.; Shribman, V.; Ben-Artzy, A.; Aizenshtein, M. Interface Phenomena and Bonding Mechanism in Magnetic Pulse Welding. *J. Mater. Eng. Perform.* **2014**, *23*, 3449–3458. [CrossRef]
4. Deribas, A.A.; Zakharenko, I.D. Surface effects with oblique collisions between metallic plates. *Combust. Explos. Shock Waves* **1974**, *10*, 358–367. [CrossRef]
5. Khaustov, S.V.; Kuz'min, S.V.; Lysak, V.I.; Pai, V.V. Thermal processes in explosive welding. *Combust. Explos. Shock Waves* **2014**, *50*, 732–738. [CrossRef]
6. Koschlig, M.; Raabe, D.; Veehmayer, M. Production of Steel-Light Metal Compounds with Explosive Metal Cladding. In Proceedings of the 3rd International Conference on High Speed Forming, Dortmund, Germany, 11–12 March 2008; Institut für Umformtechnik-Technische Universität Dortmund: Dortmund, Germany, 2008. [CrossRef]
7. Pabst, C. *Ursachen, Beeinflussung, Auswirkungen Sowie Quantifizierung der Temperaturentwicklung in der Fügezone beim Kolisionsschweißen*; Vol. Band 119, Berichte aus Produktion und Umformtechnik; Shaker Verlag: Düren, Germany, 2019.
8. Bellmann, J.; Lueg-Althoff, J.; Niessen, B.; Böhme, M.; Schumacher, E.; Beyer, E.; Leyens, C.; Tekkaya, A.E.; Groche, P.; Wagner, M.F.X.; et al. Particle Ejection by Jetting and Related Effects in Impact Welding Processes. *Metals* **2020**, *10*, 1108. [CrossRef]
9. Niessen, B.; Schumacher, E.; Lueg-Althoff, J.; Bellmann, J.; Böhme, M.; Böhm, S.; Tekkaya, A.E.; Beyer, E.; Leyens, C.; Wagner, M.F.X.; et al. Interface Formation during Collision Welding of Aluminum. *Metals* **2020**, *10*, 1202. [CrossRef]
10. Cui, J.; Ye, L.; Zhu, C.; Geng, H.; Li, G. Mechanical and Microstructure Investigations on Magnetic Pulse Welded Dissimilar AA3003-TC4 Joints. *J. Mater. Eng. Perform.* **2020**, *212*, 8. [CrossRef]
11. Raoelison, R.N.; Sapanathan, T.; Buiron, N.; Rachik, M. Magnetic pulse welding of Al/Al and Al/Cu metal pairs: Consequences of the dissimilar combination on the interfacial behavior during the welding process. *J. Manuf. Process.* **2015**, *20*, 112–127. [CrossRef]
12. Ayaz, M.; Khandaei, M.; Vahidshad, Y. Evaluating the Electromagnetic Welding Parameters for Improving the Mechanical Properties of Al–Cu Joint. *Arab. J. Sci. Eng.* **2020**. [CrossRef]
13. Pourabbas, M.; Abdollah-zadeh, A.; Sarvari, M.; Movassagh-Alanagh, F.; Pouranvari, M. Role of collision angle during dissimilar Al/Cu magnetic pulse welding. *Sci. Technol. Weld. Join.* **2020**, *25*, 549–555. [CrossRef]
14. Li, J.S.; Raoelison, R.N.; Sapanathan, T.; Hou, Y.L.; Rachik, M. Interface evolution during magnetic pulse welding under extremely high strain rate collision: Mechanisms, thermomechanical kinetics and consequences. *Acta Mater.* **2020**. [CrossRef]
15. Li, J.S.; Raoelison, R.N.; Sapanathan, T.; Zhang, Z.; Chen, X.G.; Marceau, D.; Hou, Y.L.; Rachik, M. An anomalous wave formation at the Al/Cu interface during magnetic pulse welding. *Appl. Phys. Lett.* **2020**, *116*, 161601. [CrossRef]
16. Kwee, I.; Psyk, V.; Faes, K. Effect of the Welding Parameters on the Structural and Mechanical Properties of Aluminium and Copper Sheet Joints by Electromagnetic Pulse Welding. *World J. Eng. Technol.* **2016**, *04*, 538–561. [CrossRef]
17. Psyk, V.; Scheffler, C.; Linnemann, M.; Landgrebe, D. Process analysis for magnetic pulse welding of similar and dissimilar material sheet metal joints. *Procedia Eng.* **2017**, *207*, 353–358. [CrossRef]
18. Psyk, V.; Linnemann, M.; Scheffler, C. Experimental and numerical analysis of incremental magnetic pulse welding of dissimilar sheet metals. *Manuf. Rev.* **2019**, *6*, 7. [CrossRef]
19. Psyk, V.; Hofer, C.; Faes, K.; Scheffler, C.; Scherleitner, E. Testing of magnetic pulse welded joints – Destructive and non-destructive methods. In Proceedings of the 22nd International ESAFORM Conference on Material Forming, Gasteiz, Spain, 8–10 May 2019; Galdos, L., Ed.; AIP Conference Proceedings; AIP Publishing LLC: Melville, NY, USA, 2019; p. 050010. [CrossRef]
20. Groche, P.; Becker, M.; Pabst, C. Process window acquisition for impact welding processes. *Mater. Des.* **2017**, *118*, 286–293. [CrossRef]
21. Groche, P.; Niessen, B.; Pabst, C. Process boundaries of collision welding at low energies. *Mater. Werkst.* **2019**, *50*, 940–948. [CrossRef]
22. Bargel, H.J.; Schulze, G., Eds. *Werkstoffkunde: Mit 85 Tabellen*; VDI-Buch, Springer: Berlin, Germany, 2005.

23. Groche, P.; Wagner, M.X.; Pabst, C.; Sharafiev, S. Development of a novel test rig to investigate the fundamentals of impact welding. *J. Mater. Process. Technol.* **2014**, *214*, 2009–2017. [CrossRef]
24. Akbari Mousavi, A.A.; AL-Hassani, S.T.S. Numerical and experimental studies of the mechanism of the wavy interface formations in explosive/impact welding. *J. Mech. Phys. Solids* **2005**, *53*, 2501–2528. [CrossRef]
25. Walsh, J.M.; Shreffler, R.G.; Willig, F.J. Limiting Conditions for Jet Formation in High Velocity Collisions. *J. Appl. Phys.* **1953**, *24*, 349–359. [CrossRef]
26. Cowan, G.R.; Holtzman, A.H. Flow Configurations in Colliding Plates: Explosive Bonding. *J. Appl. Phys.* **1963**, *34*, 928–939. [CrossRef]
27. Wittman, R.H. The influence of collision parameters of the strength and microstructure of an explosion welded aluminium alloy. In Proceedings of the 2nd International Symposium on Use of an Explosive Energy in Manufacturing Metallic Materials, Marianske Lazne, Czech Republic, 9–12 October 1973; pp. 153–168.
28. Raoelison, R.N.; Racine, D.; Zhang, Z.; Buiron, N.; Marceau, D.; Rachik, M. Magnetic pulse welding: Interface of Al/Cu joint and investigation of intermetallic formation effect on the weld features. *J. Manuf. Process.* **2014**, *16*, 427–434. [CrossRef]
29. Wang, K.; Shang, S.L.; Wang, Y.; Vivek, A.; Daehn, G.; Liu, Z.K.; Li, J. Unveiling non-equilibrium metallurgical phases in dissimilar Al-Cu joints processed by vaporizing foil actuator welding. *Mater. Des.* **2020**, *186*, 108306. [CrossRef]
30. Niessen, B.; Groche, P. Weld interface characteristics of copper in collision welding. In Proceedings of the 22nd International ESAFORM Conference on Material Forming, Gasteiz, Spain, 8–10 May 2019; Galdos, L., Ed.; AIP Conference Proceedings; AIP Publishing LLC: Melville, NY, USA, 2019; p. 050018. [CrossRef]
31. Kakizaki, S.; Watanabe, M.; Kumai, S. Simulation and Experimental Analysis of Metal Jet Emission and Weld Interface Morphology in Impact Welding. *Mater. Trans.* **2011**, *52*, 1003–1008. [CrossRef]
32. Emadinia, O.; Ramalho, A.M.; de Oliveira, I.V.; Taber, G.A.; Reis, A. Influence of Surface Preparation on the Interface of Al-Cu Joints Produced by Magnetic Pulse Welding. *Metals* **2020**, *10*, 997. [CrossRef]
33. Li, J.S.; Sapanathan, T.; Raoelison, R.N.; Hou, Y.L.; Simar, A.; Rachik, M. On the complete interface development of Al/Cu magnetic pulse welding via experimental characterizations and multiphysics numerical simulations. *J. Mater. Process. Technol.* **2021**, *296*, 117185. [CrossRef]
34. Lysak, V.I.; Kuzmin, S.V. Energy balance during explosive welding. *J. Mater. Process. Technol.* **2015**, *222*, 356–364. [CrossRef]

Journal of
*Manufacturing and
Materials Processing*

MDPI

Article

A Rapid Throughput System for Shock and Impact Characterization: Design and Examples in Compaction, Spallation, and Impact Welding

K. Sajun Prasad , Yu Mao, Anupam Vivek *, Stephen R. Niezgoda and Glenn S. Daehn

Department of Materials Science and Engineering, The Ohio State University, Columbus, OH 43210, USA;
kasi.4@osu.edu (K.S.P.); mao.154@buckeyemail.osu.edu (Y.M.); niezgoda.6@osu.edu (S.R.N.);
daehn.1@osu.edu (G.S.D.)
* Correspondence: vivek.4@osu.edu

Received: 15 November 2020; Accepted: 8 December 2020; Published: 10 December 2020

Abstract: Many important physical phenomena are governed by intense mechanical shock and impulse. These can be used in material processing and manufacturing. Examples include the compaction or shearing of materials in ballistic, meteor, or other impacts, spallation in armor and impact to induce phase and residual stress changes. The traditional methods for measuring very high strain rate behavior usually include gas-guns that accelerate flyers up to km/s speeds over a distance of meters. The throughput of such experiments is usually limited to a few experiments per day and the equipment is usually large, requiring specialized laboratories. Here, a much more compact method based on the Vaporizing Foil Actuator (VFA) is used that can accelerate flyers to over 1 km/s over a few mm of travel is proposed for high throughput testing in a compact system. A system with this primary driver coupled with Photonic Doppler Velocimetry (PDV) is demonstrated to give insightful data in powder compaction allowing measurements of shock speed, spall testing giving fast and reasonable estimates of spall strength, and impact welding providing interface microstructure as a function of impact angle and speed. The essential features of the system are outlined, and it is noted that this approach can be extended to other dynamic tests as well.

Keywords: impact deformation; vaporizing foil actuator; powder compaction; spallation; inclined collision welding

---

1. Introduction

Equation of state (EOS) describes the states and thermodynamic properties of the material, and its measurement is of immediate interest in military, automobile, aerospace, and other several manufacturing applications [1]. Numerous approaches have been developed for obtaining the EOS of a variety of materials through dynamic shock wave compression [2], static compression [3] and static coupled with dynamic compression techniques [4]. To investigate material behavior under dynamic loading, dedicated facilities such as gas guns, rail guns, shock guns, electrothermal gun accelerators, laser driven flyer impacts and even explosives have been used effectively [5–10]. The present work attempts to formalize and standardize new techniques that can make dynamic studies more accessible and improve throughput dramatically. Vivek et al. [11] have developed the Vaporizing Foil Actuator (VFA) technique for many impulse and impact processes, such as impact welding [12,13], forming and embossing [14], shearing [15], and high-energy rate metalworking [16]. In this method, the pressure created from the electrically driven rapid vaporization of a thin aluminum (Al) foil is used to generate an immense transient pressure that can be used to launch projectiles to velocities well in excess of 1 km/s. Flyer speeds can be accurately measured with a laser-based technique, Photonic Doppler Velocimetry (PDV), the working principles for which can be found elsewhere [17]. In the present work, three critical dynamic material phenomena are shown as examples of the utility of this system: powder compaction, spallation, and impact welding.

Dynamic powder compaction (DPC) can reach higher densities than static compaction [18] and offers near-net-shape part fabrication by which subsequent finishing operations are minimized or eradicated, resulting in cost-effective manufacturing [19]. All consolidation processes including conventional hydraulic pressing, forging, hot/cold isostatic pressing, triaxial compaction [20] retain residual porosity, which during the sintering process leads to dimensional changes. Several DPC techniques using explosive compaction, gun-type units, hydraulic impact, spring-loaded hammers, and electromagnetic impulse units have been reported to improve the green density of compacts [21]. Green compacts with low defects and uniform density are essential for the fabrication of a high strength with little and uniform shrinkage. For DPC, higher densification is ascribed to the passage of the shock wave through the powder when the impact source hits the powder mass. Sethi et al. [22] demonstrated that DPC yields better green strength of the compact compared to the conventional compaction technique. Wang et al. [23] carried out DPC by high-energy impact delivered by a hammer with a mass of 135 kg traveling at speeds of 7–10 m/s. The result showed that the green density and bending strength of compacts increase with an increase in impact velocity due to an increase in shock wave pressure. Sometimes, multiple impacts have also been used to increase green density [24,25] In the past, particular interest has been given to DPC of metallic powders using explosives [26,27]. Though very high densities have been achieved, there have been issues with reflected shock waves producing cracking after initial consolidation [26,27].

Spallation is one of the significant modes of material failure during high-velocity impact. During impact, the interaction of the incident and reflected shock waves inside the material causes an internal fracture and separation of the outer surface. Knowledge of the spall strength of the material is essential in designing armor, military vehicles, and aircraft structures [28–30]. In this regard, Forbes [31] indicated that the Hugoniot Elastic Limit (HEL) is dependent on the thickness of the specimen and ultimately affects the spall strength of the component. Apart from these, it was concluded that the spall strength depended on several factors, including the processing route of material, anisotropy of the material and directional distribution of the second phases. In recent years, significant advances have been made to generate high tensile stress at high strain rates, which led to the generation of the database for known strategic materials [32–34]. Copper has been used as a common model material to study incipient spall strength and its correlation to microstructure and crystal orientation [35,36]. Hence, considering the vast amount of data available, the polycrystalline copper material (Cu110) was chosen for the present work.

Impact welding is a high strain rate, solid-state joining process. It is complex and involves high pressure and high local strains. These cause removal of the initial surfaces by jetting and the high pressure produces intimate contact. High transient temperatures can also cause melting and the formation of brittle intermetallic compounds. Furthermore, the process can cause wave formation dominated by fluid dynamics. Several groups have studied these phenomena and supporting microstructural representation can be found in the literature by Lee et al. [37]. Kuzmin and Lysak [38] experimentally proved that successful collision welds depend not only on the impact velocity but also on the collision angle. Zhang et al. [39] reported that successful collision welds are generally obtained when the collision velocity is between 150 and 1500 m/s and the collision angle is between 5° and 20°. With optimal welding parameters, melting can be favorably suppressed and localized or eliminated, creating minimal to no intermetallic formation. Metals that are difficult or impossible to join using conventional welding methods are therefore able to be joined using collision welding technique. In this regard, different welding technologies viz. explosive welding [40,41], magnetic pulse welding [42], underwater explosion [43] and laser impact welding [44] have been extensively used in the past, and the limitations of these techniques are mentioned by Ngaile et al. [45]. Recently, vaporizing foil actuator welding (VFAW) techniques have been extensively used to weld the target with varying oblique angles for a different class of materials including AA1100-O to AISI1018 [46], 15-5

PH SS [47], Cu110 to CP-Ti [48,49], OFHC copper to Grade 2 Ti [50] and many more combinations of exotic materials. A versatile tool based on the vaporizing foil actuator is described here, and the first results from the aforementioned three experiments are presented.

2. Materials and Methods

In the present work, the VFA technique assisted projectile acceleration and the impact drove impulse-based dynamic characterization experiments in powder compaction, spallation, and collision welding of inclined target samples. A similar approach using electric gun technology at Lawrence Livermore Laboratory discusses problems of interest to shock-wave researchers [51,52]. For this purpose, a chamber for dynamic processing and characterization was designed and fabricated at the Impulse Manufacturing Laboratory at The Ohio State University (http://iml.osu.edu). A schematic CAD model and fabricated chamber with details are depicted in Figure 1. The chamber is equipped with an automated pneumatic control pedestal system to transmit and control energy. The base structure of the operational level inside of the chamber is electrically insulated using Garolite G-10 material (glass-fiber-reinforced epoxy) material. The chamber has ports to allow entry to focusing probes from a time-domain multiplexed 16-channel PDV system. This PDV system is built with standard arrays of 4×4 probes that allow measurement of workpiece velocity, position, and acceleration and its essential construction is described in greater depth elsewhere [17]. The 16-channel PDV system allows for a more accurate and higher resolution measurement of deformation fields generated by VFA launch or impact.

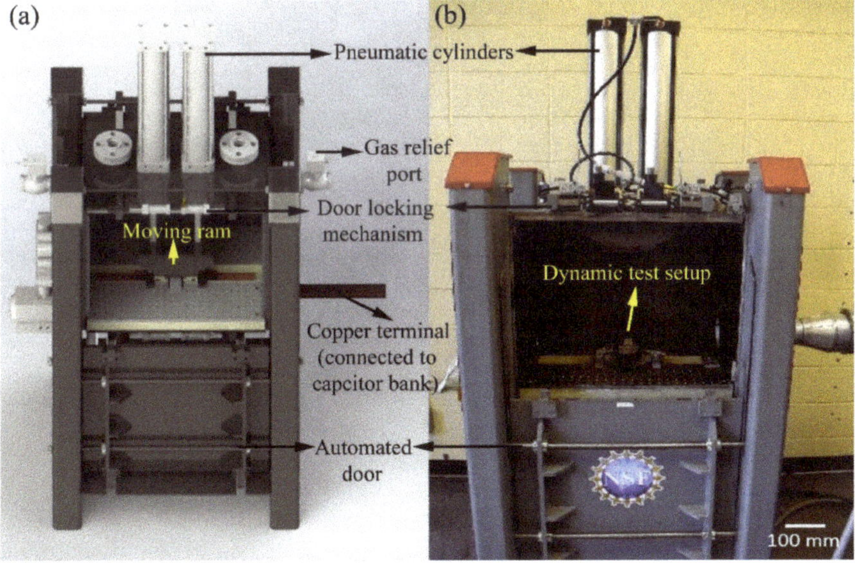

Figure 1. (**a**) A rendered CAD-model schematic and (**b**) actual photo of the chamber.

The measurement of impactor speed is carried out separately from material characterization experiments. Figure 2a depicts the schematic of the dynamic experimental setup with the PDV probe for the measurement of impact velocity. A tool steel (S7 steel) barrel with a total length of 30 mm and an inner bore diameter of 12.8 mm was used to house the projectile. Aluminum (Al 6061-T6) projectiles with 12.5 mm diameter with different lengths of 10 mm and 5 mm were used to understand its effect on the impact velocity. Additionally, a composite projectile consisting of a 1-mm-thick steel impactor (AISI 1018) backed by a 4-mm-thick aluminum puck (Al 6061-T6) was used in order to increase the impact pressure. An industrial adhesive (J-weld) was used to join the impactor and puck. Furthermore, graphite-based lubricant or motor oil was used to reduce friction from the bore wall and increase projectile speed. A thin piece of polyvinyl chloride (PVC) and an aluminum sheet (6061-T6) of 1 mm was placed at the bottom of the barrel to increase the stability and efficiency of the plasma-driven acceleration of the projectile, separating the projectile from two-layered aluminum foils. Two layers of aluminum foils (Al1100) with a thickness of 0.0762 mm were connected to the copper terminals. The copper terminals are connected to the terminals of the capacitor bank.

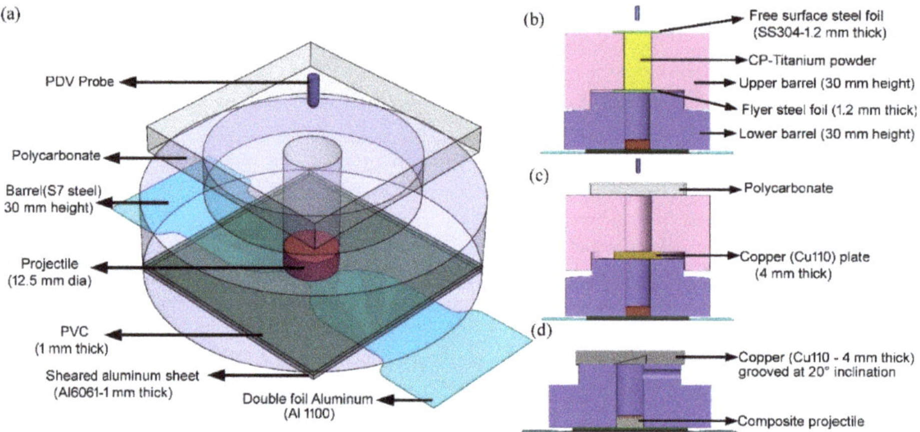

Figure 2. A schematic showing: (**a**) the experimental setup for the measurement of projectile velocity using the Photonic Doppler Velocimetry (PDV) probe, (**b**) powder compaction of commercial pure titanium (CP-Ti), (**c**) spalling of copper (Cu110) plate, and (**d**) 20° inclined collision welding between copper (Cu110) and steel (AISI 1018).

For all experiments, a current with an amplitude around 200 kA was rapidly discharged from the capacitor bank through the terminals to the aluminum foils, causing nearly instant vaporization of the foil, as has been detailed elsewhere [11,53]. The generated impulse plasma pressured and sheared the PVC and aluminum sheets through the barrel, accelerating the projectile to a high velocity traveling along the barrel. The input energy was set to be 14 kJ. The capacitor bank is a 16-kJ commercial Magneform system with a total capacitance of 462 μF, short circuit current rise time of 12 μs, an inductance of 100 nH, and resistance of 10 mΩ. All the conductive parts of the fixture were insulated with Kapton^{TM} tape to avoid alternate current paths other than through the foil. A piece of transparent polycarbonate was placed at the top of the barrel, and a steel backing block with a hole pattern was positioned on top of the polycarbonate to secure and protect the PDV probe. A 25 mm focuser probe was used to measure the projectile velocity and a collimator was used to measure the velocity of free surfaces.

After determining projectile velocity, the same experimental setup was used for powder compaction, spallation, and collision welding of inclined target samples as shown in Figure 2b–d, respectively. For carrying out powder compaction experiments, commercially pure titanium (CP-Ti) powder was filled inside the upper barrel as depicted in Figure 2b. Detailed material description of the used CP-Ti powder can be found elsewhere [15]. Thin steel foil (SS304) just above the lower barrel was used to hold the powder, and the steel foil (SS304) at the top was used to measure the free surface velocity of the compacted powder. For spallation experiments, a 4 mm thick copper target (Cu110) was kept at the top of the lower barrel and backed by an upper barrel as presented in Figure 2c. Polycarbonate was kept at the top of the barrel to capture the spalled fragments. In order to study impact welding, a specific angle of 20° was machined into a 4mm thick copper target plate as shown in Figure 2d. This inclined target provides the collision angle required for impact welding. Finally, the PDV data collected was exported to MATLAB to analyze the speed resulting from initial projectile velocity and subsequent experiments. The voltage and current traces from the corresponding experiments were also collected using a 1000:1 BK PR28A voltage divider and a 50 kA: 1 V Rogowski coil, respectively.

3. Result and Discussion

3.1. Effect of Projectile Configuration on Impact Velocity

As discussed earlier, VFA experiments using different projectile lengths and masses were used to understand their effect on impact velocity. The output of time-based evolution of current, voltage, and velocity using an aluminum projectile with 10 mm length at 14 kJ energy input is shown in Figure 3a. The voltage profile showed a sudden pop-in trace around 15 µs, and the instance is usually referred to as the shock burst out time. This kink in voltage is due to foil vaporization into high-pressure plasma [15]. Subsequently, an increase in voltage and a decrease in current was observed. At the same instance, the projectile started to gain its momentum, and the projectile approached a stagnation velocity of 310 m/s at about 40 µs. The velocity trace of different projectile configuration with 14 kJ input energy is shown in Figure 3b. It can be observed that the impact velocities were 310 m/s, 387 m/s and 372 m/s for an aluminum projectile with a 10 mm length, 5 mm length, and a composite projectile, respectively. Similarly, it is evident from the figure that the higher mass in the composite projectile led to a lower velocity compared to the 5 mm aluminum projectile. Repeatability was confirmed by testing two specimens at each condition.

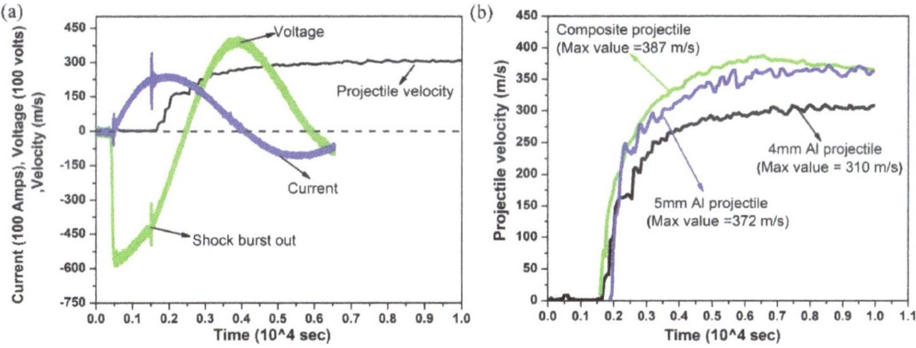

Figure 3. (**a**) Time-based evolution of voltage, current and velocity using an aluminum projectile with a 10 mm length, and (**b**) velocity comparison with different projectile configurations.

3.2. Effect of Lubrication on Projectile Velocity

Graphite-based lubricant or motor oil was applied between projectile and barrel surfaces to reduce the sliding friction. Prior to each experiment, the barrel was cleaned with extreme care using acetone to avoid any external resistance from the scale produced from the previous experiment. It was observed that the usage of different lubrication had no mitigating difference in impact velocity. However, motor oil was used for further analysis due to consistency in results. Figure 4 presents the comparison of projectile velocity for both unlubricated and lubricated cases. The maximum impact velocity using lubricated 5 mm aluminum and the composite projectile showed an approximate increase by 2.5 and 1.7 times with respect to the unlubricated condition. Lubricated conditions had a significant effect on impact velocity compared to experiments without lubrication. Repeatability was ensured by carrying out two experiments at each conditions. Henceforth, projectiles with motor oil lubrication were considered for the different case studies discussed in the subsequent sections.

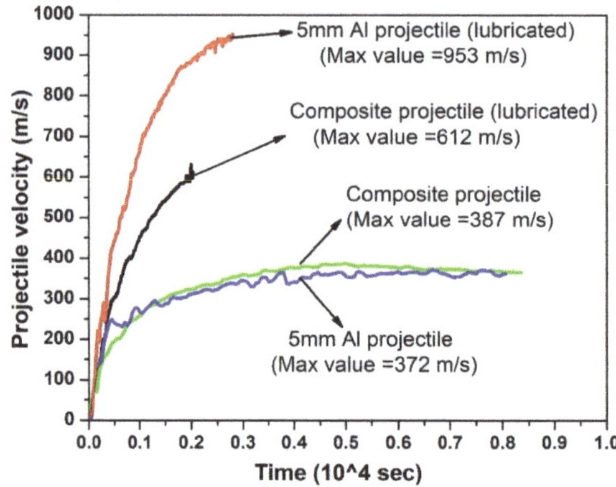

Figure 4. Graph showing the comparison of projectile velocity using lubrication and without lubrication for 5 mm aluminum and composite projectiles.

3.3. Case Studies

3.3.1. Powder Compaction of CP-Ti

An example of using this system for powder compaction and its relevant diagnostics follow. The CP-Ti powder was loaded into the 30 mm barrel and found to be compressed to 11.4 mm and 8.7 mm lengths when the 5 mm aluminum and the composite projectiles was used, respectively. The corresponding compacts are shown in Figure 5a. The setup data and results of the two experiments are consolidated in Table 1. Subsequently, the densities of both of the compacts were determined using the Archimedes principle. It can be seen that the compact obtained using the composite projectile showed a higher relative density of 88% compared to the 74% dense compact using the 5 mm aluminum projectile. The higher densification is postulated to be because a higher pressure is imparted into the powder body by the composite projectile which as a steel tip, which has a higher shock impedance [54]. A similar result was found for compacting CP-Ti powder using a VFA assisted sheared flyer [15].

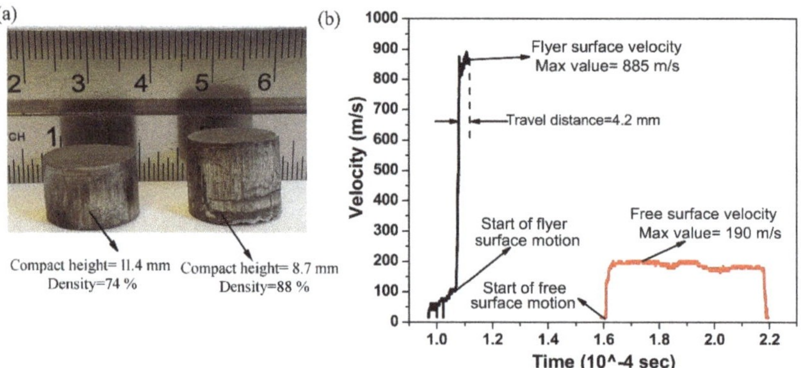

Figure 5. (**a**) Powder compaction results in terms of compact height (mm), relative density (kg/m^3), and microhardness (HV) and (**b**) flyer and free surface velocity for estimation of shock velocity.

Table 1. Experimental setup parameters and the result obtained using different projectile configurations.

Projectile Configuration	Initial Tap Density (g/cm³)	Plunger Impact Speed (m/s)	Plunger Kinetic Energy (J)	Final Density (g/cm³)
5 mm Al projectile	1.32	953	777	3.33
5 mm composite projectile	1.32	612	442	3.96

DPC involves propagation of a compressive shock wave through a powder to cause its densification as mentioned by Vogler and coworkers [18,55,56]. The work details planar shock experiments on porous powders consolidation on a stepped die provided by a gun-driven impactor. The experiment is further identified with shock arrival time measurements on the sample impact and rear surfaces using velocity interferometry (VISAR) probes to obtain an accurate measurement of shock velocity and, consequently, determine the EOS or shock Hugoniot state. A similar approach using PDV was adopted to estimate flyer velocity, free surface velocity, and shock velocity through composite projectile impact as depicted in Figure 5b. The maximum flyer velocity and free surface velocity were found to be around 885 m/s and 190 m/s, respectively. The reduction in velocity is due to the energy expended for powder compaction. The movement of the flyer starts at a time of 1.06×10^{-4} s, whereas the free surface motion was encountered around 1.6×10^{-4} s. From this, the average shock speed was estimated to be approximately 555 m/s. By measuring the shock velocity, the impact pressure can be determined for a material for which EOS is known [57,58]. Moreover, by repeating this procedure for different impact velocities and measuring shock pressures using gauges such as manganin [59] and attenuator material [60], empirical for EOS can be determined for unknown material.

3.3.2. Spallation in Copper

In the present work, the copper sample resulting in spall with a large spalled top surface with one piece along with several smaller shards was collected as presented in Figure 6a. Figure 6b shows the PDV velocity traces of the composite projectile and free surface near the shock wave breakout. The impactor speed before colliding with the target was is estimated at 612 m/s. The resulting initial breakout speed of the spalled sample was found to be approximately 527 m/s. The spall speed being lower than impactor speed is expected given the lower density of the steel impactor compared to the copper target. The voltage traces obtained from the experiments confirm the initiation of impactor movement and spall. The velocity profile shows the clear spall signatures that consist of a release into tension represented by velocity pullback; furthermore, the occurrence of spall at the velocity minimum as presented in Figure 6b. In the same instance, recompression waves propagate in both directions from near the spall plane denoted by a small spike in the velocity profile. Subsequently, recompression appears as a velocity increase after spall, and it is followed by ringing in the spall scab. The deceleration in the curve observed after the first shock denotes continuing damage evolution until the spall scab completely separates from the bulk copper plate.

The data collected from the spall experiment was used to estimate the spall strength as per the commonly used procedure [33,61] and Equation (1).

$$\sigma_{spall} = \frac{\rho C (V_{max} - V_{Spall-max})}{2} \tag{1}$$

where ρ denoted the density of copper (8950 kg/m³) and C is bulk sound speed in copper material (4600 m/s). The strain rates were calculated by applying a classical acoustic approximation as given by Equation (2) [62].

$$\dot{\varepsilon} = \frac{0.5 (V_{max} - V_{Spall-max})}{C \Delta t} \tag{2}$$

where Δt is the time difference between V_{max} and $V_{Spall-max}$, and was obtained from the velocity spectrogram plot as shown in Figure 6b. The average spall strength for the 4 mm thick copper sample was estimated to be approximately 2.26 GPa and the average strain rate was around -2×10^4 s^{-1}. The obtained spall strength and the average strain rate was in the range mentioned by Turley and coworkers [33,61]. Extensive studies by Remington et al. [62] concluded that decreasing sample thickness results in higher strain rates by almost 10 times and an increase in spall strength by 30% as compared to that of its counterparts. Furthermore, materials shocked by a high power laser show a rapid increase in the spall strength with the strain rate at about 10^7 s^{-1} [63]. Thus, a dependence of spall strength on strain rate and sample thickness for an unknown material can be established using this tool.

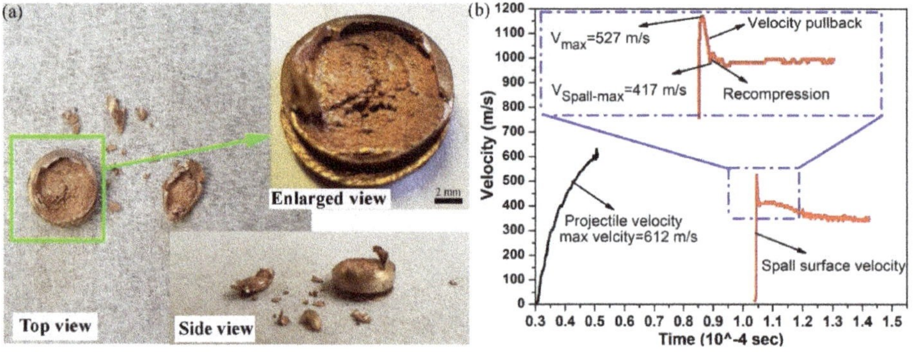

Figure 6. (a) Copper target showing large spalled top surface with one piece along with several smaller shards, and (b) velocity traces of composite projectile and copper free surface near shock wave breakout.

3.3.3. Collision Welding of Inclined Target

For collision welding, the lubricated composite projectile was launched at 14 kJ input energy level with an impact speed of 612 m/s (refer Section 3.2) to the copper target. The kinetic energy of steel tip of composite projectile was estimated to be 186.7 J. The copper target grooved with 20° inclination angle is shown in Figure 7a. The steel part became detached from the composite projectile and welded to the copper target as shown in Figure 7b. The detachment of steel from the aluminum might have occurred due to the generated reversal wave that had higher strength than the bonding strength of the material. Furthermore, the material was sectioned along an X–X contour for optical microscopy as shown in Figure 7c. A clear wave profile signifying a strong impact weld is visible even with unaided eyes. A work by Ngaile et al. [45] where a copper flyer mass of 14.82 gm was launched towards steel target using a chemically produced hydrogen energy-based impact technique. A similar wavy weld profile at the steel–copper interface was observed with an impact speed and kinetic energy of 637 m/s and 3000 J, respectively.

Figure 7d shows that an ungapped intimate contact weld interface was obtained over the entire length imaged using optical microscopy. The large consistent waves going along most of the welded interface are a testament to a constant impact angle during the welding process, which has not been previously obtained with VFAW [64]. The waviness of the interface starts a few mm after the start of welding. The weld's bonding strength can be estimated using the micro-tensile test [65]. A similar phenomenon of delay in wave initiation was observed during explosive welding [66]. Szecket et al. demonstrated that the strong waves are commonly allied with the best-welded joints, although they are not a prerequisite [67].

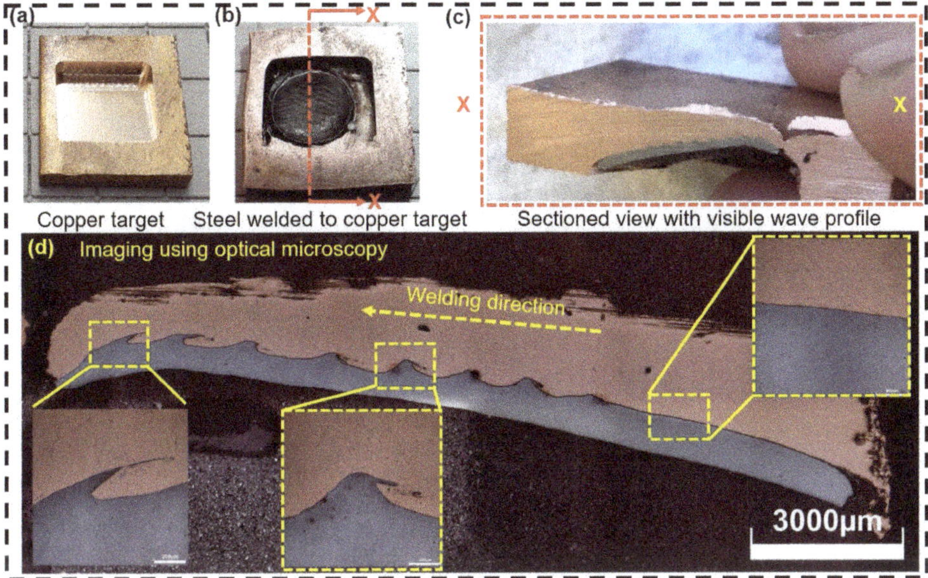

Figure 7. (**a**) Copper target grooved with 20° inclination, (**b**) steel welded to copper target, (**c**) sectional view of the welded part, and (**d**) optical image showing complete welding profile.

4. Summary

This work's objective was to present the new and rapid throughput experimental setup to do shock physics and impulse manufacturing experiments. Three applications, namely powder compaction, spallation, and impact welding experiments, were presented using the proposed facility. The first one deals with dynamic compaction of titanium powder and conventional measurement of free surface velocity to evaluate shock velocity through the compact loaded barrel. In the second part, the spall strength and strain rate of polycrystalline copper were estimated. The third application relates to metallurgical bonding between the steel and copper plate with a 20° inclined groove. The experimental results substantiate the potential of a powerful new tool to study dynamic material behavior and related processes.

Author Contributions: Conceptualization, K.S.P., Y.M., A.V.; methodology, K.S.P., Y.M., A.V.; software, Y.M.; formal analysis, K.S.P., Y.M.; investigation, K.S.P., Y.M.; resources, A.V., G.S.D.; data curation, K.S.P., Y.M., A.V.; writing—original draft preparation, K.S.P.; writing—review and editing, K.S.P., Y.M., A.V., S.R.N., G.S.D.; visualization, A.V., S.R.N., G.S.D.; supervision, A.V., G.S.D.; project administration, G.S.D. All authors have read and agreed to the published version of the manuscript.

Funding: This research received no external funding.

Acknowledgments: The authors are thankful to National Science Foundation (NSF) under a Major Research Instrument grant number: 1531785 for the funding received for carrying out this research work. The authors acknowledge Impulse Manufacturing Laboratory (IML) members for their valuable help in carrying out these experiments.

Conflicts of Interest: The authors declare no conflict of interest.

References

1. Davison, L.; Graham, R.A. Shock compression of solids. *Phys. Rep.* **1979**, *55*, 255–379. [CrossRef]
2. Wang, H.; Zhang, W.; Wan, Y.; Chu, G.; Chen, D. Study on equation of state and spall strength of sintered Nd-Fe-B magnets. *J. Magn. Magn. Mater.* **2020**, *497*, 166014. [CrossRef]

3. Mao, H.; Hemley, R.J. Ultrahigh-pressure transitions in solid hydrogen. *Rev. Mod. Phys.* **1994**, *66*, 671. [CrossRef]

4. Kimura, T.; Ozaki, N.; Okuchi, T.; Mashimo, T.; Miyanishi, K.; Endo, T.; Jitsui, T.; Hirose, A.; Ikoma, M.; Kakeshita, T.; et al. Static compression experiments for advanced coupling techniques of laser-driven dynamic compression and precompression target. *J. Phys. Conf. Ser.* **2010**, *215*, 12152. [CrossRef]

5. Banishev, A.A.; Shaw, W.L.; Bassett, W.P.; Dlott, D.D. High-Speed Laser-Launched Flyer Impacts Studied with Ultrafast Photography and Velocimetry. *J. Dyn. Behav. Mater.* **2016**, *2*, 194–206. [CrossRef]

6. Curtis, A.D.; Banishev, A.A.; Shaw, W.L.; Dlott, D.D. Laser-driven flyer plates for shock compression science: Launch and target impact probed by photon Doppler velocimetry. *Rev. Sci. Instrum.* **2014**, *85*, 043908. [CrossRef]

7. Moshe, E.; Eliezer, S.; Henis, Z.; Werdiger, M.; Dekel, E.; Horovitz, Y.; Maman, S.; Goldberg, I.B. Experimental measurements of the strength of metals approaching the theoretical limit predicted by the equation of state. *Appl. Phys. Lett.* **2000**, *76*, 1555–1557. [CrossRef]

8. Okada, A.; Aso, Y.; Hamada, K.; Yasunaga, K.; Kiritani, M. Ultra-high-speed deformation by impact of a projectile flying at speeds on the order of km s^{-1}. *Mater. Sci. Eng. A* **2003**, *350*, 86–91. [CrossRef]

9. Field, J.E.; Walley, S.M.; Proud, W.G.; Goldrein, H.T.; Siviour, C.R. Review of experimental techniques for high rate deformation and shock studies. *Int. J. Impact Eng.* **2004**, *30*, 725–775. [CrossRef]

10. Han, R.; Zhou, H.; Liu, Q.; Wu, J.; Jing, Y.; Chao, Y.; Zhang, Y.; Qiu, A. Generation of Electrohydraulic Shock Waves by Plasma-Ignited Energetic Materials: I. Fundamental Mechanisms and Processes. *IEEE Trans. Plasma Sci.* **2015**, *43*, 3999–4008. [CrossRef]

11. Vivek, A.; Hansen, S.; Liu, B.C.; Daehn, G.S. Vaporizing foil actuator: A tool for collision welding. *J. Mater. Process. Technol.* **2013**, *213*, 2304–2311. [CrossRef]

12. Kapil, A.; Lee, T.; Vivek, A.; Bockbrader, J.; Abke, T.; Daehn, G. Benchmarking strength and fatigue properties of spot impact welds. *J. Mater. Process. Technol.* **2018**, *255*, 219–233. [CrossRef]

13. Lee, T.; Mao, Y.; Gerth, R.; Vivek, A.; Daehn, G. Civilized explosive welding: Impact welding of thick aluminum to steel plates without explosives. *J. Manuf. Process.* **2018**, *36*, 550–556. [CrossRef]

14. Vivek, A.; Brune, R.; Hansen, S.R.; Daehn, G. Vaporizing foil actuator used for impulse forming and embossing of titanium and aluminum alloys. *J. Mater. Process. Technol.* **2014**, *214*, 865–875. [CrossRef]

15. Vivek, A.; DeFouw, J.D.; Daehn, G. Dynamic compaction of titanium powder by vaporizing foil actuator assisted shearing. *Powder Technol.* **2014**, *254*, 181–186. [CrossRef]

16. Vivek, A.; Daehn, G. Vaporizing Foil Actuator: A Versatile Tool for High Energy-rate Metal Working. *Procedia Eng.* **2014**, *81*, 2129–2134. [CrossRef]

17. Johnson, J.R.; Taber, G.; Vivek, A.; Zhang, Y.; Golowin, S.; Banik, K.; Fenton, G.K.; Daehn, G.S. Coupling Experiment and Simulation in Electromagnetic Forming Using Photon Doppler Velocimetry. *Steel Res. Int.* **2009**, *80*, 359–365. [CrossRef]

18. Vogler, T.; Lee, M.; Grady, D. Static and dynamic compaction of ceramic powders. *Int. J. Solids Struct.* **2007**, *44*, 636–658. [CrossRef]

19. Yamamoto, Y.; Kiggans, J.; Clark, M.B.; Nunn, S.D.; Sabau, A.S.; Peter, W.H. Consolidation Process in Near Net Shape Manufacturing of Armstrong CP-Ti/Ti-6Al-4V Powders. *Key Eng. Mater.* **2010**, *436*, 103–111. [CrossRef]

20. Nagarathnam, K.; Renner, A.; Trostle, D.; Kruczynski, D.; Massey, D. Development of 1000-Ton combustion-driven compaction press for materials development and processing. *Adv. Powder Metall. Part. Mater.* **2007**, *1*, 3.

21. Skoglund, P.; Kejzelman, M.; Hauer, I. High density P/M components by high velocity compaction. *Adv. Powder Metall. Part. Mater.* **2002**, *4–85*. Available online: https://www.hoganas.com/globalassets/download-media/technical-papers/pm/highdensitypmcomponentsbyhighvelocitycompaction.pdf (accessed on 9 December 2020).

22. Sethi, G.; Hauck, E.; German, R. High velocity compaction compared with conventional compaction. *Mater. Sci. Technol.* **2006**, *22*, 955–959. [CrossRef]

23. Wang, J.; Qu, X.; Yin, H.; Yi, M.; Yuan, X. High velocity compaction of ferrous powder. *Powder Technol.* **2009**, *192*, 131–136. [CrossRef]

24. Yi, M.-J.; Yin, H.-Q.; Wang, J.-Z.; Yuan, X.-J.; Qu, X.-H. Comparative research on high-velocity compaction and conventional rigid die compaction. *Front. Mater. Sci. China* **2009**, *3*, 447–451. [CrossRef]

25. Grigoriev, S.N.; Dmitriev, A.M.; Dmitriev, A.M.; Fedorov, S.V. A Cold-Pressing Method Combining Axial and Shear Flow of Powder Compaction to Produce High-Density Iron Parts. *Technologies* **2019**, *7*, 70. [CrossRef]

26. Raming, T.P.; Van Zyl, W.E.; Carton, E.P.; Verweij, H. Sintering, sinterforging and explosive compaction to densify the dual phase nanocomposite system Y_2O_3-doped ZrO_2 and RuO_2. *Ceram. Int.* **2004**, *30*, 629–634. [CrossRef]

27. Johnson, K.; Murr, L.E.; Staudhammer, K. Comparison of residual microstructures for 304 stainless steel shock loaded in plane and cylindrical geometries: Implications for dynamic compaction and forming. *Acta Met.* **1985**, *33*, 677–684. [CrossRef]

28. Lenihan, D.; Ronan, W.; O'Donoghue, P.E.; Leen, S.B. A review of the integrity of metallic vehicle armour to projectile attack. *Proc. Inst. Mech. Eng. Part L J. Mater. Des. Appl.* **2018**, *233*, 73–94. [CrossRef]

29. Li, W.; Yao, X. The spallation of single crystal SiC: The effects of shock pulse duration. *Comput. Mater. Sci.* **2016**, *124*, 151–159. [CrossRef]

30. Montgomery, J.S.; Wells, M.G.H.; Roopchand, B.; Ogilvy, J.W. Low-cost titanium armors for combat vehicles. *JOM* **1997**, *49*, 45–47. [CrossRef]

31. Forbes, J.W. *Shock Wave Compression of Condensed Matter: A Primer*; Springer Science & Business Media: Berlin/Heidelberg, Germany, 2013.

32. Vignjevic, R.; Bourne, N.K.; Millett, J.C.F.; Devuyst, T. Effects of orientation on the strength of the aluminum alloy 7010-T6 during shock loading: Experiment and simulation. *J. Appl. Phys.* **2002**, *92*, 4342–4348. [CrossRef]

33. Turley, W.D.; Fensin, S.J.; Hixson, R.S.; Jones, D.R.; La Lone, B.M.; Stevens, G.D.; Thomas, S.A.; Veeser, L.R. Spall response of single-crystal copper. *J. Appl. Phys.* **2018**, *123*, 055102. [CrossRef]

34. Millett, J.C.F.; Whiteman, G.; Bourne, N.K. Lateral stress and shear strength behind the shock front in three face centered cubic metals. *J. Appl. Phys.* **2009**, *105*, 033515. [CrossRef]

35. Xie, P.; Wang, Y.; Shi, T.; Wang, X.; Hu, C.; Hu, J.; Zhang, F. Damage evolution and spall failure in copper under complex shockwave loading conditions. *J. Appl. Phys.* **2020**, *128*, 055111. [CrossRef]

36. Khomskaya, I.V.; Razorenov, S.V.; Garkushin, G.V.; Shorokhov, E.V.; Abdullina, D.N. Dynamic Strength of Submicrocrystalline and Nanocrystalline Copper Obtained by High-Strain-Rate Deformation. *Phys. Met. Met.* **2020**, *121*, 391–397. [CrossRef]

37. Lee, T.; Nassiri, A.; Dittrich, T.; Vivek, A.; Daehn, G. Microstructure Development in Impact Welding of a Model System. *SSRN Electron. J.* **2019**, *178*, 203–206. [CrossRef]

38. Kuzmin, S.V.; Lysak, V.I. Main regularities of transfer to waveless modes of joint formation in explosive welding. In *Explosive Welding and Properties of Welded Joints*; Inter-Departmental TransactionVolgograd Polytechnic Institute: Volgograd, Russia, 1991; pp. 29–38.

39. Zhang, Y.; Babu, S.S.; Prothe, C.; Blakely, M.; Kwasegroch, J.; Laha, M.; Daehn, G.S. Application of high velocity impact welding at varied different length scales. *J. Mater. Process. Technol.* **2011**, *211*, 944–952. [CrossRef]

40. Malakhov, A.Y.; Saikov, I.V.; Denisov, I.V.; Niyezbekov, N.N. AlMg6 to Titanium and AlMg6 to Stainless Steel Weld Interface Properties after Explosive Welding. *Metals* **2020**, *10*, 1500. [CrossRef]

41. Carvalho, G.; Galvão, I.; Mendes, R.; Leal, R.; Loureiro, A. Aluminum-to-Steel Cladding by Explosive Welding. *Metals* **2020**, *10*, 1062. [CrossRef]

42. Kapil, A.; Sharma, A. Magnetic pulse welding: An efficient and environmentally friendly multi-material joining technique. *J. Clean. Prod.* **2015**, *100*, 35–58. [CrossRef]

43. Manikandan, P.; Lee, J.O.; Mizumachi, K.; Mori, A.R.; Raghukandan, K.; Hokamoto, K. Underwater explosive welding of thin tungsten foils and copper. *J. Nucl. Mater.* **2011**, *418*, 281–285. [CrossRef]

44. Wang, K.; Wang, H.; Zhou, H.; Zheng, W.; Xu, A. Research Status and Prospect of Laser Impact Welding. *Metals* **2020**, *10*, 1444. [CrossRef]

45. Ngaile, G.; Löhr, P.; Lowrie, J.; Modlin, R. Development of chemically produced hydrogen energy-based impact bonding process for dissimilar metals. *J. Manuf. Process.* **2014**, *16*, 518–526. [CrossRef]

46. Vivek, A.; Hansen, S.R.; Benzing, J.; He, M.; Daehn, G. Impact Welding of Aluminum to Copper and Stainless Steel by Vaporizing Foil Actuator: Effect of Heat Treatment Cycles on Mechanical Properties and Microstructure. *Met. Mater. Trans. A* **2014**, *46*, 4548–4558. [CrossRef]

47. Liu, B.; Palazotto, A.; Vivek, A.; Daehn, G.S. Impact welding of ultra-high-strength stainless steel in wrought vs. additively manufactured forms. *Int. J. Adv. Manuf. Technol.* **2019**, *104*, 4593–4604. [CrossRef]

48. Nassiri, A.; Vivek, A.; Abke, T.; Liu, B.; Lee, T.; Daehn, G. Depiction of interfacial morphology in impact welded Ti/Cu bimetallic systems using smoothed particle hydrodynamics. *Appl. Phys. Lett.* **2017**, *110*, 231601. [CrossRef]

49. Vivek, A.; Liu, B.; Hansen, S.R.; Daehn, G. Accessing collision welding process window for titanium/copper welds with vaporizing foil actuators and grooved targets. *J. Mater. Process. Technol.* **2014**, *214*, 1583–1589. [CrossRef]

50. Gupta, V.; Lee, T.; Vivek, A.; Choi, K.S.; Mao, Y.; Sun, X.; Daehn, G. A robust process-structure model for predicting the joint interface structure in impact welding. *J. Mater. Process. Technol.* **2019**, *264*, 107–118. [CrossRef]

51. Osher, J.E.; Barnes, G.; Chau, H.H.; Lee, R.S.; Lee, C.; Speer, R.; Weingart, R.S. Operating characteristics and modeling of the LLNL 100-kV electric gun. *IEEE Trans. Plasma Sci.* **1989**, *17*, 392–402. [CrossRef]

52. Chau, H.H.; Dittbenner, G.; Hofer, W.W.; Honodel, C.A.; Steinberg, D.J.; Stroud, J.R.; Weingart, R.C.; Lee, R.S. Electric gun: A versatile tool for high-pressure shock-wave research. *Rev. Sci. Instrum.* **1980**, *51*, 1676–1681. [CrossRef]

53. Hahn, M.; Tekkaya, A.E. Experimental and Numerical Analysis of the Influence of Burst Pressure Distribution on Rapid Free Sheet Forming by Vaporizing Foil Actuators. *Metals* **2020**, *10*, 845. [CrossRef]

54. Sethi, G.; Myers, N.S.; German, R. An overview of dynamic compaction in powder metallurgy. *Int. Mater. Rev.* **2008**, *53*, 219–234. [CrossRef]

55. Borg, J.P.; Vogler, T.J. Aspects of simulating the dynamic compaction of a granular ceramic. *Model. Simul. Mater. Sci. Eng.* **2009**, *17*, 45003. [CrossRef]

56. Gluth, J.W.; Hall, C.A.; Vogler, T.J.; Grady, D.E. *Dynamic Compaction of Tungsten Carbide Powder*; No SAND2005-1510; Sandia National Laboratories: Albuquerque, NM, USA, 2005.

57. Meyers, M.A. *Dynamic Behavior of Materials*; John Wiley & Sons: Hoboken, NJ, USA, 1994.

58. Kerley, G.I. The linear us-up relation in shock-wave physics. *arXiv* **2013**, arXiv:1306.6916.

59. Tarver, C.M.; Forbes, J.W.; Garcia, F.; Urtiew, P.A. Manganin Gauge and Reactive Flow Modeling Study of the Shock Initiation of PBX 9501. *AIP Conf. Proc.* **2002**, *620*, 1043–1046.

60. Watson, R.W. Gauge for Determining Shock Pressures. *Rev. Sci. Instrum.* **1967**, *38*, 978–980. [CrossRef]

61. Turley, W.D.; Stevens, G.D.; Hixson, R.S.; Cerreta, E.K.; Daykin, E.P.; Graeve, O.A.; La Lone, B.M.; Novitskaya, E.; Perez, C.; Rigg, P.A.; et al. Explosive-induced shock damage in copper and recompression of the damaged region. *J. Appl. Phys.* **2016**, *120*, 085904. [CrossRef]

62. Remington, T.; Hahn, E.N.; Zhao, S.; Flanagan, R.; Mertens, J.; Sabbaghianrad, S.; Langdon, T.G.; Wehrenberg, C.; Maddox, B.; Swift, D.; et al. Spall strength dependence on grain size and strain rate in tantalum. *Acta Mater.* **2018**, *158*, 313–329. [CrossRef]

63. Turley, W.L.; Daykin, E.P.; Hixson, R.S.; LaLone, B.M.; Perez, C., Jr.; Stevens, G.D.; Veeser, L.; Ceretta, E.; Gray, G.T.; Rigg, P.; et al. *Copper Spall Experiments with Recompression Using HE (Copper Spall Soft Recovery Experiments)*; Rep No DOE/NV/25946-2108; Nevada Test Site/National: Nye County, NV, USA, 2014.

64. Li, J.; Panton, B.; Mao, Y.; Vivek, A.; Daehn, G. High strength impact welding of NiTi and stainless steel wires. *Smart Mater. Struct.* **2020**, *29*, 105023. [CrossRef]

65. Benzing, J.; He, M.; Vivek, A.; Taber, G.A.; Mills, M.; Daehn, G.S. A Microsample Tensile Test Application: Local Strength of Impact Welds between Sheet Metals. *J. Mater. Eng. Perform.* **2017**, *26*, 1229–1235. [CrossRef]

66. Szecket, A.; Mayseless, M. The triggering and controlling of stable interfacial conditions in explosive welding. *Mater. Sci. Eng.* **1983**, *57*, 149–154. [CrossRef]

67. Szecket, A.; Inal, O.T.; Vigueras, D.J.; Rocco, J. A wavy versus straight interface in the explosive welding of aluminum to steel. *J. Vac. Sci. Technol. A* **1985**, *3*, 2588–2593. [CrossRef]

Publisher's Note: MDPI stays neutral with regard to jurisdictional claims in published maps and institutional affiliations.

Journal of
*Manufacturing and
Materials Processing*

Article

Analysis of Proximity Consequences of Coil Windings in Electromagnetic Forming

Siddhant Prakash Goyal [1,*], Mohammadjavad Lashkari [1], Awab Elsayed [2], Marlon Hahn [1] and A. Erman Tekkaya [1]

[1] Institute of Forming Technology and Lightweight Components (IUL), TU Dortmund, Baroper Strasse 303, 44227 Dortmund, Germany; Mohammadjavad.Lashkari@iul.tu-dortmund.de (M.L.); marlon.hahn@iul.tu-dortmund.de (M.H.); Erman.Tekkaya@iul.tu-dortmund.de (A.E.T.)

[2] TU Dortmund, August-Schmidt-Straße 1, 44227 Dortmund, Germany; awab.elsayed@tu-dortmund.de

* Correspondence: siddhant.goyal@iul.tu-dortmund.de; Tel.: +49-231-755-7431

Abstract: Multiturn coils are required for manufacturing sheet metal parts with varying depths and special geometrical features using electromagnetic forming (EMF). Due to close coil turns, the physical phenomena of the proximity effect and Lorentz forces between the parallel coil windings are observed. This work attempts to investigate the mechanical consequences of these phenomena using numerical and experimental methods. A numerical model was developed in LS-DYNA. It was validated using experimental post-mortem strain and laser-based velocity measurements after and during the experiments, respectively. It was observed that the proximity effect in the parallel conductors led to current density localization at the closest or furthest ends of the conductor cross-section and high local curvature of the formed sheet. Further analysis of the forces between two coil windings explained the departure from the "inverse-distance" rule observed in the literature. Finally, some measures to prevent or reduce undesired coil deformation are provided.

Keywords: electromagnetic forming; proximity effect; Lorentz forces; coil windings

Citation: Goyal, S.P.; Lashkari, M.; Elsayed, A.; Hahn, M.; Tekkaya, A.E. Analysis of Proximity Consequences of Coil Windings in Electromagnetic Forming. *J. Manuf. Mater. Process.* **2021**, *5*, 45. https://doi.org/10.3390/jmmp5020045

Academic Editor: Steven Y. Liang

Received: 9 March 2021
Accepted: 4 May 2021
Published: 8 May 2021

Publisher's Note: MDPI stays neutral with regard to jurisdictional claims in published maps and institutional affiliations.

1. Introduction

Electromagnetic forming (EMF) is a noncontact forming process involving the deformation of sheets with high electrical conductivity by the discharge of pulsed electrical power. This process comes under the broad umbrella of impulse-based or high-speed forming processes, as strain rates obtained during the process range from 10^3 to 10^4 s^{-1}. Psyk et al. [1] describe the process of electromagnetic forming as obtaining a pulsed current by discharging a capacitor bank (or many). This current is time-varying and damped sinusoidal in nature. With the time-varying magnetic field, which accompanies the current and is concentrated between the coil and the workpiece, eddy currents are induced in the workpiece. Psyk et al. [1] also mention the presence of Maxwell pressure, a magnetic pressure that acts on the workpiece and deforms it by exceeding flow stress. Workpiece velocities of several hundred meters per second have been reported by Ahmed et al. [2] and Demir et al. [3] with Thibaudeau et al. [4] observing velocities between 100 and 300 m/s. The setup for electromagnetic forming is shown in Figure 1. The figure shows a general coil geometry forming a general sheet geometry, with setups to measure the current and workpiece velocity.

The process offers many advantages over conventional sheet metal forming processes, such as higher forming limits reported by Demir et al. [3], reduction of springback reported by Gayakwad et al. [5], possibility to perform operations, such as coining and embossing reported by Psyk et al. [1].

107

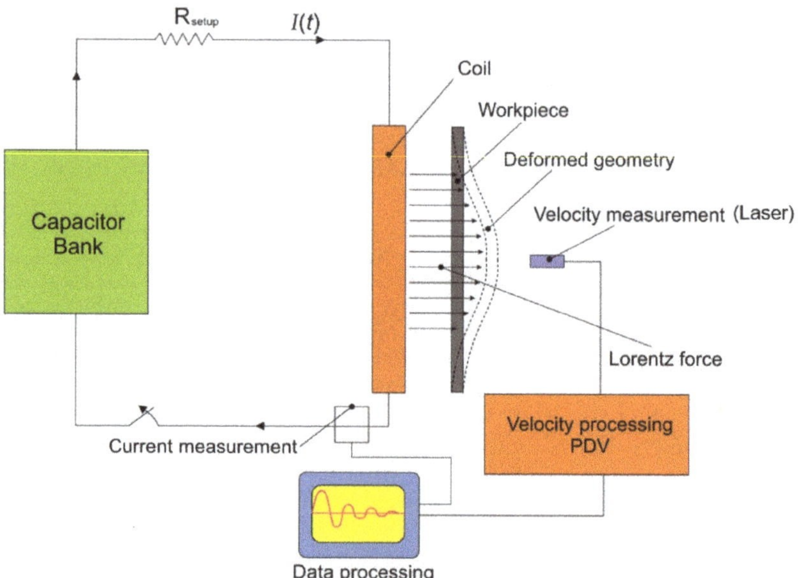

Figure 1. General experimental and measurement setup for electromagnetic forming.

1.1. Context of the Research

Extensive research has been devoted to applying the principles of electromagnetic induction to achieve compression or expansion of tubes and rings, welding of tubes and sheets, as well as cutting of sheets. In addition, the performance of the process has been improved by the design of coils, capacitor banks with shorter discharge times, and higher charging energies and high-speed measurements. Numerical modeling of the process, e.g., as introduced by Takatsu et al. [6] for a flat spiral coil, also helps visualize the process and observe the development of various unmeasurable parameters during the process.

With the understanding of the process dynamics established, the focus turned to the comprehensive study of process parameters, such as characteristics of the pulsed current, coil geometry, and tool life of the coil. Coils have been examined for their mechanical and thermal aspects, such as in the study by Gies et al. [7], in which the temperature distribution of the coil was identified to determine conclusions about the long-term usage of the coil.

This work examines another important parameter, namely, the distance between potentially multiple windings of coils. This is also known as the proximity of the coils.

1.2. Tool Life and Geometries of Electromagnetic Forming Coils

As EMF is a noncontact process, the role of the tool is taken by the coil (like a punch in deep drawing). As one of the most important parts of the setup, Golovaschenko [8] recommends that the objective for coil optimization should be the improvement in the ability to withstand many repeated discharges. Taebi et al. [9] propose that the optimization approach adjust the coils' geometrical parameters based on the deviation of the obtained workpiece geometry from the required geometry, which serves as an objective function. Thibaudeau et al. [4] state that for a given part, the coil design process should involve the decision of whether a high-pressure or a high velocity is needed. If the requirement is to have high magnetic pressure, a coil with a high number of turns must be selected, while for the latter case, a coil with a smaller number of turns must be selected, as this would lead to a shorter rise to the peak current. Coils used for processes, such as embossing and forming into shallow features, require multiturn coils with a high number of turns, as reported by

Watanabe and Kumai [10]. The use of many turns for applications with small and restricted geometries of the workpiece leads to reduced spacing and dimensions of the coil.

1.3. Skin and Proximity Effects between Conductors

The skin effect can be defined as the concentration of the current in the outer annulus of the conductor at high frequencies. It has been observed in many analyses, e.g., by Dwight [11] and Kennelly [12]. Due to this so-called crowding of the current at the outer periphery of the conductor, the alternating current (AC) resistance of the conductor is much higher than its direct current (DC) resistance, according to Reatti et al. [13]. This leads to increased power loss. Sigg et al. [14] emphasize that even though the calculation of eddy current effects and the AC resistance is essential, analytical methods for doing so are limited to simple geometries, as it is required to solve Maxwell's equations in inhomogeneous unbounded regions. Popovic [15] states that the skin depth δ can be obtained using Equation (1), where ω is the angular frequency, μ is the permittivity of the material and σ is the electrical conductivity:

$$\delta = \sqrt{\frac{2}{\omega\mu\sigma}} \tag{1}$$

The skin effect is due to the generation of eddy currents induced by the magnetic field of the primary current. In the center of the conductor, this eddy current cancels the primary current but reinforces the current flow at the periphery, according to Riba [16].

When dealing with multiple conductors or windings of the same conductor, the proximity effect comes into play because the current density of the conductors is affected by the current flowing in the neighboring conductors. This form of inductive coupling is known as the proximity effect, which becomes more pronounced at higher frequencies and small spacings between the conductors. Depending on the direction of the current flow, the proximity effect increases the current density at certain locations and reduces the same at other locations. If the current in adjacent coils is flowing in the same direction, then the current density is the highest in the furthest corners of the coils and vice versa.

Riba [16] states that the mathematical analysis of the proximity effect is even more difficult than that of the skin effect and possible only for a very few geometries. An analytical determination of the skin and proximity effects separately is attempted by Abdelbagi [17]. For two parallel plates with rectangular cross-sections, Abdelbagi [17] considers the flow of current in two plates in the same and opposite directions by solving the corresponding Helmholtz equation.

While this development is helpful in the analysis of the proximity effect in conductors with rectangular cross-sections, the underlying assumption of the nullification of the current density at the near conductor end ignores the actual effect of the spacing between the conductors. Furthermore, in electromagnetic forming, the additional presence of a conducting workpiece also modifies the current density distribution. The model to include the combined proximity effect due to the workpiece, the other coil windings, and the inherent skin effect at typical frequencies is predicted to be mathematically very complex and hardly scalable to many coils turns or complex coil geometries. Such a conclusion leads to using numerical models rather than analytical ones for solving complex electromagnetic problems. Vitelli [18] attempted to evaluate the current density and the subsequent losses in two conductors due to the proximity effect numerically.

1.4. Purpose of the Study

The objective of the present article is to evaluate the effect of spacing (proximity) between coil windings and its effect on their deformation and the forming of the workpiece in EMF. The primary objective above can be divided into three secondary objectives as follows:

- To develop and experimentally verify a numerical model to investigate the forming of the workpiece for various coil proximities and input energies;
- To evaluate the coil deformation for various discharge energies, cross-sections, and numbers of coils, so that this can be avoided in prospective applications.

2. Materials and Methods

2.1. Numerical Modeling in LS-DYNA

2.1.1. Development of the Simulation

As the EMF process is completed in a very small amount of time (approximately 200 μs), a transient numerical model provides a deeper understanding of the process evolution. The numerical model was developed employing the finite element code LS-DYNA, which provides the possibility to model the electromagnetic aspects of the problem along with the thermal and mechanical aspects in a coupled method. The methodology and theory of numerical modeling of electromagnetic forming are established by L' Éplattenier [19], with the same approach being implemented here. There, the surrounding air is modeled using the boundary element method (BEM). In this way, the air does not need to be meshed, and higher deformations of the parts are possible without continuous re-adaptation of the air mesh. The numerical model can provide information about the various parameters that cannot be measured directly, such as current densities, Lorentz forces, and the magnetic field. The simulation arrangement is presented in Figure 2.

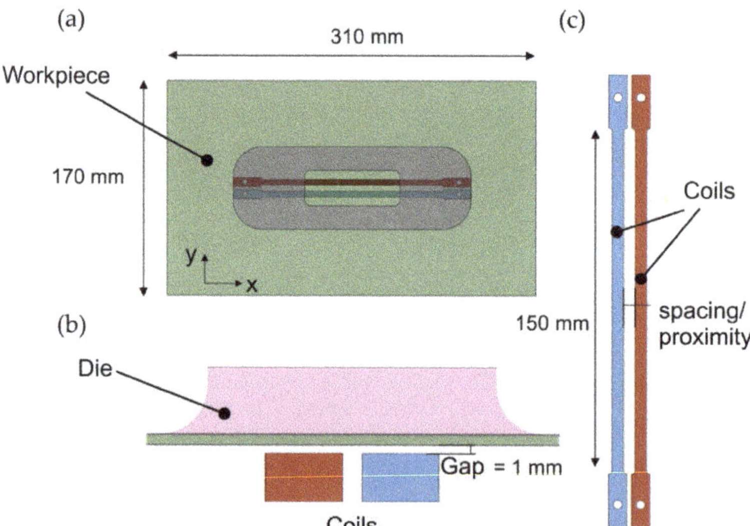

Figure 2. Numerical model in LS-DYNA. (**a**) Top view showing the dimensions of the workpiece; (**b**) front view showing the relative arrangement of the setup, (**c**) arrangement of the coils relative to one another for various values of spacing.

The arrangement consists of two (or more) identical coils, which are connected in parallel. To ensure that the coils are connected in parallel, the total input current is divided by the number of coils. This input current is obtained from the experiments, as described in Section 2.3.3. A sample input current provided to the simulation is presented in Figure 3. This current is damped sinusoidal with a reducing frequency.

The material assigned to both the coils is copper alloy CuCr1Zr, which compared to wrought pure copper alloy, has lower electrical conductivity, yet excellent mechanical strength. The holes at the ends of both the coils are fixed by constraining all displacements and rotations. The workpiece is modeled to be aluminum alloy EN AW-5083-H111, with a thickness of 1 mm. The other dimensions of the workpiece were chosen according to the requirements of the experimental setup. The gap between the workpiece and the coils was set to 1 mm. The workpiece and the coils were both modeled as deformable elastic-plastic solids. The die used for the open-window electromagnetic forming process was modeled as a rigid, nonconducting and fixed shell.

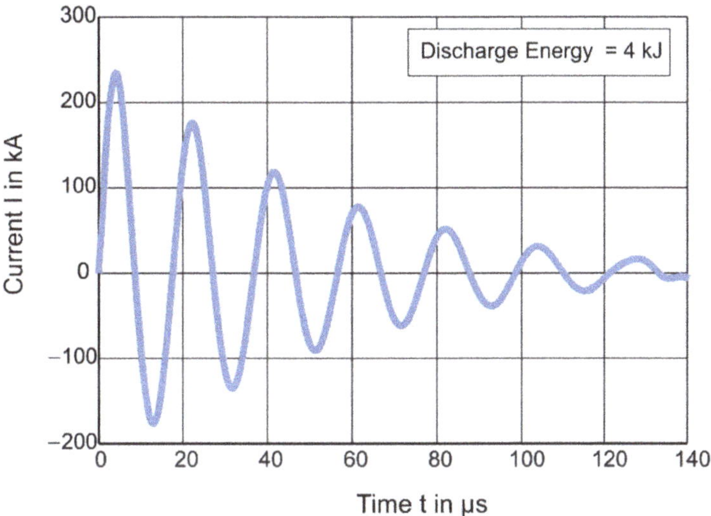

Figure 3. Input current curve for the simulations obtained for a discharge energy E_0 of 4 kJ. This current is distributed in both coils.

In coils with a workpiece above them, the current flows close to the coil's top surface. The coil induces a current in the opposite direction in the workpiece, and the currents flow as close as possible to one another due to the proximity effect, as explained in Section 1.3. This effect was also observed and modeled by Gies [20]. Similar straight coils were meshed so that the mesh density closer to the workpiece was high (thinner elements) and reduced gradually towards the opposite direction. This is done to capture the higher surface current density more precisely with a higher number of nodes in the vicinity. However, due to the very thin elements at the top, the aspect ratio of the elements is greatly skewed. In large deformations, such elements are prone to nonconvergence, erroneous results and long solving times. The mesh used for the coils is shown in Figure 4. The thickness of the topmost element is comparable to the calculated skin depth but experiences a high variation of current density within itself. Attempts with smaller thicknesses led to nonconvergence of the electromagnetic solver in LS-DYNA. The workpiece was meshed with two elements along the thickness, capturing the skin depth calculated at a frequency of 58,000 Hz (average obtained from various measurements), of EN AW-5083 sufficiently. Liu [21] stated that the deformation of the workpiece continues after the discharge of the current is finished. This happens due to the inertia of the workpiece. Furthermore, Gies [20] observed and modeled the temperature distribution in the numerical model until much longer after the process completion. The simulations were run for 0.009 s, when the deformation of the workpiece was observed to be completely stabilized for 90% of the run of the simulation.

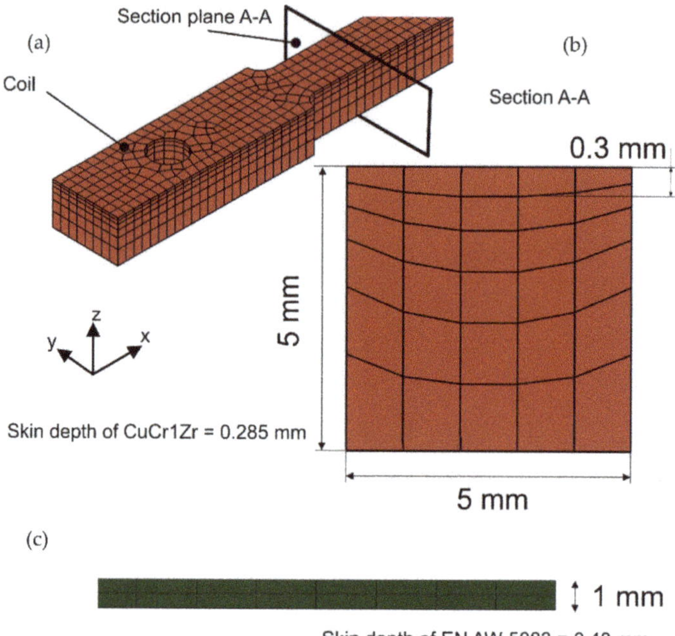

Figure 4. Meshing of the coil for the numerical model. (**a**) View of the coil end and the hole feature; (**b**) cross-sectional view showing thinner top element layers (**c**) cross-sectional view showing thickness-elements of the workpiece.

2.1.2. Material Models and Data

The properties assigned to the components can be classified into electrical, thermal and mechanical properties. The electrical conductivity (σ), thermal conductivity (k) and heat capacity (C_p) vs. temperature for CuCr1Zr and EN AW-5083 is shown in Table 1.

Table 1. Electrical and thermal properties for CuCr1Zr and EN AW-5083.

Temperature (K)	CuCr1Zr σ (MS/m)	CuCr1Zr k (W/mK)	CuCr1Zr C_p (J/kgK)	EN-AW 5083 k (W/mK)	EN-AW 5083 C_p (J/kgK)
298	52.0	343	385	126	899
350	44.8	347	392	133	919
399	39.7	350	397	138	939
448	35.3	350	401	143	959
498	31.6	348	405	149	979
548	28.8	349	408	155	999
598	26.2	347	410	159	1018
647	23.9	343	412	163	1038
698	21.9	339	414	164	1059
749	20.2	335	416	166	1079
774	19.5	334	418	167	1089

For EN AW-5083, the electrical conductivity is provided only at a room temperature of 19 MS/m. For CuCr1Zr, the plastic properties are provided as dependent on temperature. These properties were determined by hot tensile tests and were extrapolated exponentially, shown in Figure 5. The strain-rate dependency of CuCr1Zr was ignored.

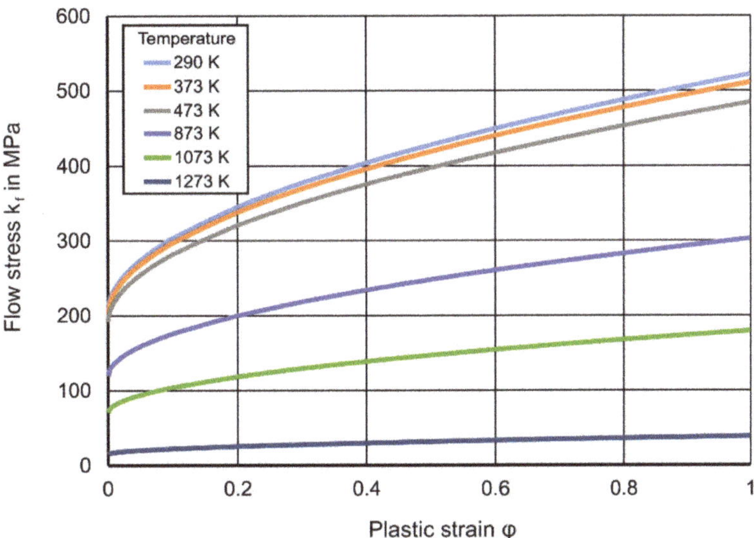

Figure 5. Temperature-dependent stress-plastic strain flow properties of CuCr1Zr from high-temperature tensile tests and extrapolation exponentially.

The Young's modulus of EN AW-5083 was taken as 70 Gpa; the Poisson's ratio was taken as 0.33. The plastic behavior of EN AW-5083 was defined to be both strain-rate- and temperature-dependent. The strain-rate sensitivity of plastic flow of aluminum alloys is assumed low until 10^3 s^{-1}, but a sharp rise is observed at strain rates between 10^3 s^{-1} and 10^4 s^{-1}. For few aluminum alloys, this dependence of initial flow stress on the strain rate is shown by Weddeling et al. [22]. Clausen et al. [23] observed that for AA-5083-H116 (different temper than the currently used alloy), the material shows negative strain-rate sensitivity for low strains (softening when the strain rate is increased). For higher strain rates (more than 1 s^{-1}), the strain-rate sensitivity is positive. Until 10^3 s^{-1}, the material data for aluminum is based on high-temperature and high strain-rate tensile tests. For strain rates higher than 10^4 s^{-1}, the flow curves are not readily available or easily measured. The sudden change in the initial flow stress for aluminum alloys is attributed to a change in deformation mechanism, according to Kabirian [24]. The mechanism for deformation at very high strain rates is the viscous drag of the dislocations. A mechanism-based constitutive relation is suggested, which is shown in Equation (2). In this equation, σ is the flow stress at a given strain rate and temperature, σ_0 is the flow stress at a strain rate of 1 s^{-1}, T_m is the melting temperature, φ presents the sigmoidal feature of the function and $\dot{\overline{\varepsilon}_p}$ is the strain rate:

$$\sigma\left(\overline{\varepsilon}_p, \dot{\overline{\varepsilon}_p}, T\right) = \begin{cases} \sigma_0\left(1 - (C_1 \exp\left(-k_2\dot{\overline{\varepsilon}_p}\right)\left(\frac{T}{T_m}\right)^{m_2}\frac{\varphi}{1+\varphi}\right) & \dot{\overline{\varepsilon}_p} \ll 1 \\ \sigma_0 \exp(k_1\dot{\overline{\varepsilon}_p}) & \dot{\overline{\varepsilon}_p} \gg 1 \\ \sigma_0 & \dot{\overline{\varepsilon}_p} = 1 \end{cases} \tag{2}$$

For the current case, the exponential relation proposed for very high strain rates is relevant. This relation was used to predict the flow stresses for strain rates of 10^4 s^{-1} and above, with σ_0 as the flow stress for lower strain rates (any lower strain rate can be chosen as the sensitivity at lower strain rates is very small). The final flow curves used in the simulations (at room temperature) are shown in Figure 6. The flow curves for strain rates of 0.001 s^{-1} and 1000 s^{-1} are almost identical due to the negligible strain-rate sensitivity of the material in this regime. As the flow curves for all strain rates in this regime are almost identical, they have not been provided in Figure 6.

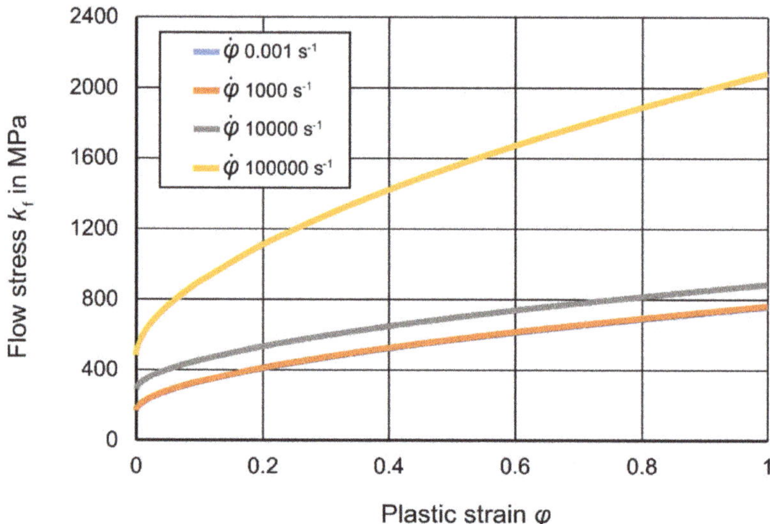

Figure 6. Implementation of the model proposed by Kabirian [24] to determine strain-rate-dependent flow curves for EN AW-5083, $\dot{\varphi}$ represents the strain rate in 1/s.

In LS-DYNA, the flow curves (experimentally measured and extrapolated both) are provided to the solver in the form of tabular data. The material model used in LS-DYNA is denoted as MAT_106, which is an elastic-viscoplastic-thermal model. The strain dependence is expressed in terms of the sum of exponential terms, and the strain-rate dependence is expressed using the Cowper–Symonds model. From the provided tabular data, the parameters of the material model are calculated by the solver.

2.2. Numerical Modeling in FEMM

Development of the Simulation

FEMM is an open-source finite element software for a harmonic 2D or axisymmetric modeling of electromagnetics. The advantage in using FEMM is the ability to quickly model a simplified version of the present problem for efficient further processing (compared to the computationally much more costly LS-DYNA model). The cross-sections of the two coils, along with the cross-section of the sheet, are modeled. The boundary condition of the zero magnetic fields is applied at a large distance away from the coils and the workpiece.

The input of the current in FEMM happens by providing amplitude and a frequency, after which a constant sinusoidal current is applied. As the actual current, as shown in Figure 3, is damped sinusoidal, only the first half-wave of the current is modeled by using the amplitude and frequency of the same. While the number of elements in the cross-section of the coil is only 30 in the LS-DYNA model, the number of cross-section elements in the FEMM model is upwards of 10,000 with only a minor impact of total solving time and no impact on convergence. Only room temperature values for the electrical conductivity for CuCr1Zr and EN AW-5083 are input into FEMM. These values are given in Section 2.1.2.

2.3. Experimental Analysis

2.3.1. Experimental Setup

The experimental setup includes the capacitor bank for the discharge of energy, the coil setup with two or more coils for the study, the EN AW-5083 workpiece and the equipment for clamping and safety. The capacitor bank used for the analysis was the SMU612 by the company Poynting. The main characteristics of this capacitor bank are given in Table 2.

Table 2. Characteristics of capacitor bank SMU612.

Characteristic	Value
Max. discharge energy E_{max}	9 kJ
Max. discharge voltage U_{max}	15 kV
Capacitance C	80 µF
Inductance L	1093 nH
Discharge frequency f	$\approx$17 kHz

When the bank is discharged, the current flows through the circuit into the coil setup and is suitably distributed into the coils, which are connected in a parallel system. The CAD model of the coil setup used for the experiments is shown in Figure 7. The coils are connected to the setup using M3 screws, which are expected to serve as fixed support. The geometry of the coil used for the experiments is shown in Figure 2. The cross-section of the coil is 5 mm × 5 mm, similar to the ones used by Gies [7]. The width of the ends was reduced to ensure closer proximities between the coils.

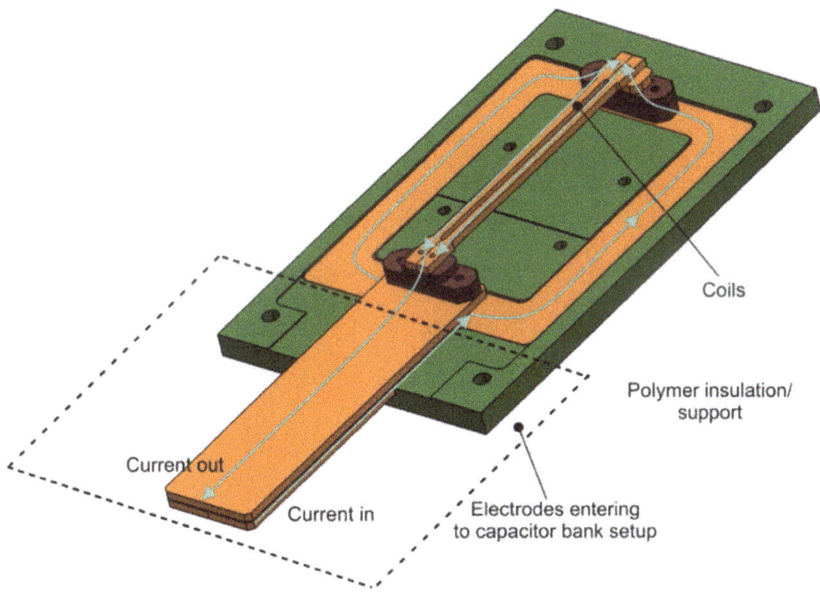

Figure 7. CAD model of the coil setup used for the experiments. The path of the current is shown in blue. Some support elements have not been shown for clarity of representation.

2.3.2. Parameters Studied in the Experiments

To study the effect of the spacing between the coils, the spacing was varied between 4 mm and 9 mm at an interval of 1 mm. Spacings smaller than 4 mm were not possible due to the geometry of the coil. Therefore, the effect of smaller intervals was investigated through the numerical models. For each value of spacing, various energies ranging from 3 kJ to 6 kJ were discharged, and the deformation of the coil and the workpiece were recorded.

For some experiments, coils of higher cross-sections were used as possibilities to reduce the coil deformation. The cross-sections investigated were 5 mm × 10 mm and 10 mm × 5 mm. The three cross-sections used are shown in Figure 8.

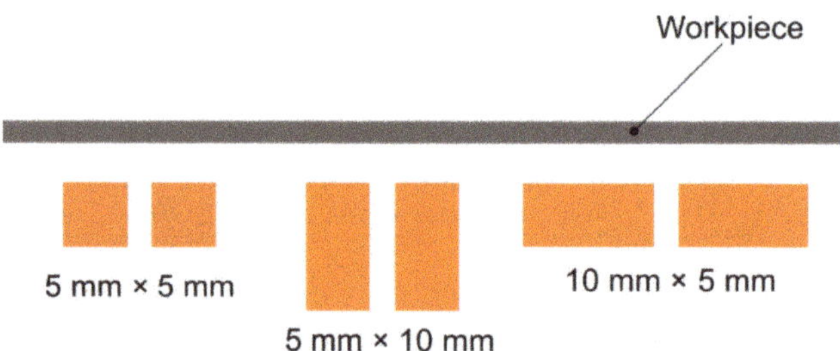

Figure 8. Three different coil cross-sections studied for the experiments.

2.3.3. Measurement Methods

The parameters measured in the experiments, along with the techniques used to measure them, are described below: The total discharge current was measured using a Rogowski coil with a conversion factor of 1.2 mV/A. Strain measurement was performed after the process is complete. Before the process, a pattern, which is a grid of dots, is created on the top surface of the EN AW-5083 sheet by selective anodizing. After the process, the deformed pattern is analyzed by the optical strain measurement system ARGUS by GOM. The parameter of interest measured through this setup is the von Mises equivalent plastic strain. The ARGUS system can measure only the two surface strains. The thickness strain is estimated using the volume constancy law in plasticity and the assumption that the strain is uniform across the thickness. The displacement of the coils was measured using ATOS by GOM also after the completion of the process. ATOS is an optical coordinate–measuring setup. The displacements were computed by comparing the original and the deformed geometries. As both the strain measurement of the workpiece and the displacement measurement of the coils could only be conducted after the experiments, the vertical velocity of the center of the specimen (middle point) was measured during the process using photonic Doppler velocimetry (PDV), as explained by Moro [25], which is a laser-based system that uses the phase shift in the reflected beam to determine the signal.

3. Results

3.1. Validation of the Numerical Model

The numerical model was verified by comparing the post-experiment strains and the apex velocity during the process.

3.1.1. Comparison of Plastic Strains

The strains measured from the experiments at the end of the process are the plastic strains on the sheet surface. They are compared to the effective plastic strains from LS-DYNA. The comparison was made along specific paths. Path 1 is the path along the coil in the middle of the two coils and the workpiece. The deformation along this path happens mostly due to inertia. Path 2 is the path across the workpiece in the y-direction. These paths are shown in Figure 9. The comparison for the numerically and experimentally determined strains is shown in Figure 10 for path 1.

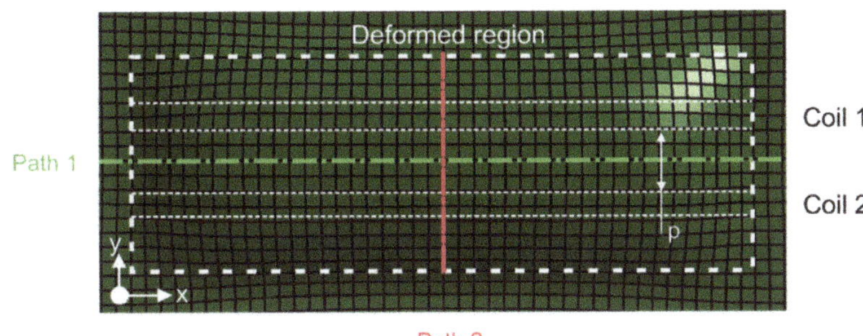

Figure 9. Two different paths chosen for comparison of plastic strains from the numerical and experimental analyses (top view).

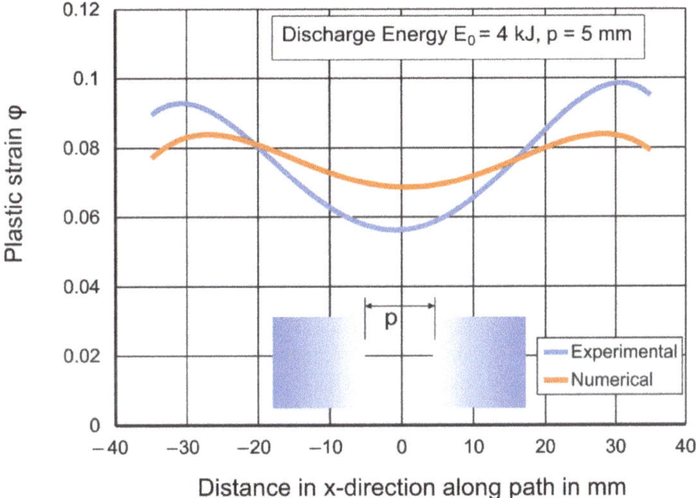

Figure 10. Comparison of numerical and experimental equivalent plastic strains.

This process was repeated for various discharge energies and spacings between the coils. Through measurements along the aforementioned paths, it was determined that the maximum average difference between the experimentally and numerically calculated values was 8%. This means that even at very high strain rates, the plastic strain comparison yielded results with acceptable accordance.

3.1.2. Comparison of Midpoint Velocity

The velocity measurement in z-direction was performed for reassurance and revalidation of the numerical LS-DYNA model to match the experimental parameters during the process. The comparison of the numerical and experimental values provides a very good agreement, as shown in Figure 11. The figure shows the comparison between simulated and measured values for two different values of spacing between the coils but the same discharge energy. The comparison for different energies and values of spacing between the coils resulted in a maximum average velocity difference of 5%. When PDV measurements were made by Demir et al. [3] to determine the average strain rate of the process, the experiments at higher energies always led to a higher deviation between the experimental and numerical values because only a quasi-static flow curve was used for the modeling of the EN AW 5083-H111 sheet.

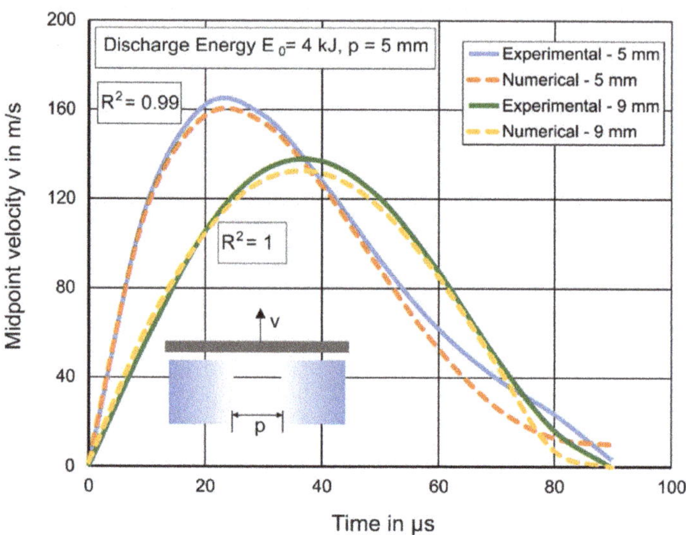

Figure 11. Comparison of the vertical velocity of the midpoint of the workpiece from numerical and experimental (PDV) analyses for two different spacings (p) between the coils.

3.1.3. Comparison of the FEMM and LS-DYNA Model

The numerical model of FEMM was verified for use in further analysis by comparison of the current density with that obtained from the numerical model in LS-DYNA. This comparison is shown in Figure 12. The values predicted by the numerical model in LS-DYNA are slightly lower because of the actual damped current input, while FEMM can only account for a constant amplitude sinusoidal current. However, the current density obtained from FEMM contains many data points, lowering the requirement for interpolation between two calculated data. The current density from LS-DYNA is discrete and has only five data points.

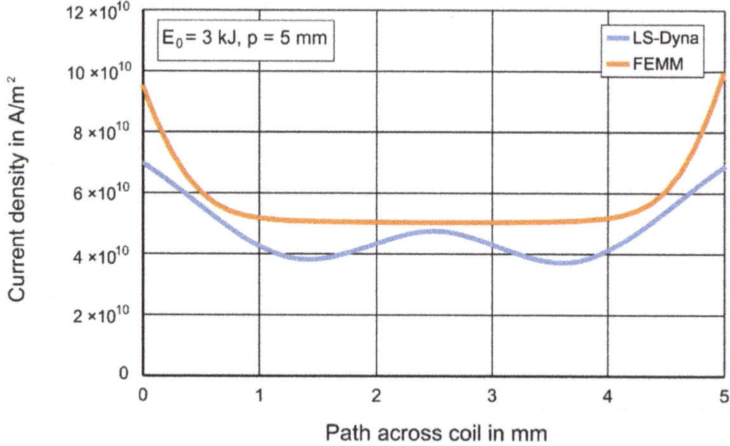

Figure 12. Comparison of the current density distribution along the path from numerical models in LS-DYNA and FEMM. The data from LS-DYNA is very discrete and not a smooth curve, as shown in the graph.

3.2. Deformation of the Coils

3.2.1. Coil Deformation for Various Discharge Energies

Even though the coils were assigned as deformable solid bodies in the numerical model, the deformation obtained from the numerical model was much smaller than the experimentally measured deformation. In this section, the experimentally measured coil deformations are addressed.

The coils were fixed to the setup at both ends, as shown in Figure 7. Apart from this constraint, the coils were permitted to deform freely in all directions. Due to the current flowing in the same direction, the coils exert an attractive force towards one another. Based on the Ampère's circuital law, the force per unit length l between two parallel infinitely long wires is given in Equation (3), where the currents in the wires are given by I_1 and I_2 and the spacing between the wires is r:

$$\frac{F}{l} = \frac{\mu_0 I_1 I_2}{2\pi r} \tag{3}$$

There is an inverse dependence of the force on the distance between the coils. For particular discharge energy, Figure 13 shows the displacement of the coils towards another. It is observed that as the spacing between the coils is increased, the displacement of the coil is reduced. This is because of the reduction of attractive Lorentz forces between the coils. For the discharge of 5 kJ, the 5 mm spacing causes the coils to displace to such an extent that a collision and repulsion of the coils occurs, leading to a different final shape. No conclusions can be made about the force between the two coils as plastic deformation of the coils takes place, and due to varying temperatures and strain rates, no direct relation between the force and displacements can be deduced. The figure helps in the qualitative understanding of the dependence of the coil displacement on the spacing between them.

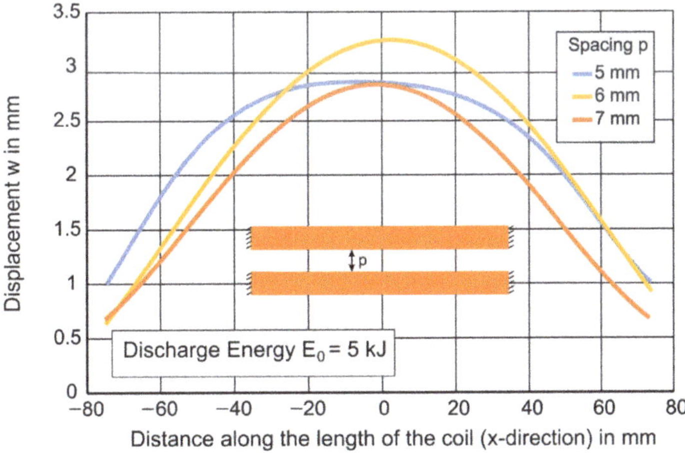

Figure 13. Experimentally measured deformation of the coil along the coil length towards one another in x-direction due to attractive Lorentz forces for three different coil spacings. The coordinate directions are shown in Figure 2.

3.2.2. Coil Deformation for Various Cross-Sections

Due to the volumetric Lorentz force, the coil experiences a distributed load, which causes bending. This bending can be reduced by increasing the cross-section of the coil, as that increases the moment of inertia of the coil. In the experimental results explained in Section 3.2.1, the coil cross-section used was 5 mm × 5 mm. Two other cross-sections, shown in Figure 8, were analyzed experimentally. Figure 14 shows the displacement of the coils towards one another for the smallest cross-section and for the 10 mm × 5 mm one. The advantage of the 5 mm × 10 mm cross-section is that the moment of inertia when bending due to the attractive force of the other coil is much higher (4 times) than the other cross-section. This leads to a reduction in coil deformation. In the wider coil, the "effective distance" between the currents is higher due to the proximity effect, as the current migrates to the extremities of the coil cross-section. Figure 14 shows the comparison between the cross-sections 10 mm × 5 mm and 5 mm × 10 mm for the same values of discharge energy and spacing.

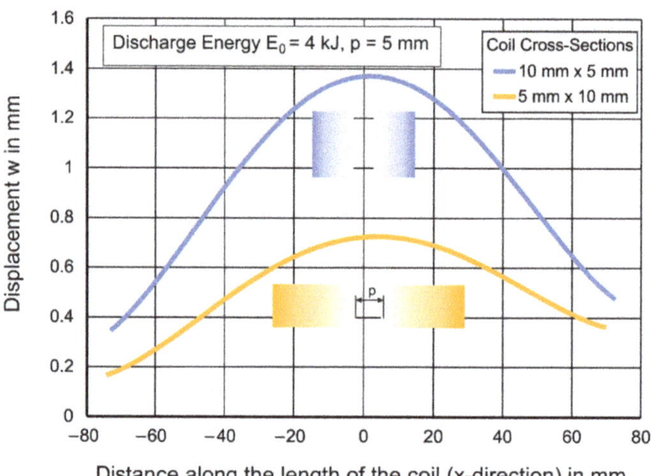

Figure 14. Experimentally measured displacement of the coils in the y-direction for two different coil cross-sections.

4. Discussion

4.1. Deformation of the Sheet

4.1.1. Variation of Displacement, Midpoint Velocity and Efficiency of the Process with Coil Spacing

The deformation of the sheet was examined using the verified numerical model. Path 2, along which the deformation was examined, is shown in Figure 9. In the simulations with two coils, the displacement of the midpoint of the workpiece happens due to the combined influence of the Lorentz force applied by the two coils and the inertia of the workpiece. To examine the effect of the coil spacing between the final workpiece geometry, the discharge energy was kept constant. For a discharge energy of 4 kJ, the impact of spacing is shown in Figure 15. It is observed that the deformation of the apex of the workpiece is reduced with increasing coil spacing.

While the current remains the same, this deformation is explained by the reduction of the magnetic field intensity and, therefore, of the Lorentz force at the midpoint. The magnetic field intensity is smaller at the midpoint as the coils are further away for larger spacings, and it is inversely proportional to the distance. Figure 16 shows the magnetic field at the midpoint of the gap for various coil spacings, including those, which were not possible in the experiments due to design limitations.

In the case of the wider spacings, due to the low magnetic pressure at the midpoint, the maximum kinetic energy of the sheet at the midpoint is also low. Selected midpoint velocities obtained from the numerical model are presented in Figure 17. When the spacing is increased, the maximum velocity of the midpoint is reduced but saturates at greater values of coil spacing. The increase in spacing causes similar maximum velocity for the apex but with smaller acceleration. The reason for the lower velocity is not any difference in the circuit properties, and the current curves were also compared and determined to be highly similar. As for the lower spacing, the midpoint is located closer to the coils; the applied Lorentz force is higher, leading to higher acceleration. As the discharge energy is the same, a higher spacing causes less energy to be transferred to the midpoint and more to the local area just above the coil. The midpoint velocity saturates as it becomes a resultant of the inertia of the sheet rather than accelerating due to direct force application.

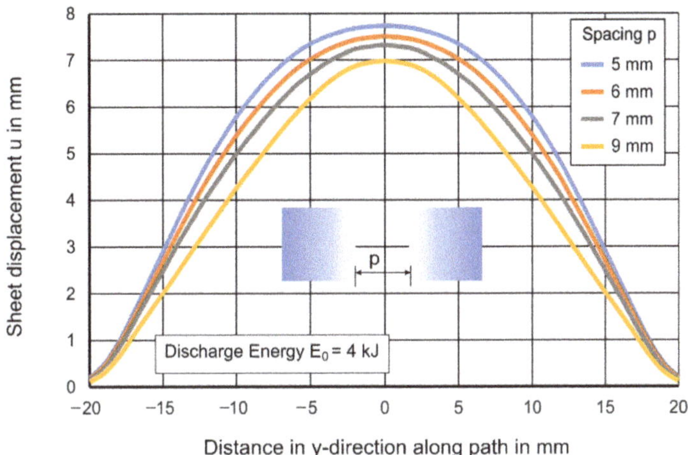

Figure 15. Experimentally measured displacement of the workpiece in z-direction along path 2.

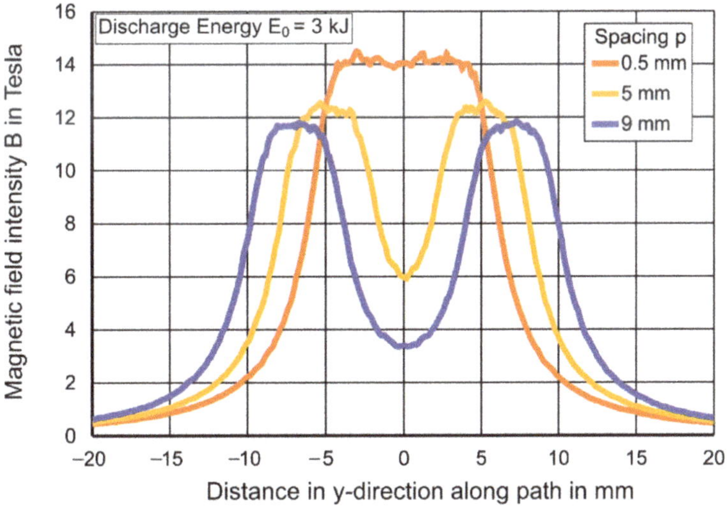

Figure 16. Numerically determined magnetic field intensity along path 2.

The efficiency of the process ($\eta_{process}$) can be calculated as the ratio of the energy used for forming the workpiece and the total energy discharge. The energy used for forming the workpiece is obtained from the simulations by integration of the plastic energy density (Equivalent strain times the flow stress) over the whole part volume at the end of the process. These efficiencies are typically very low, as more than half of the supplied energy is lost as Joule heat, according to Gies [7]. The efficiency is also dependent on the design of the setup and may be low because of the small size of the die cavity than the length of the coils. For the same setup and discharge energy, the efficiency of the process was observed to be related to the spacing between the coils. This dependency is observed in Figure 18 for different values of discharge energy.

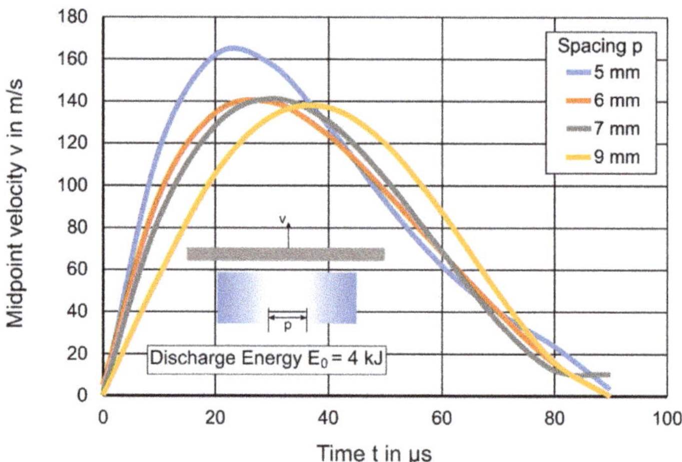

Figure 17. Comparison of experimentally measured vertical midpoint velocity for four different coil spacings.

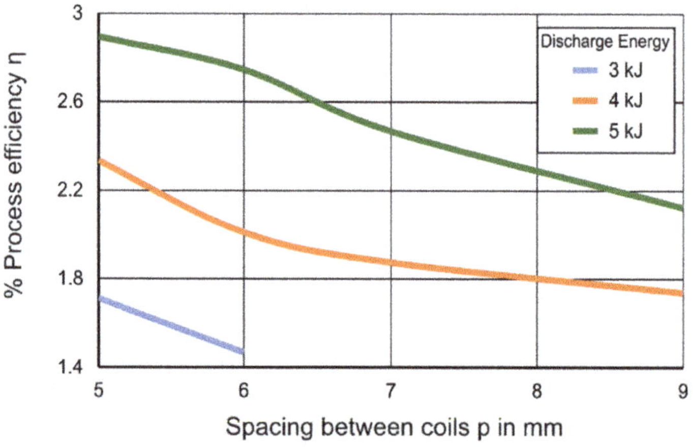

Figure 18. Reduction in process efficiency for various values of coil spacings. The overall process efficiency increases as the discharge energy increases.

When the discharge energy is the same, the reason for a lower efficiency is the improper conversion of the supplied energy into the kinetic energy of the workpiece. The maximum kinetic energy of the workpiece was investigated during the process and analyzed for varying spacings between the coils. While the velocity was analyzed for just the midpoint, the maximum kinetic energy was monitored for the entire workpiece. It was seen that for the same discharge energy, the kinetic energy of the workpiece reduces with increased spacing between the coils from 5 mm to 9 mm by 22% for a discharge of 4 kJ energy and 26% for a discharge of 5 kJ energy. As a fraction of this kinetic energy is converted into the energy to deform the workpiece plastically, this energy reduces with increased spacing as well. The kinetic energy not converted to plastic work must get dissipated finally as heat along with other losses, such as friction.

4.1.2. Change of Local Curvature with Coil Proximity

The concentration of the current at the extreme periphery of the coil due to the proximity effect affects the quality of the workpiece as well. This effect is not explicitly visible when the displacement of the sheet is plotted, as shown in Figure 19. For the same discharge energy but various values of spacing, the local curvature of the sheet was calculated along path 2, shown in Figure 9 at the end of the process. The curvature (κ) may be considered as a local bending-cum-stretching of the sheet, and it is defined as the inverse of the local radius. The curvature is calculated using Equation (4) and is plotted for two values of spacing in Figure 19. If the deformed contour of the workpiece is defined as $\mathbf{S}$ and its spatial derivatives as $\mathbf{S}'$ and $\mathbf{S}''$ is as defined in Equation (4):

$$\kappa = \frac{\mathbf{S}' \times \mathbf{S}''}{\|\mathbf{S}'\|^3} \tag{4}$$

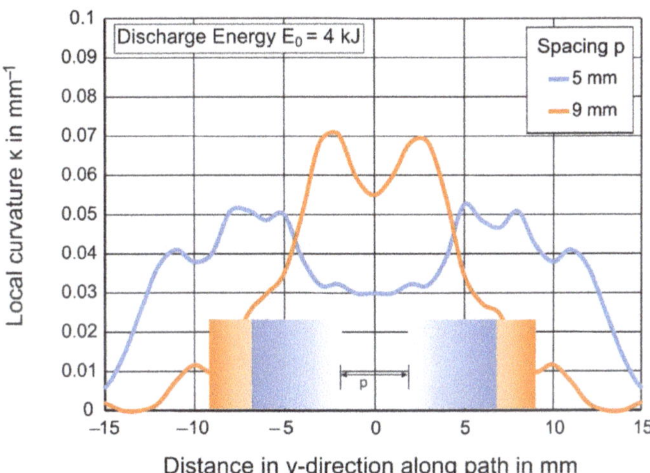

Figure 19. Change in local workpiece curvature for coil spacings of 5 mm and 9 mm.

When the coils are close to one another, as is the case is for 5 mm spacing, the current migrates to the extremes. This high current density at the extreme applies a local pressure on the sheet and causes high bending in the region around it. This effect is much lower when the spacing is higher. Due to most of the current flowing along the top surface, the Lorentz force applied by the skewed current distribution causes a high local curvature of the sheet. When the current is almost uniform on the top surface, as for the case with higher values of spacing, one observes the bending localized at the midpoint (y = 0). This happens in the absence of large localized current densities, which bend the sheet locally.

4.2. Deformation of the Coils

4.2.1. Numerical Prediction of Force between Two Coils

To predict the total force between the two coils, the FEMM simulations were used. The assumptions used here were the following:

1. No displacement of the coils and workpiece: As the force is applied, the coils begin to displace first elastically and then plastically towards one another. As the spacing between the coils is changed, it has an effect on the force as well, which will be higher than the predicted value. However, the displacement of the coils must start after a significant portion of the current is already discharged and must happen mostly due to inertia. Furthermore, according to Beerwald [26], most of the energy is transferred in

the first two large pressure pulses of the current, until which, not much displacement of the coil is expected.

2. An assumption about discharge current: FEMM can only model the discharge of a constant amplitude sinusoidal current, while the real current is damped in nature. Therefore, FEMM is used only for the calculation of the maximum force between the two coils, which would happen at the maximum value of current flowing between the two coils.

For the calculation of the Lorentz force between the two coils, FEMM uses Maxwell's equations, which provide the Lorentz force exerted per unit volume. This is then multiplied to the volume of the element, and a vector addition for all elements is performed to give the overall force exerted. For various values of spacing, the force between the two coils is shown in Figure 20. The diagram is divided into two parts based on how well the diagram conforms to Equation (3), predicting the force between the two parallel current-conducting wires. For higher values of spacing, the force between the two coils follows Equation (3), and the conductors act as isolated from one another. For smaller values of spacing, the Equation (3) rule is abandoned. The force between the coils is much smaller than predicted by the equation rule. This is due to the redistribution of the current in the cross-section. Equation (3) assumes a uniform distribution of current in the cross-section. As the spacing decreases to small values, the current locates itself in the furthest end of the cross-section of the conductor, which leads to increased "effective distance" between the currents, and therefore, to a reduction of force.

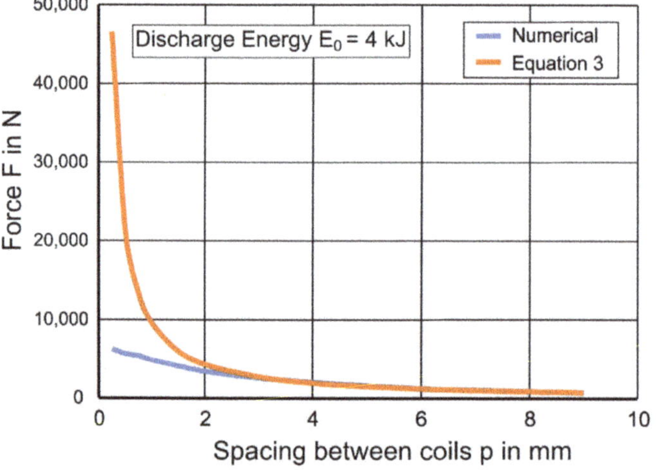

Figure 20. Comparison of total force between two coils predicted by the numerical model in FEMM and the 1/r rule.

4.2.2. Measures to Avoid Coil Deformation

The deformation of the tool coil must be avoided in a real application to increase its reuse. To reduce coil deformation, the following measures are recommended:

1. Changing the coil cross-section: Increasing the coil cross-section leads to lower deformations due to increased coil rigidity. However, the increase in dimensions of the cross-section leads to a lower current density and smaller workpiece deformation. Furthermore, if the width of the coil is increased to protect against the deformation due to the attractive force from the other coils, this will not provide much benefit against the reaction force exerted by the coil, which is orthogonal to the former force and about four times the magnitude in the geometry used in this study. It also increases the material used for the coils.

2. Providing insulation support: In some of the experiments, the coils were separated by insulation made of glass-fiber-reinforced polymer, which has excellent compressive strength and insulating properties. In such experiments, the coil deformation due to the attractive force could be completely avoided. However, this led to a higher deformation in the orthogonal direction (the direction where the coil deformed due to the reaction force from the workpiece). With a good insulation design, the deformation of the coil can be completely avoided. Before implementation, however, the damping characteristics of the polymer must also be investigated.

3. Innovative coil design: Gies [27] proposed using hybrid conductors for the reduction of coil deformation and suggested the methodology for the design as well. With the appropriate design of the steel support, plastic deformation of the coil can be avoided for some discharge energies. Other ideas for an improvement of the coil design could be the addition of cooling possibilities in the coil so that thermal softening of the material is prevented.

5. Conclusions

In this study, the spacing between the coil and the workpiece and the coil windings themselves was examined from multiple perspectives. Our objective was to understand and quantify the various phenomena when EMF is performed using conductors having multiple discrete turns. For this study, numerical and experimental methods were employed. The effect of spacing between coil turns on the overall workpiece displacement, workpiece midpoint velocity, efficiency and local curvature were observed. It was found that due to the proximity effect, the local curvature of the sheet is high for smaller spacings between the coils. Increasing the spacing reduces the process efficiency. For the coils, it was observed that the (undesired) deformation and the corresponding force between them both increase with reducing the spacing.

Generally, for parts with larger forming heights and high desired process efficiency, it is recommended to reduce the proximity between the coils. However, as the proximity between the coils is reduced, the forces between the coils increase, potentially leading to coil failure. Thus, there exists a tradeoff between process efficiency and tool life. The tool life may be increased by using coils with larger cross-sections and appropriate orientation.

As the spacing between the coils is an input parameter, it is essential to understand these phenomena and implement them in the setup design to obtain reliable results from EMF. Other results, such as the possible deformation of the coils, must be considered for a specific setup. For general safety of the tool coil, some recommendations, such as supporting the coil using rigid insulation and the potential addition of steel supports, are provided. Future studies in the prevention of tool coil deformation will focus on such hybrid coils.

Author Contributions: Conceptualization, S.P.G., M.L. and A.E.T.; methodology, S.P.G. and M.L.; software, S.P.G.; validation, S.P.G., A.E. and M.L.; formal analysis, S.P.G. and M.L.; investigation, S.P.G., A.E. and M.L.; resources, A.E.T. and M.H.; data curation, S.P.G. and M.H.; writing—original draft preparation, S.P.G.; writing—review and editing, S.P.G., M.H. and A.E.T.; visualization, S.P.G.; supervision, M.H. and A.E.T.; project administration, M.H. and A.E.T.; funding acquisition, M.H. and A.E.T. All authors have read and agreed to the published version of the manuscript.

Funding: This research was funded by the Deutsche Forschungsgemeinschaft (DFG) under project number 259797904. The authors express their gratitude and appreciate their commitment to support research.

Data Availability Statement: Not applicable.

References

1.   Psyk, V.; Risch, D.; Kinsey, B.L.; Tekkaya, A.E.; Kleiner, M. Electromagnetic forming—A review. *J. Mater. Process. Technol.* **2011**, *211*, 787–829. [CrossRef]

2. Ahmed, M.; Panthi, S.K.; Ramakrishnan, N.; Jha, A.K.; Yegneswaran, A.H.; Dasgupta, R.; Ahmed, S. Alternative flat coil design for electromagnetic forming using FEM. *Trans. Nonferrous Met. Soc. China* **2010**, *21*, 618–625. [CrossRef]
3. Demir, O.K.; Goyal, S.P.; Hahn, M.; Tekkaya, A.E. Novel Approach and Interpretation for the Determination of Electromagnetic Forming Limits. *Materials* **2020**, *13*, 4175. [CrossRef] [PubMed]
4. Thibaudeau, E.; Kinsey, B.L. Analytical design and experimental validation of uniform pressure actuator for electromagnetic forming and welding. *J. Mater. Process. Technol.* **2015**, *215*, 251–263. [CrossRef]
5. Gayakwad, D.; Dargar, M.K.; Sharma, P.K.; Purohit, R.; Rana, R.S. A Review on Electromagnetic Forming Process. *Procedia Mater. Sci.* **2014**, *6*, 520–527. [CrossRef]
6. Takatsu, N.; Kato, M.; Sato, K.; Tobe, T. High-Speed Forming of Metal Sheets by Electromagnetic Force. *JSME Int. J.* **1988**, *31*, 142–148. [CrossRef]
7. Gies, S.; Tekkaya, A.E. Effect of workpiece deformation on Joule heat losses in electromagnetic forming coils. *Procedia Eng.* **2017**, *207*, 347– 346. [CrossRef]
8. Golovashchenko, S.F. Material Formability and Coil Design in Electromagnetic Forming. *J. Mater. Eng. Perform.* **2007**, *16*, 314–320. [CrossRef]
9. Taebi, F.; Demir, O.K.; Stiemer, M.; Psyk, V.; Kwiatkowski, L.; Brosius, A.; Blum, H.; Tekkaya, A.E. Dynamic forming limits and numerical optimization of combined quasi-static and impulse metal forming. *Comput. Mater. Sci.* **2012**, *54*, 293–302. [CrossRef]
10. Watanabe, M.; Kumai, S. High-Speed Deformation and Collision Behavior of Pure Aluminum Plates in Magnetic Pulse Welding. *Mater. Trans.* **2009**, *50*, 2035–2042. [CrossRef]
11. Dwight, H.B. Skin effect in tubular and flat conductors. *Trans. Am. Inst. Electr. Eng.* **1918**, *XXXVII-2*, 1379–1403. [CrossRef]
12. Kennelly, A.E.; Laws, F.; Pierce, P. Experimental researches on skin effect in conductors. *Trans. Am. Inst. Electr. Eng.* **1915**, *34*, 1953–2021. [CrossRef]
13. Reatti, A.; Grasso, F. Solid and Litz-wire winding non-linear resistance comparison. In Proceedings of the 43rd IEEE Midwest Symposium on Circuits and Systems (Cat.No.CH37144), Lansing, MI, USA, 8–11 August 2000; pp. 466–469.
14. Sigg, H.J.; Strutt, M.J.O. Skin effect and proximity effect in polyphase systems of rectangular conductors calculated on an RC network. *IEEE Trans. Power App. Syst.* **1970**, *89*, 470–477. [CrossRef]
15. Popovic, Z.; Popovic, B.D. *Introductory Electromagnetics*; Prentice Hall: Upper Saddle River, NJ, USA, 2000.
16. Riba, J.R. Calculation of the ac to dc resistance ratio of conductive nonmagnetic straight conductors by applying FEM simulations. *Eur. J. Phys.* **2000**, *36*. [CrossRef]
17. Abdelbagi, H.E. Skin and Proximity Effects in Two Parallel Plates. 2007. Available online: https://corescholar.libraries.wright.edu/etd_all/18419 (accessed on 30 November 2020).
18. Vitelli, M. Numerical evaluation of 2-D proximity effect conductor losses. *IEEE Trans. Power Deliv.* **2004**, *19*, 1291–1298. [CrossRef]
19. L'Eplattenier, P.; Cook, G.; Ashcraft, C.; Burger, M.; Imbert, J.; Worswick, M. Introduction of an Electromagnetism Module in LS-DYNA for Coupled Mechanical-Thermal-Electromagnetic Simulations. *Steel Res. Int.* **2009**, *80*, 351–358. [CrossRef]
20. Gies, S.; Tekkaya, A.E. Analytical prediction of Joule heat losses in electromagnetic forming coils. *J. Mater. Process. Technol.* **2017**, *246*, 102–115. [CrossRef]
21. Liu, X.; Huang, L.; Su, H.; Ma, F.; Li, J. Comparative Research on the Rebound Effect in Direct Electromagnetic Forming and Indirect Electromagnetic Forming with an Elastic Medium. *Materials* **2018**, *11*, 1450. [CrossRef] [PubMed]
22. Weddeling, C. Electromagnetic Form-Fit Joining. Ph.D. Dissertation, Universität Dortmund, Dortmund, Germany, 2014.
23. Clausen, A.H.; Børvik, T.; Hopperstad, O.S.; Benallal, A. Flow and fracture characteristics of aluminium alloy AA5083–H116 as function of strain rate, temperature and triaxiality. *Mater. Sci. Eng. A* **2004**, *364*, 260–272. [CrossRef]
24. Kabirian, F.; Khan, A.S.; Pandey, A. Negative to positive strain rate sensitivity in 5xxx series aluminum alloys: Experiment and constitutive modeling. *Int. J. Plast.* **2014**, *55*, 232–246. [CrossRef]
25. Moro, E.A. New Developments in photon Doppler velocimetry. *J. Phys. Conf. Ser.* **2014**, *500*, 142023. [CrossRef]
26. Beerwald, C. Grundlagen der Prozessauslegung und -Gestaltung Bei der Elektromagnetischen Umformung. Ph.D. Dissertation, Universität Dortmund, Dortmund, Germany, 2005.
27. Gies, S.; Tekkaya, A.E. Design of Hybrid Conductors for Electromagnetic Forming Coils. In Proceedings of the 8th International Conference on High Speed Forming, Columbus, OH, USA, 14 May 2018. [CrossRef]

Journal of
*Manufacturing and
Materials Processing*

Article

Numerical and Experimental Investigation of the Impact of the Electromagnetic Properties of the Die Materials in Electromagnetic Forming of Thin Sheet Metal

Björn Beckschwarte [1,2,*], **Lasse Langstädtler** [1,2], **Christian Schenck** [1,2,3], **Marius Herrmann** [1,2,3] and **Bernd Kuhfuss** [1,2,3]

1   Bremen Institute for Mechanical Engineering, University of Bremen, Badgasteiner Str. 1,
    28359 Bremen, Germany; langstaedtler@bime.de (L.L.); schenck@bime.de (C.S.); herrmann@bime.de (M.H.);
    kuhfuss@bime.de (B.K.)
2   University of Bremen, Bremen 28359, Germany
3   Mapex Center for Materials and Processing, Postbox 330440, 28334 Bremen, Germany
*   Correspondence: beckschwarte@bime.de; Tel.: +49-421-218-64805

Abstract: In electromagnetic forming of thin sheet metal, the die is located within the effective range of the electromagnetic wave. Correspondingly, a current is induced not only in the sheet metal, but also in the die. Like the current in the workpiece, also the current in the die interacts with the electromagnetic wave, resulting in Lorentz forces and changes of the electromagnetic field. With the aim to study the influence of different electromagnetic die properties in terms of specific electric resistance and relative magnetic permeability, electromagnetic simulations were carried out. A change in the resulting forming forces in the sheet metals was determined. To confirm the simulation results, electromagnetic forming and embossing tests were carried out with the corresponding die materials. The results from simulation and experiment were in good agreement.

Keywords: impulse forming; bulge forming; permeability; conductivity

Citation: Beckschwarte, B.; Langstädtler, L.; Schenck, C.; Herrmann, M.; Kuhfuss, B. Numerical and Experimental Investigation of the Impact of the Electromagnetic Properties of the Die Materials in Electromagnetic Forming of Thin Sheet Metal. *J. Manuf. Mater. Process.* **2021**, 5, 18. https://doi.org/10.3390/jmmp5010018

Received: 7 December 2020
Accepted: 9 February 2021
Published: 12 February 2021

Publisher's Note: MDPI stays neutral with regard to jurisdictional claims in published maps and institutional affiliations.

1. Introduction

Electromagnetic forming is an energy-based process that can be used for high speed forming of tubes and sheets [1]. In this process, a current pulse causes a strong magnetic field at a coil. The field induces eddy currents in an electrically conductive workpiece. Lorentz forces arise by interaction of magnetic field and eddy currents that are used for forming [2], cutting [3], joining [4] and embossing operations [5]. Various advantages in contrast to quasi-static forming methods such as changes in the plastic material behavior [6] and one-sided simple die [7] were investigated. Furthermore, in electromagnetic forming wrinkling [8] and springback [9] can be reduced. Starting with the first patent related to electromagnetic tube forming mentioned by Harvey and Brower [10], an in-depth forming investigation was carried out for sheets and tubes with a thickness greater than 1 mm [11]. In contrast to thick sheet metal, thin sheet metal are completely penetrated by the electromagnetic field. They have only been examined sparsely starting with the investigations of Haiping and Chunfeng [12]. Due to the insufficient shielding of thin sheet metal and the resulting current flow in the die, the consideration of the die influence is extended by the aspect of electromagnetic influences. Basic considerations of the die during electromagnetic forming refer to the mechanical interaction between sheet metal and die. It is known, for example, that electromagnetic embossing leads to increased adhesion and sheet-die welding [5]. In addition, the mechanical properties of the die influence the forming process. The die strength is determined by mechanical properties, but process-specific effects such as the rebound effect caused by high impact velocities influences also the forming result [13]. Furthermore, not only the process is influenced by the die material, but also the process application as well as the workpiece and die manufacture. Therefore, it is necessary to be able to describe the mechanical and electromagnetic influences. The relevant influences increase with the use of thin sheet metal.

The increase of electromagnetic influences is explained by the insufficient shielding of the magnetic field by thin metal sheets, which allows field interaction at the backside of the workpiece. Therefore, based on the workpiece thickness in relation to penetration depth of the magnetic field the electromagnetic forming process can be divided into two cases [14]. The first case occurs, when the penetration depth is lower than the workpiece thickness. Therefore, a current flow is only generated in a layer of the workpiece [15]. The second case is characterized by a current flow in the complete volume and a magnetic body force, which is a result of the high penetration depth in relation to a thin workpiece. In this case, a magnetic field acts behind the workpiece and a current is inducted in the die [16]. A scale for the penetration depth of a magnetic field is the skin depth δ, which is dependent on the mathematical constant π, the specific electrical resistance ρ, permeability of vacuum μ_0, relative magnetic permeability μ_r and discharge frequency f of the current pulse:

$$\delta = \sqrt{\frac{\rho}{\pi \cdot \mu_r \cdot \mu_0 \cdot f}} \tag{1}$$

Therefore, the penetration depth in the workpiece and die is only dependent on the material (ρ, μ_r) and the discharge frequency f. The specific resistance ρ and the relative magnetic permeability μ_r of the die influence the change of shielding in contrary directions. In case of a real forming setup, the discharge frequency is set by the inductivity of the machine and the tool coil. Due to this inductivity, the discharge frequency is limited which limits the reduction of the penetration depth. The impact of an occurring current in a forming die is shown by Cao [17] by simulation of two different tool coil setups and specific resistance of the die. The work based on the usage of thick sheet metal, with sufficient shielding of the magnetic field, so a die current is generated by a setup of two tool coils. For the setup with one coil without a resulting current in the die, the specific resistance value of the die did not change the geometry of the forming results. In the other setup with two tool coils, the second tool coil is used to induce a current in the forming die. Through this, Cao found that a higher current density in the die raised the draw-in and the forming depth. In contrast to the investigations related to the forming die, investigations regarding the workpiece show influences on the plastic material behavior by an electric current density (electroplastic effect—EPE) [18]. In this context, an increase in plasticity was shown for both long-term [19] and short-term [20] current duration time. This influence is attributed to the movement of free electrons, which lead to an increased (local) temperature of the workpiece due to the Joule heating caused by the electron movement [21]. But the comparison of electrically assisted forming with forming at different workpiece temperatures still shows differences regarding the plastic material behavior and grain microstructure [22,23]. In addition to the influences on the material behavior, an influence of the current density and its distribution on the force during electromagnetic forming can be observed. This can be explained by the influence of the current density on the Lorentz force, whereby the electric and magnetic components are influenced (see Equation (2)). As a result, the thesis can be formulated that an influence on the electromagnetic forming can be attributed to a change in the electric charge density ρ_c, the electric field strength E and magnetic field strength H or magnetic flux density B. Whereby the electric charge density ρ can be expressed by the current density j related to the velocity of the charge v (see Equation (2)). Depending on these variables in this study, the influence of the electromagnetic properties of the die is to be investigated by simulations:

$$\vec{F} = \iiint_V \rho_c \cdot \left(\vec{E} + \vec{v} \times \vec{B} \right) dV$$
$$\vec{F} = \iiint_V \frac{\vec{j}}{\vec{v}} \cdot \vec{E} + \vec{j} \times \vec{B} \, dV \tag{2}$$

The process effect is quantified by the impulse as an integrated value of the impact. As the die cavity influences the magnetic field, a plane geometry was used. Consequently, no forming was considered in the simulation model. In the model, both electromagnetic parameters were varied in the range of real materials values. Beyond that, the influence of the die material was shown by experiments with different real die materials. These experiments were based on the simulation model, whereby the experiments enabled a validation of the simulated results.

2. Materials and Methods

2.1. Die Material

The ranges of the simulated electromagnetic die properties were chosen that they include real material properties. The specific resistance value was varied from 1×10^{-8} $\Omega \cdot$m to 1 $\Omega \cdot$m, the relative permeability ranged from 0.9 to 300. A change in relative permeability μ_r and specific conductivity ρ due to a change in temperature is not considered in the simulation.

During experiments, the variation of the electromagnetic properties was done by different die materials. The selection included common tool steels (90MnCrV8) with high specific resistance and high relative permeability. In the simulation part, the ferromagnetic properties of the tool steel were neglected, and a linear behavior was applied. Further, an austenitic chrome-nickel steel (X5CrNi18-10) with high specific resistance and a relative permeability of 1 was used. In contrast to the high specific resistance of ferrous based materials, copper and aluminium with a low specific resistance were also applied as die material. The applied specific resistance and the relative permeability for the used die materials and for the used aluminium sheet metal for simulation is shown in Table 1 [24].

Table 1. Electromagnetic properties of the used die and workpiece materials from the ANSYS database.

	Material	Specific Resistance ρ [$\Omega \cdot$m]	Relative Permeability μ_r [-]
	austenitic chrome-nickel steel X5CrNi18-10/AISI 304	9.0909×10^{-7}	1.000000
	cold work tool steel 90MnCrV8/AISI O2	5×10^{-7}	~300
die	copper E-Cu57/C11000/ETP	1.7241×10^{-8}	0.999991
	aluminium AlCuMgPb/AA2007	2.6316×10^{-8}	1.000021
workpiece	aluminium Al99.5/AA1050A	2.6316×10^{-8}	1.000021

2.2. Simulation Model

A two-dimensional planar electromagnetic simulation model within the software environment ANSYS Maxwell 2D 19.1.0 was used in this work. The magnetic transient model consisted of four elements; coil, workpiece, die and surrounding air, see Figure 1a. The maximum mesh element length for the sheet metal was set to 0.02 mm. For all other parts of the model the maximum mesh element length was set to 0.5 mm. A triangular mesh with an adaptive mesh algorithm was applied. As tool coil, a single straight conductor was used to receive a plane symmetric magnetic field, which can be described in the planar area. Mutual interactions between the magnetic field and both the changing geometry during forming and the die cavity were avoided by applying a flat die. Thus, a coupled mechanical model of the electromagnetic forming was not necessary. Due to the use of a single straight conductor with low inductance, the discharge is largely determined by the electrical properties of the capacitor system. Changes to the coil-workpiece-die system therefore play a subordinate role. Following a constant discharge was used in simulation. During all simulations, the input current in Figure 1b was used. The current based on a charging energy E_C = 1200 J discharge was modelled as a damped sine function with a frequency of f = 14.5 kHz, a maximum current amplitude of I_{max} = 37 kA and a damping factor of 5600. All simulations were evaluated in 1 μs time steps. According to Equation (1), the corresponding skin depth in aluminium (Al99.5) was 678 μm. During all simulations, sheet metal with a thickness of s_0 = 200 μm was used. As a result, the initial electromagnetic field was not shielded completely by the sheet metal and hence, a current was generated in the die depending on the die material. To specify the impact, the impulse J of the sheet metal was calculated by integrating the body force F_{LN} in -Z direction over the whole volume of the workpiece, see Equation (3) and over four current periods T of the oscillating circuit described by the damped sine oscillation:

$$J = \int_0^{4 \cdot T} F_{LN,x} \mathrm{d}t \qquad (3)$$

(a) (b)

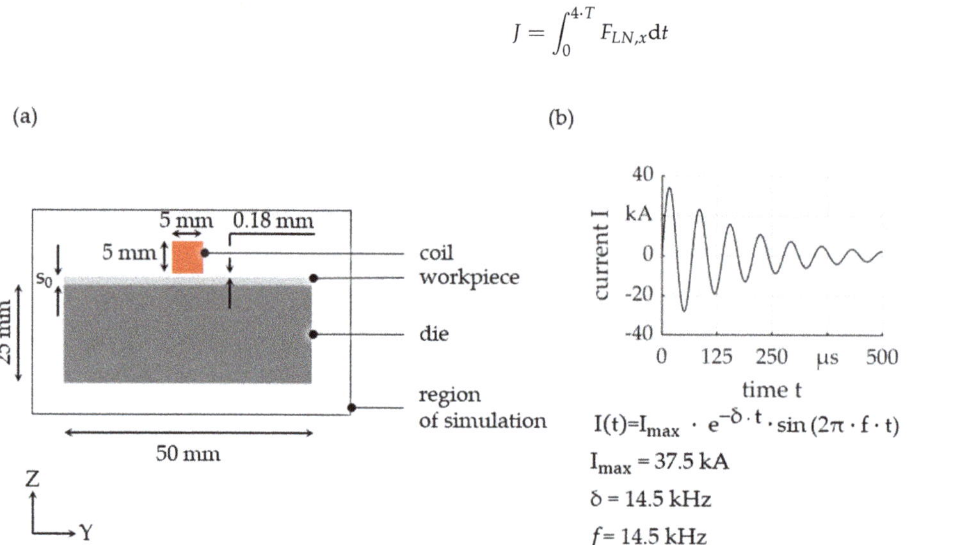

Figure 1. Simulation model; (**a**) schematic; (**b**) input current (tool coil).

2.3. Experimental Setup

During the experiments, a capacitor bank (capacity of C = 100 μF) with a maximum charging energy of E_C = 1800 J at maximum charging voltage of U_0 = 6 kV was used, see Figure 2a. Forming experiments were carried out with charge energy of 1200 J. Switching was done by a single ignitron (NL508/NL508A, National Electronics, LaFox, IL, USA). The inductance of the circuit including the tool coil was approximately L = 0.6 μH. As tool coil, a single straight conductor with a rectangular cross-section of 5 × 5 mm² was applied (Figure 2b). The tool coil was made of copper (E-Cu57) by milling. It was embedded in a polylactide carrier structure and insulated from the workpiece with 195 μm thick polyimide foil.

Al99.5 sheets with a thickness of s_0 = 200 μm dimensioned 50 × 50 mm² were used during the experiments. In combination with the single straight conductor coil and capacitor bank used in the tests, the skin depth in the aluminium sheet metal was 678 μm. Therefore, the die is in the range of the electromagnetic field during the experiments. For validation of the simulation results, four dies for free bulging were used, see Figure 3a.

During the experiments, five workpieces were formed with each die. The convex side of the formed sheet metal geometry was measured by a laser confocal microscope (VXK-210, Keyence, Itasca, IL, USA). With these measurements, the die specific bulge height h was determined. The bulge height was defined as the maximum distance of the convex formed sheet geometry and the undeformed flange geometry (see Figure 3b).

Furthermore, embossing experiments were carried out on raised circular structures, regarding the transfer of the results to replication processes, (see Figure 4a). These experiments were carried out with Al99.5 sheets with a thickness of 50 µm and 800 J charge energy. Austenitic chrome-nickel steel (X5CrNi18-10) and copper (E-Cu57) dies were used. The corner radius r achieved along and perpendicular to the effective tool coil area was determined as a measure for the embossing result (see Figure 4b). The measurements were carried out with a laser confocal microscope (Keyence VXK-210).

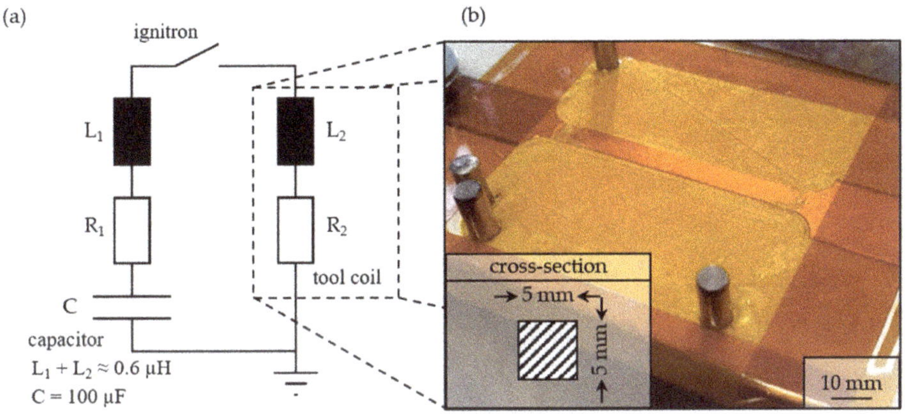

Figure 2. Experimental setup; (**a**) equivalent circuit diagram; (**b**) tool coil.

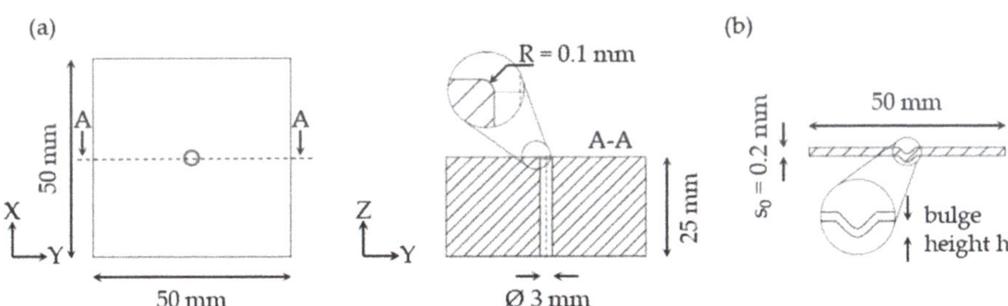

Figure 3. Forming experiment; (**a**) free forming die geometry; (**b**) resulting workpiece geometry.

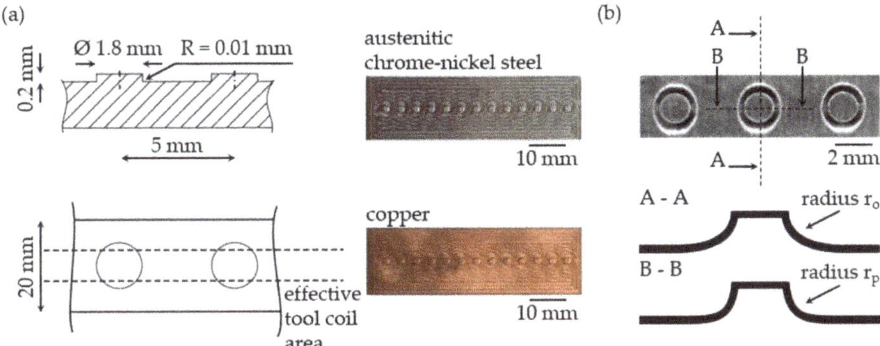

Figure 4. Embossing experiment; (**a**) die geometry; (**b**) resulting workpiece geometries.

3. Results

3.1. Simulation Results

In a first step, the simulated impulses J on the workpiece were plotted against the resulting skin depth δ of the die material (see Equation (1)). The impulse J could not be correlated to the skin depth δ and the maximum values can be achieved with different combinations of specific resistance ρ and relative permeability μ_r values (see Figure 5). In contrast to influence of sheet metal thickness in electromagnetic forming, the electromagnetic die properties must influence the impulse on the sheet metal independently.

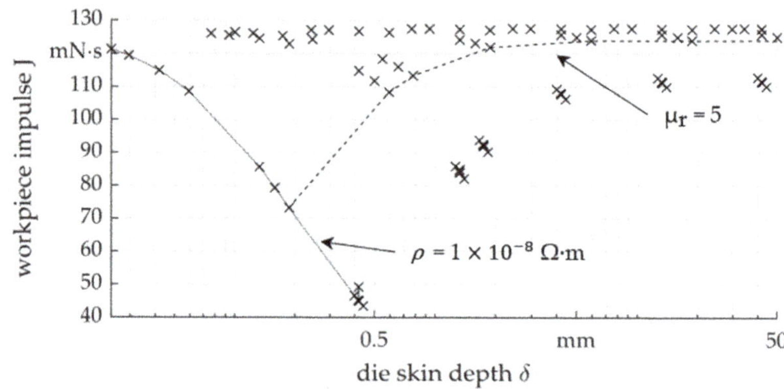

Figure 5. Simulated impulse as a function of the resulting skin depth for the parameter combination of specific resistance ρ and relative permeability μ_r values.

The influence of relative permeability μ_r and specific resistance ρ on the impulse J of the workpiece is shown in Figure 6. The impulse J rises significantly with the specific resistance ρ until a maximum value is reached at about $1 \times 10^{-5}\ \Omega\cdot m$. Further increasing specific resistance ρ does not affect the impulse J. With an increase of the relative permeability μ_r, the impulse J also increases. Thus, the die material must be chosen carefully for efficient electromagnetic forming of thin sheets with the demand of minimum specific resistance ρ and high relative permeability μ_r.

The process effect of the electromagnetic force depends on the current density j, the magnetic flux density B and electric field strength E in the workpiece (see Equation (2)). Based on this, the die material must have an influence on the current density j, magnetic flux density B, magnetic field strength H or the electric field strength E in the workpiece. Figure 7 shows the dependence of the maximum current density j_{max} in the sheet volume as a function of the die material properties. An increase in the current density j is caused by high specific resistance ρ and relative permeability μ_r values. Thus, the increased pulse could be explained by an increased current density j.

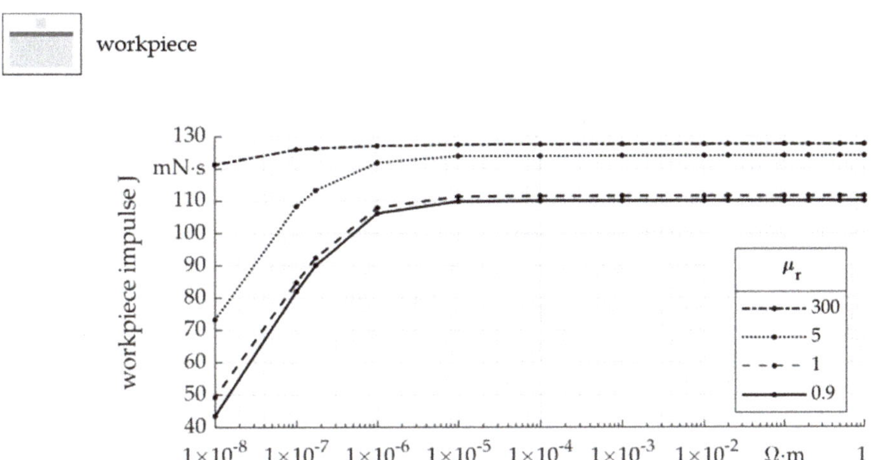

Figure 6. Simulated impulse J on thin aluminium sheets for different electromagnetic properties of the die material.

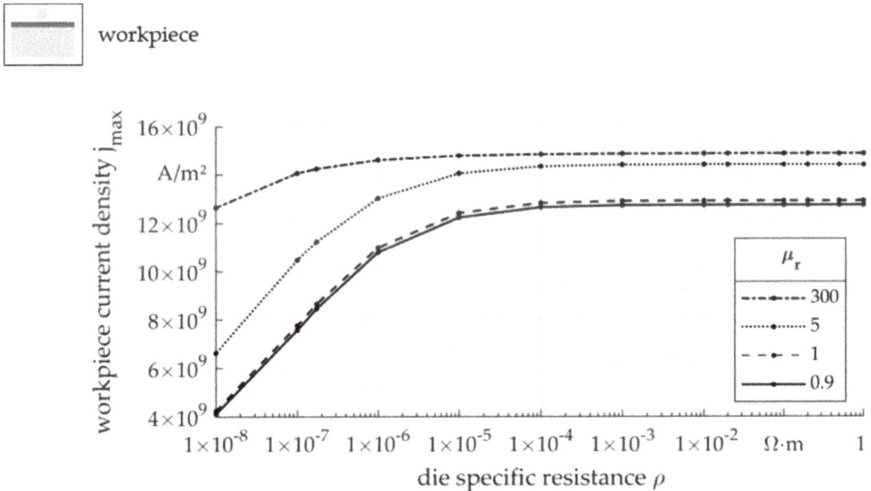

Figure 7. Maximum current density j_{max} in the workpiece volume in respect of the specific resistance ρ of the die material.

To analyse the effect of the electromagnetic field penetrating the die, the electric charge q of the die is calculated, by the integral current density j in the die y-z-surface over time t. This value enables the investigation of the energy which is transferred in the die. The electric charge q is reduced by an increased relative permeability μ_r and specific resistance ρ (see Figure 8), so a current flow in the die is suppressed by the material properties.

According to a current flow in the die, an electromagnetic field is induced from the die. This field causes a force to act against the forming direction, which reduce or can even result in a negative impulse J_N. Corresponding to the current flow in the die, it is induced for low specific resistances ρ and low relative permeability μ_r (see Figure 9). So, the impulse J during electromagnetic forming is decreased though the acting of the inducted die current density j.

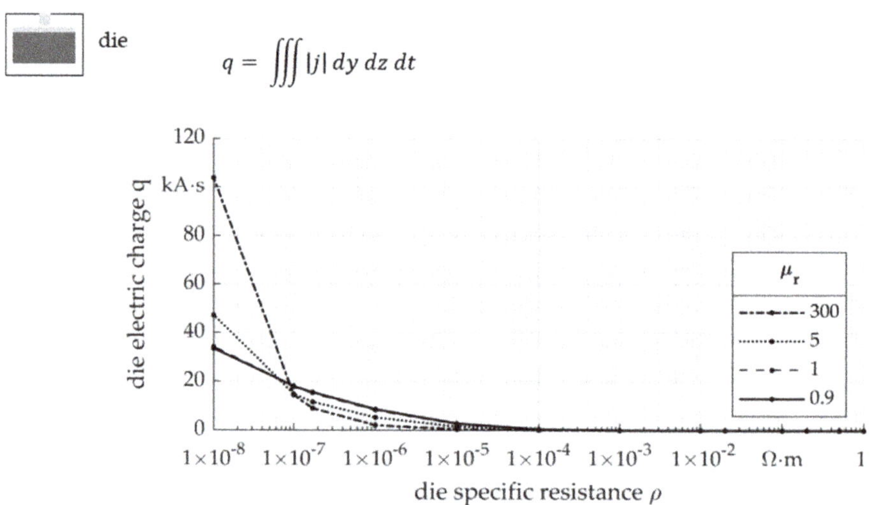

Figure 8. Integral of the current density j in the die volume over time in respect of the specific resistance ρ of the die material.

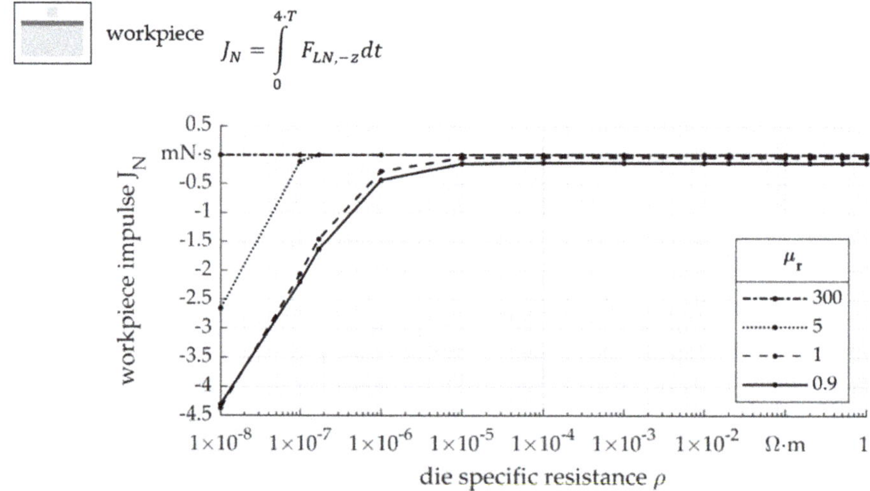

Figure 9. Integral of the negative force components over time in respect of the specific resistance ρ of the die material.

In addition to the influence of the die material on the current density j in the workpiece, the field changes in strength and shape by the different material properties. Figures 10 and 11 shows this change in the field regarding the material properties, each highest and lowest specific resistance ρ and relative permeability μ_r. For dies with high relative permeability μ_r within the die volume, a low field strength H is present. However, this is followed by a field concentration within the sheet volume which leads to a higher impulse J. A decrease of the field strength H is observed by the increase of the specific resistance ρ. However, the field is more concentrated in the die for a low specific resistance ρ, thus the impulse J is lower for the workpiece due to the higher effect in opposite direction.

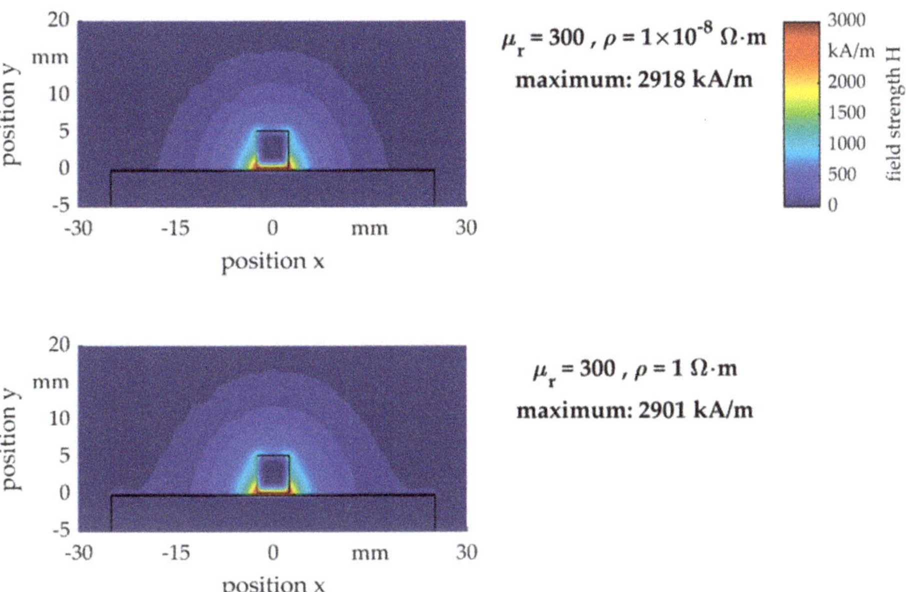

Figure 10. Distribution of the magnetic field with high relative permeability μ_r of the die during the maximum current density in the tool coil.

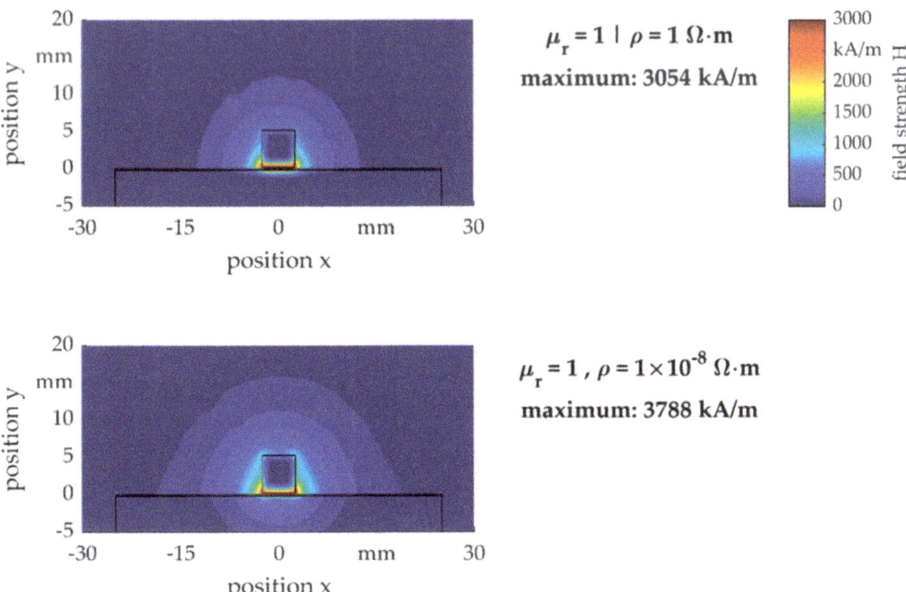

Figure 11. Distribution of the magnetic field with low relative permeability μ_r of the die during the maximum current density in the tool coil.

For a further comparison of the influence of the die material properties on the magnetic field, a normal distribution of the field is assumed. The average magnetic field strength H_a is calculated as the average in the workpiece volume at the time step of the maximum tool coil current. In respect of the electromagnetic die properties the results are shown in Figure 12. The average field strength H_a in the workpiece can be increased with high relative permeability μ_r and high specific resistance ρ of the die material.

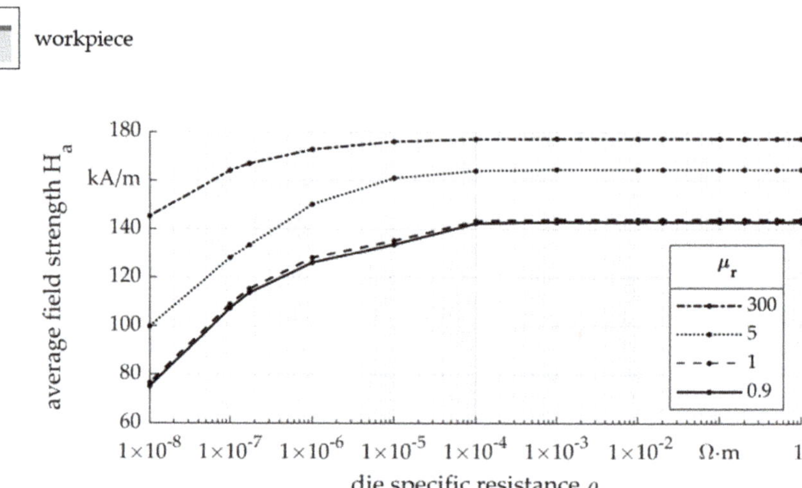

Figure 12. Average magnetic field strength H_a in the workpiece in respect of the specific resistance ρ of the die material.

To show the influence of the die material properties on the distribution of the field strength H which penetrates the workpiece, the deviation of the distribution of the field strength H_d in the workpiece volume during maximum tool coil current is shown in Figure 13. Once more, a high relative permeability μ_r and a high specific resistance ρ increase the spread of the field strength H in the workpiece. Thus, the spread or the effective range of the electromagnetic field of the tool coil is changed by the die material properties.

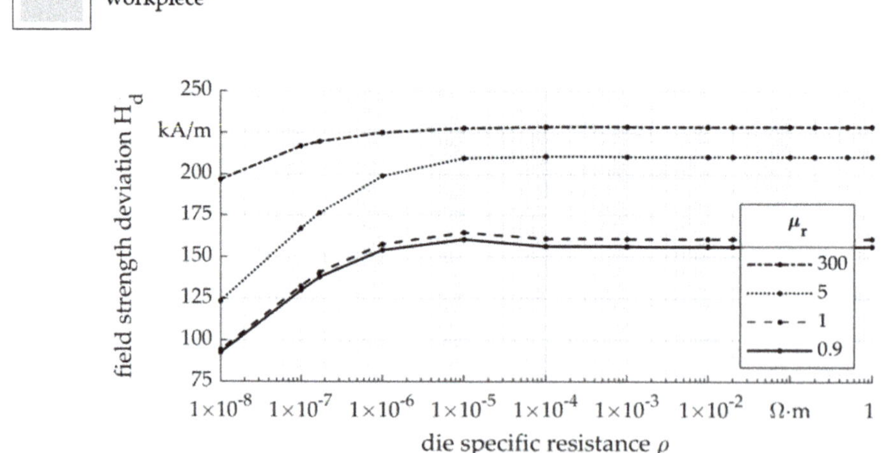

Figure 13. Field strength deviation H_d (standard deviation of the magnetic field strength) in the workpiece in respect of the specific resistance of the die material.

3.2. Experimental Results

Based on the electromagnetic simulation, an initial qualitative confirmation of the results can be carried out by comparison with experiments. Due to the dependence of the impulse J on the current density j (see Equation (2)), which in turn influences the plastic material behavior, a clear validation by the forming result is not possible with the actual model. The correlation in Figure 14 of the impulse to the forming result, which represented by the bulge height h, shows a linear dependence with good agreement. The results of the simulation model are confirmed under the boundary condition that more impulse causes a greater forming result. In contrast to this are the effects from the change of the plastic material behavior, which can be traced back to the die material. First, the current density in the workpiece is changed, which results in an influence on the plastic material properties by exceeding the limiting current density described in the literature (EPE) [21]. At the same time in dependence of the die material, current flow and forming lead to heating, which affect the material properties during forming. Likewise, it cannot be ruled out that different temporal and local force effects, which are not considered by the impulse, lead to a change in the strain rate and strain state, as well their material property dependencies. Regarding the differentiated consideration into influence of the relative permeability μ_r, the influence on the impulse can be confirmed by the comparison of the iron-based dies with different relative permeability μ_r and comparable specific resistance ρ. The impact of specific resistance ρ is shown by comparing the results of aluminium or copper-based dies with austenitic chrome-nickel steel-based dies, which have comparable relative permeability μ_r.

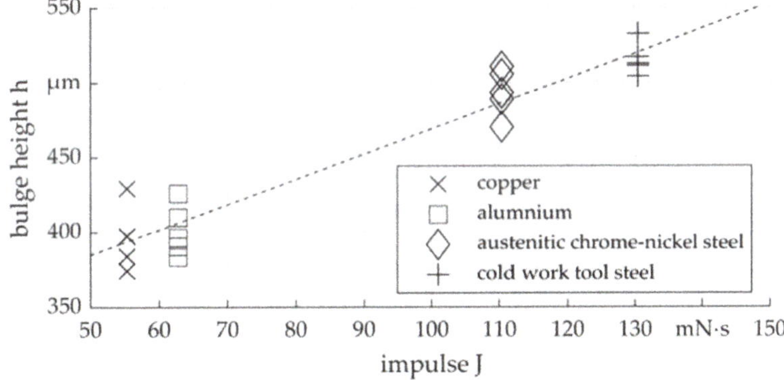

Figure 14. Corresponding experimental bulge height h for the simulated impulse J for different die materials with linear correlation line for a 1250 J forming experiment.

Regarding the embossing by electromagnetic forces, the performance of electromagnetic embossing can be increased by the influence of the die material. The influence of the die material is shown in Figure 15 by comparing the achieved inner corner radii r. Austenitic chrome-nickel steel-based dies achieve smaller radii r than copper-based tools, with lower specific resistance ρ for both directions considered. This makes embossing with steel-based dies more efficient in terms of die properties.

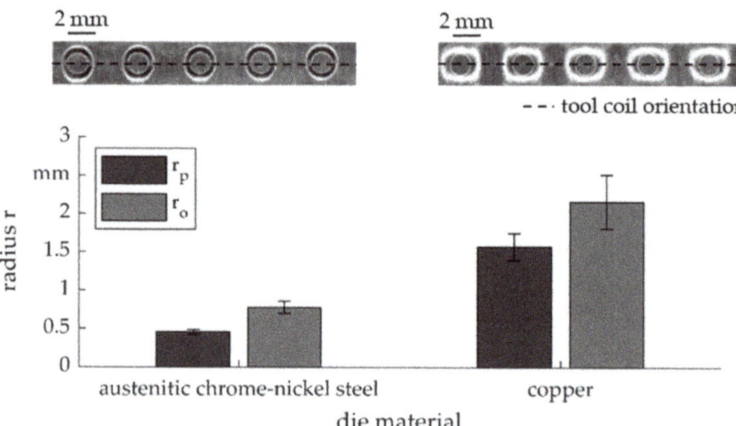

Figure 15. Comparison of the inner corner radius in two directions and the embossing result of two embossing dies made of austenitic chrome-nickel steel and copper, cf. [25].

4. Conclusions

In the case that the die is within the effective range of the electromagnetic wave, an influence of electromagnetic properties of the die material on the forming process can be proven in simulation and experiments. The specific resistance ρ and relative permeability μ_r influence the transmitted impulse J during the forming process. Both a high specific resistance ρ and a high relative permeability μ_r of the die material cause a higher impulse. The following conclusions can be drawn:

- Below a specific resistance ρ of around 10^{-5} Ω·m, a decrease in the workpiece impulse during electromagnetic forming can be expected. To increase forming efficiency, the die material should have a higher specific resistance ρ.
- The relative permeability influences the impulse up to a relative permeability μ_r of 5, whereby the influence decreases with higher relative permeability μ_r. A high relative permeability μ_r is positive regarding process efficiency.
- The influence of the die material on the impulse J is due to the current density j in the workpiece. A high specific resistance ρ and relative permeability μ_r cause higher current densities j, which increases the impulse J on the workpiece.
- The properties of the die material influence the field, which can be seen in the change of the distribution and level of the field strength H. A high specific resistance ρ and relative permeability μ_r lead to a magnetic field concentration.
- The results of the simulation confirmed by forming experiments using free forming of bulge structures.
- The results of the simulation and free forming could be transferred to electromagnetic embossing and it could be shown that an improvement of the impression is possible due to the die material influence.

From the consideration of the influence of the electromagnetic properties of the die material it can be concluded that the die with material properties and geometry by local changes influences the process. Consequently, when designing the process for electromagnetic forming thin sheet metal, it is necessary to consider the repercussions of the die properties and the cavity. In combination with the dependency of thin sheet metal on the discharge frequency, the die, workpiece, tool coil and impulse forming system must be considered. In addition to this additional effort in the process design, the influence of the die material offers a possibility to improve the performance of the forming process by local adaptation via the die. With this aim, the influence of the die on the produced geometry and the conditions prevailing during the process must be determined. Thus, the impulse is changed during the forming process, but beyond that, influences regarding workpiece temperature and strain rate are expected. According to this thesis, influences on the plastic material behaviour of the workpiece following, so that the changed generated geometry cannot be entirely explained by different impulses.

Author Contributions: Conceptualization, B.B. and L.L.; methodology, B.B.; software, B.B.; validation, B.B. and L.L.; formal analysis, B.B.; investigation, B.B. and L.L.; resources, B.K.; data curation, B.B.; writing—original draft preparation, B.B., C.S., L.L. and M.H.; writing—review and editing, B.B., C.S., L.L., M.H. and B.K.; visualization, B.B.; supervision, B.K.; project administration, B.K.; funding acquisition, L.L., C.S., M.H. and B.K. All authors have read and agreed to the published version of the manuscript.

Funding: This research was funded by the Deutsche Forschungsgemeinschaft (German Research Foundation—DFG), grant number 395821503.

Data Availability Statement: The data presented in this study are available on request from the corresponding author.

Acknowledgments: The authors gratefully acknowledge the support by the German Research Foundation DFG for the project "Electromagnetic embossing of optical microstructures".

Conflicts of Interest: The authors declare no conflict of interest.

References

1. Kleiner, M.; Beerwald, C.; Homberg, W. Analysis of Process Parameters and Forming Mechanisms within the Electromagnetic Forming Process. *CIRP Ann.* **2005**, *54*, 225–228. [CrossRef]
2. Psyk, V.; Kurka, P.; Kimme, S.; Werner, M.; Landgrebe, D.; Ebert, A.; Schwarzendahl, M. Structuring by electromagnetic forming and by forming with an elastomer punch as a tool for component optimisation regarding mechanical stiffness and acoustic performance. *Manuf. Rev.* **2015**, *2*, 23. [CrossRef]
3. Kuhfuss, B.; Schenck, C.; Wilhelmi, P.; Langstädtler, L. Magnetic pulse cutting of micro metal foils. In Proceedings of the 8th International Conference of Micromanufacturing, ICOMM, Victoria, BC, Canada, 25–28 March 2013.
4. Jimbert, P.; Pérez, I.; Eguia, I.; Daehn, G. Straight Hemming of Aluminum Sheet Panels Using the Electromagnetic Forming Technology: First Approach. *Key Eng. Mater.* **2007**, *344*, 365–372. [CrossRef]
5. Kamal, M.; Shang, J.; Cheng, V.; Hatkevich, S.; Daehn, G.S. Agile manufacturing of a micro-embossed case by a two-step electromagnetic forming process. *J. Mater. Process. Technol.* **2007**, *190*, 41–50. [CrossRef]
6. Taebi, F.; Demir, O.K.; Stiemer, M.; Psyk, V.; Kwiatkowski, L.; Brosius, A.; Blum, H.; Tekkaya, A.E. Dynamic forming limits and numerical optimization of combined quasi-static and impulse metal forming. *Comput. Mater. Sci.* **2012**, *54*, 293–302. [CrossRef]
7. Langstädtler, L.; Schönemann, L.; Schenck, C.; Kuhfuss, B. Electromagnetic embossing of optical microstructures. *J. Micro Nano Manuf.* **2016**, *4*, 1–4. [CrossRef]
8. Li, H.W.; Yao, X.; Yan, S.; He, J.; Zhan, M.; Huang, L. Analysis of forming defects in electromagnetic incremental forming of a large-size thin-walled ellipsoid surface part of aluminum alloy. *J. Mater. Process. Technol.* **2018**, *255*, 703–715. [CrossRef]
9. Iriondo, E.; Gutiérrez, M.A.; González, B.; Alcaraz, J.L.; Daehn, G.S. Electromagnetic impulse calibration of high strength sheet metal structures. *J. Mater. Process. Technol.* **2011**, *211*, 909–915. [CrossRef]
10. Harvey, G.W.; Brower, D.F. General Dynamics Corp. Metal Forming Device and Method. U.S. Patent 2,976,907, 28 March 1961.
11. Psyk, V.; Risch, D.; Kinsey, B.L.; Tekkaya, A.E.; Kleiner, M. Electromagnetic forming—A review. *J. Mater. Process. Technol.* **2011**, *211*, 787–829. [CrossRef]
12. Yu, H.; Li, C. Effects of current frequency on electromagnetic tube compression. *J. Mater. Process. Technol.* **2009**, *209*, 1053–1059. [CrossRef]
13. Risch, D.; Beerwald, C.; Brosius, A.; Kleiner, M. On the Significance of the Die Design for Electromagnetic Sheet Metal Forming. In Proceedings of the International Conference on High Speed Forming, ICHSF, Dortmund, Germany, 31 March–1 April 2004; pp. 191–200. [CrossRef]
14. Thibaudeau, E.; Kinsey, B.L. Analytical design and experimental validation of uniform pressure actuator for electromagnetic forming and welding. *J. Mater. Process. Technol.* **2015**, *215*, 251–263. [CrossRef]
15. Paese, E.; Geier, M.; Homrich, R.P.; Rossi, R. A coupled electric–magnetic numerical procedure for determining the electromagnetic force from the interaction of thin metal sheets and spiral coils in the electromagnetic forming process. *Appl. Math. Model.* **2015**, *39*, 309–321. [CrossRef]
16. Langstädtler, L.; Schenck, C.; Kuhfuss, B. Effective electromagnetic forces in thin sheet metal specimen. *MATEC Web Conf.* **2015**, *21*, 11002. [CrossRef]
17. Cao, Q.; Li, Z.; Lai, Z.; Li, Z.; Han, X.; Li, L. Analysis of the effect of an electrically conductive die on electromagnetic sheet metal forming process using the finite element-circuit coupled method. *Int. J. Adv. Manuf. Technol.* **2019**, *101*, 549–563. [CrossRef]
18. Kravchenko, V.Y. Effect of directed electron beam on moving dislocations. *Sov. Phys. JETP* **1967**, *24*, 1135–1142.

19. Wang, X.; Xu, J.; Wang, C.; Sánchez Egea, A.J.; Li, J.; Liu, C.; Wang, Z.; Zhang, T.; Guo, B.; Cao, J. Bio-Inspired Functional Surface Fabricated by Electrically Assisted Micro-Embossing of AZ31 Magnesium Alloy. *Materials* **2020**, *13*, 412. [CrossRef] [PubMed]
20. Unger, J.; Stiemer, M.; Walden, L.; Bach, F.; Blum, H.; Svendsen, B. On the effect of current pulses on the material behavior during electromagnetic metal forming. In Proceedings of the 2nd International Conference on High Speed Forming, Dortmund, Germany, 20–21 March 2006; pp. 23–32.
21. Gallo, F.; Satapathy, S.; Ravi-Chandar, K. Plastic deformation in electrical conductors subjected to short-duration current pulses. *Mech. Mater.* **2012**, *55*, 146–162. [CrossRef]
22. Sánchez Egea, A.J.; Peiró, J.J.; Signorelli, J.W.; González Rojas, H.A.; Celentano, D.J. On the microstructure effects when using electropulsing versus furnace treatments while drawing inox 308L. *J. Mater. Res. Technol.* **2019**, *8*, 2269–2279. [CrossRef]
23. Wang, X.; Sánchez Egea, A.J.; Xu, J.; Meng, X.; Wang, Z.; Shan, D.; Guo, B.; Cao, J. Current-Induced Ductility Enhancement of a Magnesium Alloy AZ31 in Uniaxial Micro-Tension Below 373 K. *Materials* **2018**, *12*, 111. [CrossRef] [PubMed]
24. ANSYS. *Material Database*; ANSYS: Canonsburg, PA, USA, 2019.
25. Langstädtler, L. Elektromagnetisches und Elektrohydraulisches Umformen in der Mikroproduktion. Ph.D. Thesis, Universität Bremen, Bremen, Germany, 2020. [CrossRef]

Journal of
*Manufacturing and
Materials Processing*

Article

Examples of How Increased Formability through High Strain Rates Can Be Used in Electro-Hydraulic Forming and Electromagnetic Forming Industrial Applications

Gilles Avrillaud *, Gilles Mazars, Elisa Cantergiani, Fabrice Beguet, Jean-Paul Cuq-Lelandais and Julien Deroy

Bmax, 30 Boulevard Thibaud, 31100 Toulouse, France; gilles.mazars@icube-research.com (G.M.);
elisa.cantergiani@icube-research.com (E.C.); fabrice.beguet@bmax.com (F.B.);
jean-paul.cuq-lelandais@icube-research.com (J.-P.C.-L.); julien.deroy@icube-research.com (J.D.)
* Correspondence: gilles.avrillaud@bmax.com; Tel.: +33-5-3461-1660

Abstract: In order to take up some challenges in metal forming coming from the recent environmental stakes, Electromagnetic Forming and Electro-Hydraulic Forming processes have been developed at the industrial scale, using the advantages of high strain rates. Such progress has been possible in particular thanks to the emergence of strongly coupled simulation tools. In this article, some examples have been selected from some industrial applications in deep forming, postforming, embossing, and complex shapes forming. It shows how in particular, the increase in formability can bring benefits to solve customer issues in the automotive, luxury packaging, aeronautic, and particles accelerator sectors. Some simulation results are presented to explain how this highly dynamic forming occurs for each of these applications.

Keywords: electromagnetic forming; magnetic pulse forming; electro-hydraulic forming; high strain rates; lightweight; high pulsed power; formability; simulation

Citation: Avrillaud, G.; Mazars, G.; Cantergiani, E.; Beguet, F.; Cuq-Lelandais, J.-P.; Deroy, J. Examples of How Increased Formability through High Strain Rates Can Be Used in Electro-Hydraulic Forming and Electromagnetic Forming Industrial Applications. *J. Manuf. Mater. Process.* **2021**, *5*, 96. https://doi.org/10.3390/jmmp5030096

Academic Editor: Steven Y. Liang

Received: 2 July 2021
Accepted: 18 August 2021
Published: 1 September 2021

Publisher's Note: MDPI stays neutral with regard to jurisdictional claims in published maps and institutional affiliations.

1. Introduction

Many industries are facing new regulations concerning greenhouse gas emissions and are urged to find efficient solutions to reduce CO2 emissions. Lighter structures are one of the solutions for the transportation industry, provided that equivalent safety levels are maintained. These mass reduction policies often impel the use of lighter metallic materials or with increasingly high performances. However, in most cases, the mechanical properties of these materials and the associated thickness reductions lead to forming issues that are hard to solve using current industrial processes. Forming limits are often too low, higher springback appears, making it difficult to meet dimensional tolerances, and these new parts require increased press tonnages. High strain rates forming processes make it possible to overcome some of these limitations at low costs by inducing interesting material behaviors. The most interesting ones are an increase in elongation at break and the ability to reduce springback significantly, in particular for parts made of Aluminum and high strength steel. Regarding springback, when the impact velocity between the sheet and die or the pressure applied is controlled, the compressive stresses generated through the thickness can eliminate most of the elastic tensile stresses happening during forming and, therefore, produce a part that is geometrically much closer to the shape of the die than with conventional cold forming processes. Furthermore, in the automotive industry, this trend to reduce mass must be satisfied by keeping the same expectations on outer panel designs for marketing reasons. High impact velocity processes bring an additional competitive advantage, allowing to get complex shapes and very fine details like the ones obtained when stamping a coin. All these abilities are of interest for many other industries such as consumer electronics, luxury packaging, watches, etc.

One of the key contributing factors to the recent success of these decades-old processes is the ability to rely on new improvements in multiphysics simulation tools. These allow for a much better understanding of underlying physical phenomena, the design of long lifetime tools, and a significant increase in the predictability and control of the complex high-speed forming parameters, thereby extending their fields of industrial application. Flow stress tends to increase with strain rate for materials with positive strain rate sensitivity. For a given part velocity, the smaller the details are to form, the higher are the strain rates, which tends to increase the difference in behavior with quasi-static properties. Therefore, dedicated material characterizations, including constitutive laws and forming limits dependent on the strain rate, are often required to get predictive simulations. With these inputs, simulations can evaluate how the thickness of the sheet is distributed along with the formed shape and identify possible risks of tearing.

This paper focuses on the Electro-Hydraulic Forming (EHF) and Electromagnetic Forming processes (EMF), although other high strain rate forming processes exist, using explosives, vaporizing foil actuators, or laser shocks. For more information about EMF, a complete review has been proposed by Psyk et al. [1]. In this paper, we first discuss the increase in formability at high strain rates. Then, even though most of the industrial parts cannot be disclosed for confidentiality reasons, several parts formed by Bmax are presented for low and high-volume production. For each of them, details are given, such as the industry involved, part geometry, material, and advantages compared to conventional processes. A particular focus is made on simulation results in order to better visualize the type of deformations specific to high-speed forming. Finally, a point is made regarding some of the main limitations of these processes from an industrial point of view.

2. Formability at High Strain Rate

Formability improvement at room temperature in pulsed forming is a phenomenon that has been observed by multiple researchers. This benefit is discussed, for example, in reviews by Daehn et al. [2], Jenab et al. [3], and Demir et al. [4].

Nonetheless, when a sheet is deformed in free-forming into an open cavity without a die, some researchers observed almost no improvement like Golovashchenko [5] for an Al 6111-T4 and an Al 5754. This can be explained as in an unarrested expansion, if the part is aimed at remaining unfailed, it must come back to zero velocity so ending by a quasi-static forming. Nevertheless, Renaud [6] observed improvements in open cavities, higher in plane strain than in uniaxial or biaxial tensions, and getting improvements in plane strain of 8% for mild steel, 50% for an Al 5052, and 70% for an Al 3004.

In free displacement of the sheet, if the speed stays high, it is generally accepted that an increase in formability can already happen due to inertial necking resistance and strain rate hardening beyond a threshold strain rate and material-dependent (typically in the order of 10^3 s^{-1} to 10^4 s^{-1}), both working against strain localization. It can be explained by a change in deformation mechanism compared to quasi-static strain rates, for example, twinning formation, dislocation drag, reduced dislocations motion, and void growth. However, although this formability improvement before impact is a feature of many materials like Aluminum, it is not a general feature for all of them. For example, Kim [7] followed high-speed free-forming deformations in time by using a fast camera and found no improvement in forming limit diagrams (FLDs) for CQ and DP590 steels.

Nevertheless, high strain rates in forming applications generally induced an impact of the sheet on a die, inducing several other phenomena that need to be considered to fully understand the cases of extraordinary increase in formability.

One of these phenomena is the ironing effect when the impact velocity, above several tens of meters per second, creates high compression stresses through the thickness of the part. As explained, for example, by Jenab [3], the compressive stresses are also transmitted downstream of the impact, i.e., in the not yet impacted zone. This compressive stress state could decrease or suppress void nucleation and growth, and therefore, prevent or at least delay the formation of damages in the following phases of deformation. Additionally, at high impact speed, a flow of material is generated downstream of the impacted zone, tending to thicken the part in the not impacted portion. This can compensate for thinning coming from bulging of the part and also prevent or postpone tearing that could happen due to excessive stretching. This ironing effect is shown in Figure 1, where a 50 mm diameter hemisphere made from an Al 5182-O is formed by EHF at different energy levels. Figure 1a demonstrates the increase in formability when the forming velocity is high enough, and Figure 1b illustrates how the ironing takes place against the die, with an arrow giving the direction of the evolution of the impact.

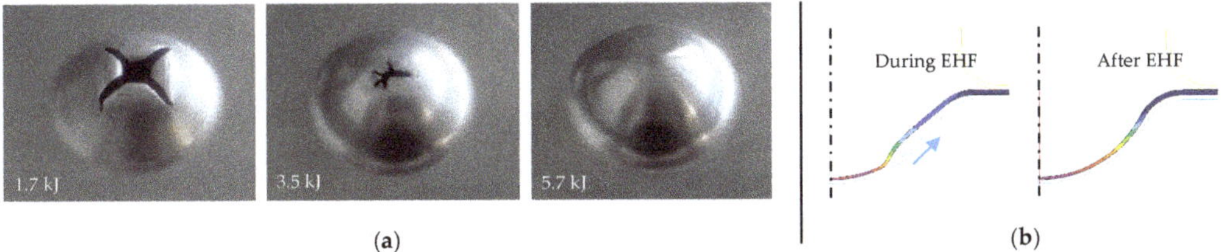

Figure 1. EH Forming of a 50 mm hemisphere: (**a**) experimental results at different energy levels; (**b**) related simulation at 5.7 kJ during and after forming.

Similar to the observations from Allwood [8], who showed an increase in formability from 7% to 300%, Demir et al. [4] suggest that another possible reason for the increase in formability comes from the out of plane shear stress due to the type of loading in EMF where Lorentz forces act perpendicular to the sheet. It could be the same in the EHF process.

Another possible positive effect on formability can be due to the simultaneous stretching and bending while forming. Emmens [9] has shown an increased elongation before rupture from 25% up to 430% in a quasi-static tensile test combined with bending. Additionally, this could be similar to what has been described by Duroux [10] in conventional stamping at the entry radius of a blank drawn inside a die. These two processes generate compressive stresses in a portion of the thickness due to bending that can delay damage formation even if this compression does not occur through the whole thickness of the sheet. Similarly, in high-speed forming, if both part velocity and distance between two contact points of the part with the die are important enough, the motion of accelerated parts tends to be flat before impact. This is illustrated in simulation plots of Figure 2 inspired from Yamada [11] for an Aluminum strip launched with an initial vertical speed. With an initial velocity of 18 m/s (a), the shape downstream of the impact with the die is curved, while it is almost flat at 120 m/s (b). Colors represent the pressure (Pa) in the material. The detail in Figure 2c shows with a white arrow the compression (red zone) in part of the thickness downstream of the impact point due to bending. This forming particularity should therefore contribute sometimes to the observed increase in formability due to high speed.

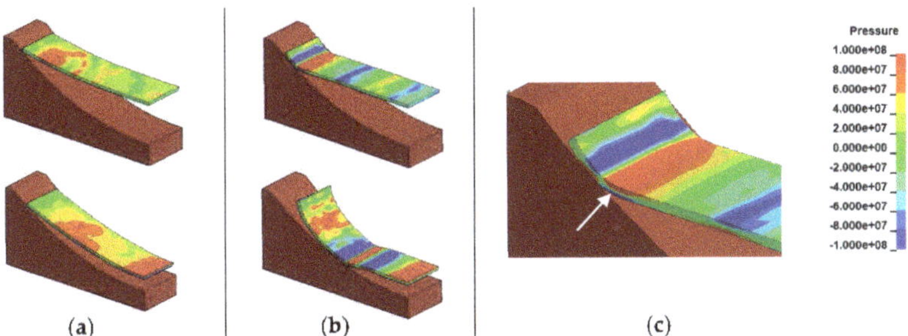

Figure 2. Pressure in an Aluminum strip launched at 18 m/s (**a**), 120 m/s (**b**), with a zoom at impact location for 120 m/s (**c**).

Based on all the above-mentioned considerations, the forming process should be designed in such a way that large strains occur to take advantage of these positive effects on formability. For example, forming local features in relatively small areas induces large strains and very high strains rates (>10^4 s^{-1}). This makes it possible to perform sharp radii without tearing in particular for Aluminum, to form and engrave metallic parts with very fine details, to use stronger materials with too low formability in quasi-static forming in order to reduce the thickness of the sheet, sometimes to eliminate heat treatments by enabling to form parts directly from sheets already in their final tempered state, and more.

Regarding FLDs at high strain rates coming from experimental results published in the literature, they should be used with caution. Indeed, forming limit curves are valid for a linear deformation path, and when the full thickness is in tension, this is not always the case when forming complex shapes at high strain rates, like sometimes also with particular slow forming processes. Moreover, when an important ironing effect takes place, the thickness of a sheet reduces, and the final measured extension at the surface can sometimes no longer be due to tensile stresses alone but also through thickness compression that deforms the part in-plane due to conservation of volume. Nevertheless, analyzing deformations from simulations in FLDs before impact remains of interest to evaluate the risks of tearing.

3. Deep Drawing and High Strain Rates

Generally speaking, forming a deep cavity requires sheet metal drawing from the flange to limit deformation. High strain rates forming are generally too fast to give enough time to the material to flow extensively under the blank holder. Therefore, deep cavity parts can be formed either by multiple pulses or by preforming operations on a press, as discussed in multiple publications reviewed by Psyk et al. for EMF [1].

The multi-pulse approach is expected to be more applicable when the cost of the tool is a more significant factor than the cycle time. The hybrid forming approach, combining possibly in the same operation quasi-static forming and pulse restriking or forming, provides a substantially shorter cycle time but requires more sophisticated tooling.

Regarding EHF, the hybrid processes can be performed by combining it either with hydroforming [12–15] or with stamping [16]. To demonstrate the latter, a part was designed to be formed by a cylindrical punch in which an Electro-Hydraulic chamber is integrated to form the central shape. The principle is shown in Figure 3.

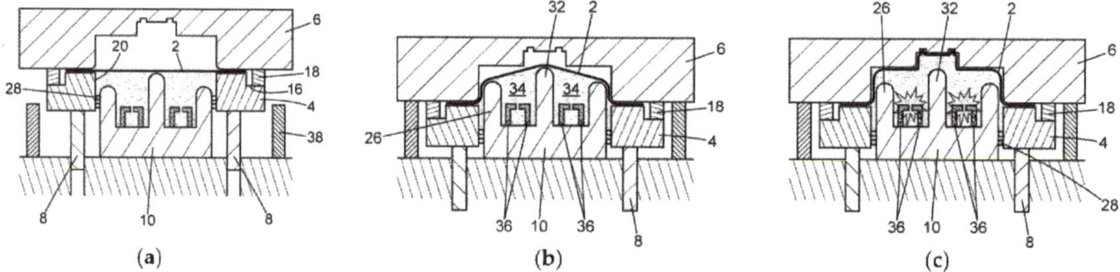

Figure 3. Principle of the process combining stamping and EHF. (**a**) Initial position; (**b**) after stamping; (**c**) after EHF. 2: blank; 4: chamber; 6: die; 8: guiding columns; 10: electrodes support; 18: kiss block; 26, 32: punch; 28: sealing; 34: water; 36: electrodes system; 28: stopper.

The demonstration part was made from a blank in Al 6061-T4, 1.5 mm thick. The cylindrical cavity was 200 mm in diameter and 80 mm deep. As shown on Autoform® simulation results presented in Figure 4, it is designed so that it cannot be formed using conventional stamping in one step. Indeed, deformations generated by the smaller central punch reported on a forming limit diagram (Figure 4c) show red spots way above the Forming Limit Curve in plane strain.

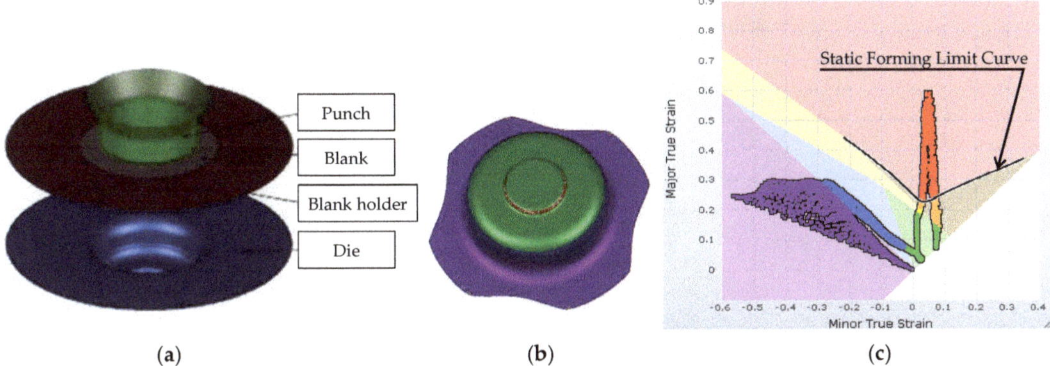

Figure 4. Conventional stamping simulation details using an outer punch first and then a central one: (**a**) tooling shapes; (**b**) formed part; (**c**) deformations plotted in the Forming Limit Diagram.

To simulate the hybrid process combining quasi-static stamping and EHF, the LS-Dyna® finite element software has been used. To accurately predict the forming, the evolution of the electrical power as a function of time has been taken into account with a model dependent on the parameters of the high voltage generator, as presented by Deroy [17]. This electrical current in the arc generates pressure waves inside an Eulerian mesh, representing the volume of water, while the metallic sheet to be formed is represented by a Lagrangian mesh. At each time step, the two meshes are coupled so that the pressure generated in the water is transferred as a force to the sheet being formed. The dynamic constitutive behavior of the material depending on strain rate, as presented by Jeanson [18,19], is also required to correctly predict the proper strain and stresses distribution in materials. The simulation results of the combined process are shown in Figure 5.

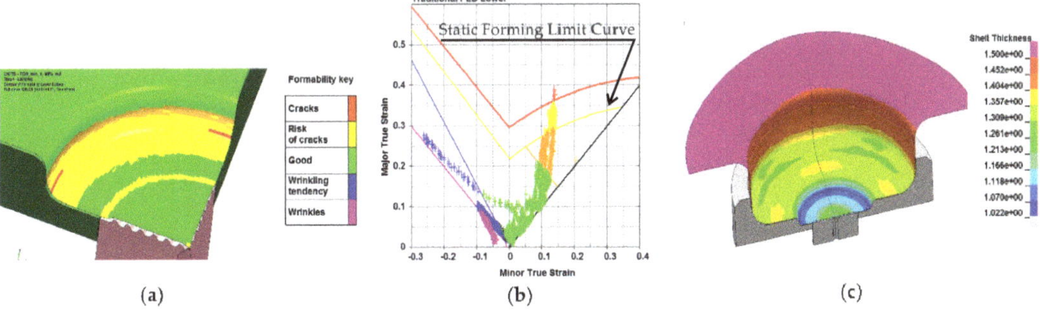

Figure 5. Simulation results of a combined stamping and EHF process using an outer punch first: (**a**) formability keys reported on the part; (**b**) deformations plotted in the Forming Limit Diagram; (**c**) thicknesses map.

As seen in Figure 5b, contrary to stamping, deformations of the central portion with EHF have positive minor deformations and so are not in plane strain. Nevertheless, some areas are still above the static forming limit curve in yellow (Figure 5c).

The EHF generator used to produce the parts has the ability to store energy up to 25 kJ and is connected to one set of electrodes located inside the punch. It has been specially developed for mass production with a lifetime of insulators relevant for industrial applications. Typical designs of EHF electrodes are reported, for example, by Golovashchenko [13] or by Felts [20] without membrane and by Brejcha [21] with membrane.

Here, the hybrid tooling is integrated into a 400-ton press, as shown in Figure 6a. The bottom of the punch is closed by a membrane in its center to avoid contact of the formed part with water and, more importantly, to avoid time loss in water filling, allowing high production rates. For the drawing phase, a blank holder is used with the proper retaining force to avoid any wrinkle of the part when flowing inside the die.

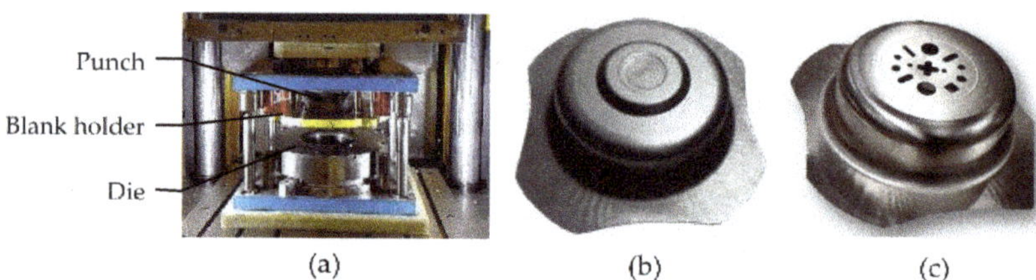

Figure 6. Pictures of tooling and parts formed in 1 stroke and 1 EHF discharge: (**a**) setup integrated in the press; (**b**) part formed with details in the center; (**c**) part formed with cutting in the center.

Despite deformations above the static forming limit curve (Figure 5b), Figure 6b demonstrates a part formed without tearing thanks to high strain rates and the ironing effect. In addition, fine details have been formed in the center. Different shapes can be proposed by changing the central insert in the die, making it possible to perform forming as well as cutting operations (see Figure 6c). Since no punch is used in the central area with EHF, it is easy to customize parts without time-consuming dies adjustment phases.

This application is an example presenting the interest to couple conventional deep drawing and EHF with a membrane in the same step, possibly at high production rates, in order to get shapes not feasible with mechanical stamping.

4. Hood Panel Postforming by EMF

For lightweighting purposes, integrating an EMF coil in a conventional press line gives the possibility to restrike character lines of an outer panel in order to get the radii on Aluminum as small as the ones obtained on steel panels, without tearing or visual defects.

Here is the case of the postforming of sharp edges in a Lamborghini Aluminum hood performed for design purposes as presented by Held [22]. The requirements for the application were the forming of character lines with an edged V design 500 mm long on a hood panel made of a 1.2 mm thick Al 6016-T4, with a size of 1 m × 2 m. The idea was to produce a unique lightweight concept with sharp edges to participate in the impression of a highly dynamic car. The difficulty for a conventional stamping process was to get small radii due to the limited formability of the Aluminum at low strain rates.

The main objectives of the project were to show the ability of the EMF process to achieve this specification without creating any surface defects, which is crucial for outer panels, to integrate the tool into a conventional production line, and to show the environmental sustainability of the process due to very low energy consumption. A cut-view of representative tooling is given in Figure 7.

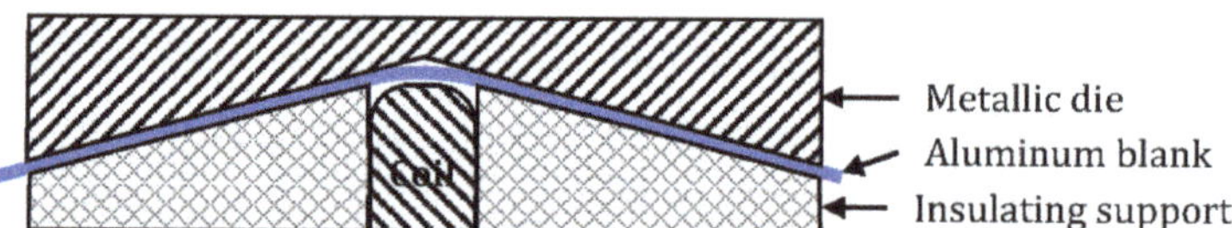

Figure 7. Principle of the EMF postforming process.

For this implementation, the process took place in a specific operation OP35 (see Figure 8), between the trimming and the first flanging step of a press line initially comprising four operations.

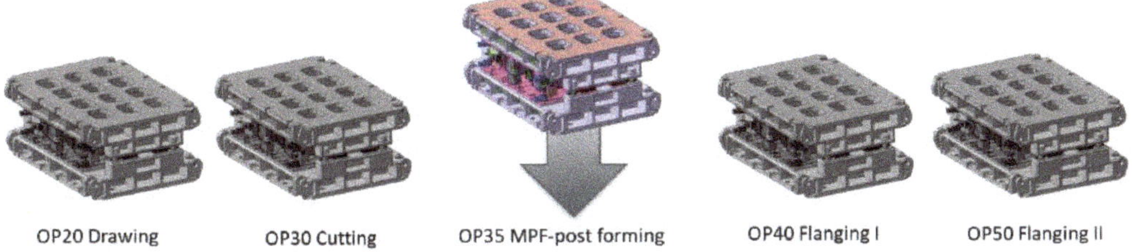

Figure 8. View of the five tools integrated into the stamping line and localization of the EMF operation.

The design of the EMF tool was performed using the strongly coupled LS-Dyna simulation code, taking into account Lorentz forces, dynamic behavior at high strain rates, and heating, which remained reasonable despite the high current involved (hundreds of kA). To generate appropriate Lorentz forces in order to get a uniform final radius along the character line while respecting acceptable stresses in the whole one-turn coil, 2D and 3D simulations were performed with specific attention to the slot region of the coil where the current enters and leaves the coil. To get the forming there, the magnetic field was kept above a certain value (see Figure 9). The required energy to get the final part was 15 kJ delivered in one discharge.

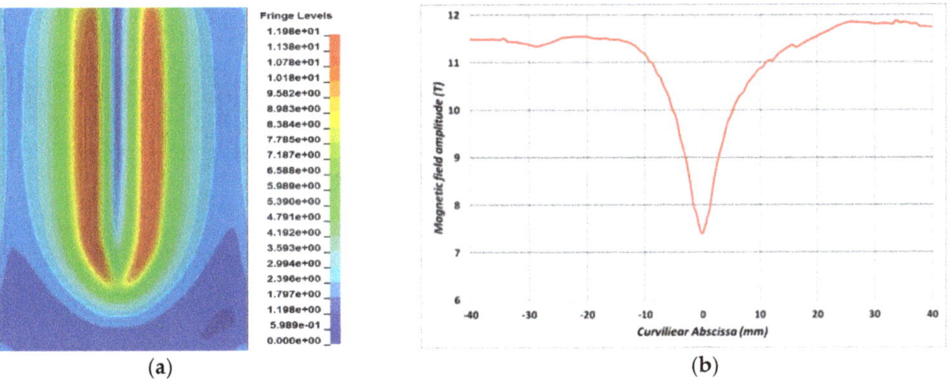

(**a**) (**b**)

Figure 9. Top view of the magnetic field along the working area of the coil close to the slot location (**a**) and the associated amplitudes (**b**).

The process robustness and reproducibility are mainly ensured by the precision of the charging voltage, the part positioning, a peak current high enough to form whatever the material batches, and the coil insulation material. An important phase to avoid any surface defects on the part is the spotting operations to adjust the surfaces of the non-conductive support around the coil. Such operation in conventional stamping is much more time-consuming, as both metallic die and punch surfaces need to be adjusted with extreme precision by manual grinding, honing, and polishing. Figure 10b,c shows the ability of the tool to create very sharp features. In these pictures, a sharper than needed radius was produced, with a final measured radius of 0.1 mm, starting from 4 mm.

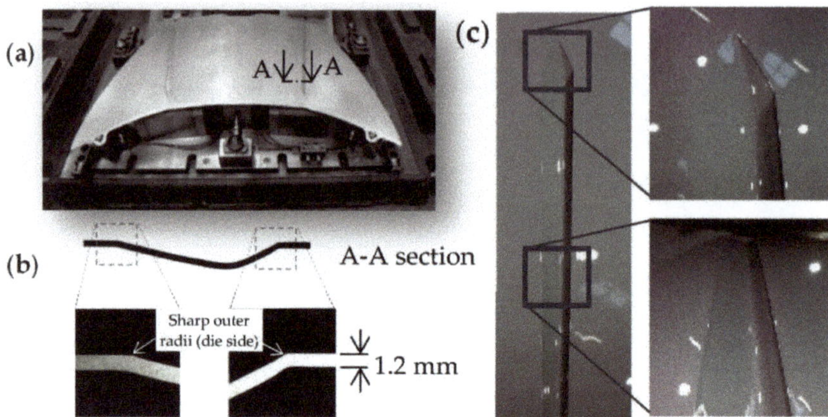

Figure 10. Hood after EMF calibration (**a**) Hood installed on the tooling, (**b**) microscopy images of the cross-section of the sharp radii, (**c**) hood details after painting.

At the end of the process, the material flow seems closer to forging than forming and shows how high strain rates make it possible to form sharp corners with high local deformation without rupture. However, radii of about 1 mm are more generally used by car manufacturers for optimum painting and pedestrian safety issues.

This on-site, small series production made it possible to qualify the technology in terms of integration in a press line and safety. Since then, in other projects involving Renault [23], such EMF tools were integrated directly in the last flanging step and used for larger series production. A cooling system was integrated into the tool to ensure that the production rate of the press line could be maintained, and a specific automated connection system was developed to respect the tooling changing time requirement of press lines. Typical times to change all the tools, such as those in Figure 8, in order to start a new series production, are between 4 and 20 min depending on the car factory. The EMF process and produced parts were fully qualified, including production rate, coil lifetime, surface quality, painting quality, as well as parts microstructure and mechanical strength.

5. Embossing by EHF for Luxury Packaging

To show the ability of EHF to create embossing, the example of the "J'Adore L'Or" perfume bottle cap is taken. This design received the Packaging of Perfume Cosmetics and Design award at the PCD Paris 2018 event.

To manufacture perfume caps in metal for premium products rather than metalized plastics, leading brands generally use Zamak casting. An alternative to this zinc-based alloy is the use of Aluminum processed by EHF.

The blanks are made of Aluminum alloy 5657 with a thickness of 0.8 mm. As for the EHF application shown previously, the process was modeled by a fluid-structure interaction and a time-dependent energy deposition to generate the correct pressure waves resulting from the discharge, and therefore, to properly predict the forming. This is illustrated in the example in Figure 11, where a parallel between simulation and experiment during the forming sequence of the cap is presented. Pressure contour plots are shown at four different timesteps, revealing the dynamic forming of the blank that impacts the die with peak velocities above 100 m/s.

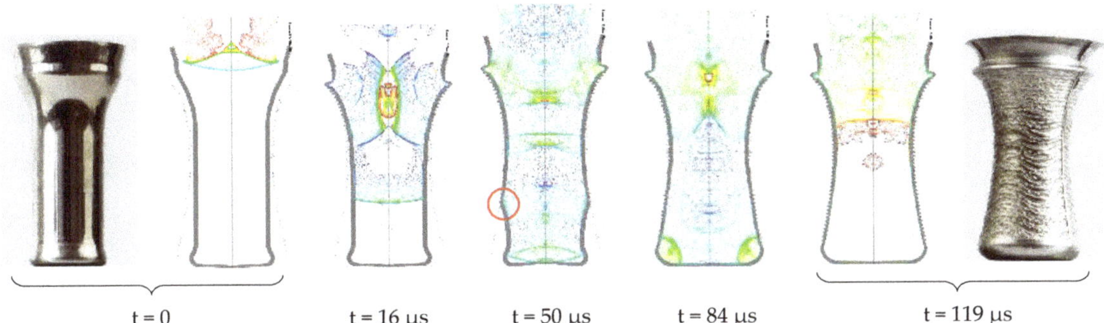

Figure 11. Forming simulation with pressure contours of propagating shock waves for the "J'adore L'Or" upper cap at different time steps.

In static, the uniaxial strain to failure of the Al5657 was 17%. Extrapolating this data and building the theoretical forming limit curve in biaxial expansion by the Storen-Rice formula, one finds 20% in major strain and 20% in minor strain. In the case of the cap, the deformation in the top corner reaches 40% in major strain and 20% in minor strain without rupture. Ironing and bending while forming phenomena that explain part of this increase in formability can be seen at the impact location with the die in the simulation plot of Figure 11 at time t = 50 μs (red circle). Regarding the details, they are properly formed once the shape of the bottom corner is complete and thanks to a reflected wave with increased pressure moving up, coming from the bottom of the part (see simulation plot at t = 119 μs).

Other relevant features of the "J'adore L'Or" perfume caps are the sharp angles and very fine details achieved shown in Figure 12b,c. The different manufacturing steps are also shown in Figure 12a. The cap is made by assembling two parts, with one remaining on the glass bottle. The blanks are formed with a conventional progressive stamping process by the G. Pivaudran company, as well as the final cutting, anodization, and assembly. The whole EHF step was developed and is performed by Bmax.

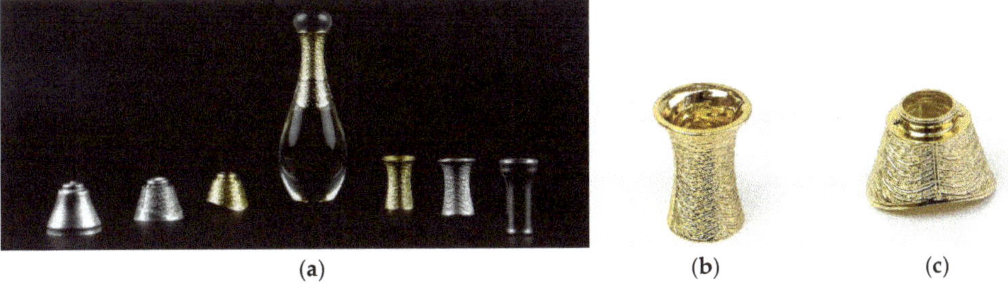

Figure 12. Manufacturing steps (**a**) and parts detail (**b**,**c**) of the "J'adore L'Or" perfume caps.

For the first time using the EHF process, mass production has been performed with a fully automated production line. Parts production of this new design had been performed over seven months for the launch of the product and continued for the following years. Figure 13 shows a single soundproof integrated EHF machine located around a pooled conveying system able to produce such a part. Its production capability in three shifts is 1 million parts per year, using only one electrode system.

Figure 13. Fully automated EHF production system for small parts.

Several EHF machines like the one in Figure 13 can be placed around this conveying system in order to increase the production rate or to produce different parts as the conveyor is equipped with a handling robot capable of optically detecting the type of parts that are being fed.

Thus, the increase in formability of this Al 5657 due to high strain rates gave the possibility to use Aluminum instead of Zamak casting, offering finer engravings, a larger color palette thanks to anodization, a durable product, and a substantially greener process. This is another example of how ultra-high-speed forming can push the current design limitations.

6. Radio-Frequency Half-Cavities Formed by EHF

EHF has been explored for the production of several Superconducting Radio Frequency half cavities for CERN of 400 MHz, 800 MHz, and 1300 MHz. The main results have been described by Atieh [24] and Cantergiani [25,26]. The 400 MHz frequency cavities made of electronic oxygen-free electronic (OFE) copper are used for the large hadron collider (LHC) and will also be used a priori for the future circular collider (FCC) that was initially foreseen at 700–800 MHz using high purity bulk niobium. Traditionally, such cavities are fabricated by the electron-beam welding of half-cells obtained through sheet metal forming techniques such as deep-drawing, spinning [24], or hydroforming [27]. These traditional shaping methods show several drawbacks. The first two of them leave 100–200 μm of the inner surface damaged layer due to friction with the tools, and this layer needs to be removed by buffered chemical polishing to improve RF performances. For deep drawing producing complex shapes with high accuracy, it is difficult, and for large half-cells, such as the 400 MHz ones, it would require high-tonnage hydraulic presses. Furthermore, springback causes significant deviations from the die geometry, in particular in the central area—the iris—[28] despite die compensation. Thus, for large components, a combination of spinning and machining to get the proper geometry was initially preferred. However, this also presents some disadvantages as it requires several spinning steps, including intermediate annealing, to avoid necking [29]. Full cavities without weld at the equator can also be obtained through hydroforming from a tube [27]. The springback is also an issue for this process, and the precision of the internal RF surface is highly dependent on the thickness variation due to the stretching and tolerance of the starting tube. Thus, geometrical tolerances remain large here also. Moreover, if Niobium is used, tube procurement is an issue financially and from a supply point of view.

These drawbacks associated with traditional techniques can be overcome or limited by EHF, making it possible to get tight geometrical tolerances due to the reduced springback and to avoid intermediate annealing due to the increase in formability.

For such large parts, multiple shots are required to get the final geometry. OFE copper and Niobium constitutive laws have been shown to be very sensitive to strain rates for a few 10^3 s^{-1} [26]. Thus, proper material models are mandatory as input in simulations to optimize the EHF process parameters, such as the energy level, the number of electrode systems, and the number of successive discharges. At the end of the simulation for a shot, the deformed sheet with its residual stress-strain distribution is used as a starting blank to simulate the next forming shot. An example of 3 simulated forming shots for a 700 MHz half-cell is shown in Figure 14, where the evolution of deformation can be seen. As the initial diameter of the blank is the same as the external diameter of the blank holder, one can see that shot after shot, the material is drawn inside the die.

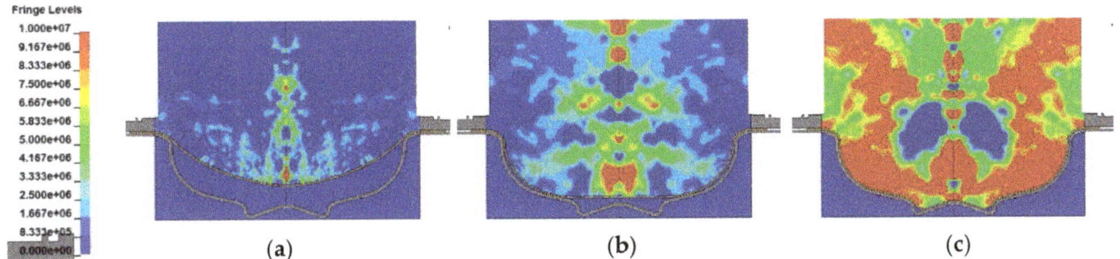

Figure 14. Two-dimensional-axisymmetric EHF simulations of a 700 MHz copper cavity showing the pressure fringes (Pa) inside the discharge chamber during three successive shots (**a**–**c**) and showing the deformation almost completed after each shot.

Figure 15a shows the EHF set-up used to produce the 400 MHz half cells of Figure 15b, and examples of 700 MHz half-cells are given made of OFE copper in Figure 15c and niobium in Figure 15d.

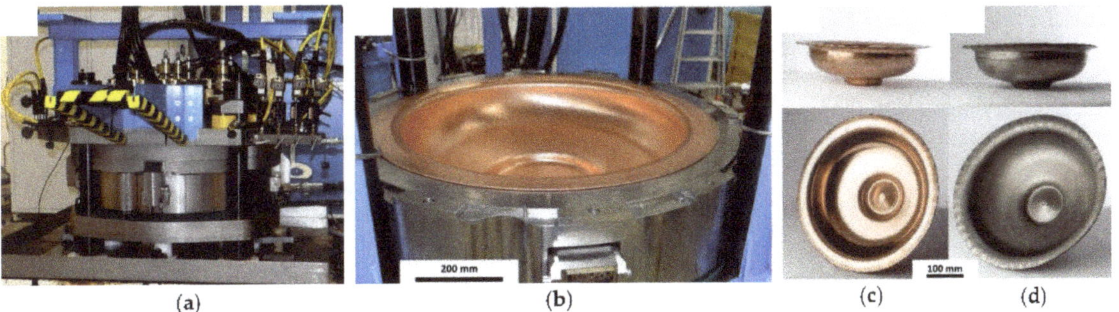

Figure 15. Experimental EHF set-ups: (**a**) closed 400 MHz tooling; (**b**) 400 MHz OFE copper half-cell; (**c**) 700 MHz OFE copper half-cell; (**d**) 700 MHz niobium half-cell.

The shape accuracy of 400 MHz half-cells formed by EHF was compared at CERN with the shape accuracy of 400 MHz half cells obtained by spinning, including an intermediate annealing step and complete machining of the inner surface [30]. The latter were obtained starting from a blank of 4 mm in order to have enough material for the final machining step to achieve the required inner shape.

EHF was performed on copper sheets of 3 mm thickness without any intermediate annealing or machining. The results obtained by geometry control are shown in Table 1, and the roughness values are shown in Table 2.

From the values, one notices that EHF gives results in agreement with the precision required and is more precise than spinning and machining. Moreover, as shown in Table 2, the value of roughness before and after EHF is preserved both for copper and niobium half-cells. The thicknesses of the formed 400 MHz half-cell were extracted for both EHF and spinning and are shown in Figure 16.

Table 1. Shape accuracy at iris and equator for 400 MHz copper half cells formed by EHF and spinning + machining [26].

Shape Accuracy	EHF	Spinning + Machining
Diameter at the Equator [mm]	687.64	687.81
Circularity at the equator [mm]	0.03	0.63
Thickness at the Equator [mm]	2.99 ± 0.02	2.30 ± 0.14
Diameter at the Iris [mm]	299.91	300.02
Circularity at the iris [mm]	0.08	0.77
Thickness at the Iris [mm]	2.05 ± 0.05	2.35 ± 0.07

Table 2. Surface roughness (R_a) measured for 400 MHz copper half-cells, 700 MHz niobium, and copper half-cells [26].

Half-Cell	R_a before EHF [μm]	R_a before Spinning + Machining [μm]	R_a after EHF [μm]	R_a after Spinning + Machining [μm]
400 MHz copper	0.17	0.25	0.48	0.76
700 MHz copper	0.2	-	0.2	-
700 MHz niobium	0.8–0.9	-	0.9–1	-

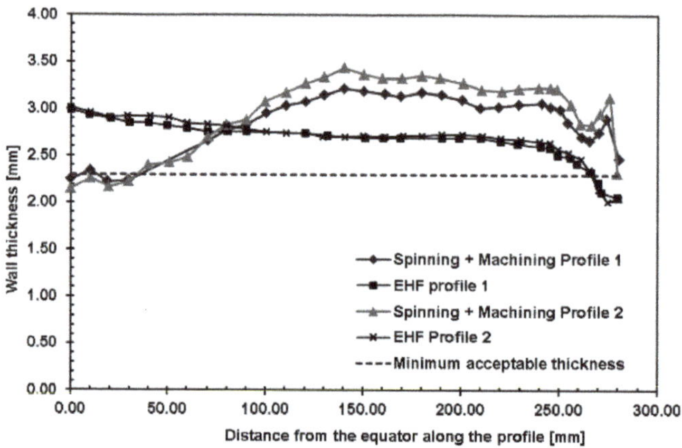

Figure 16. Thickness distribution for 400 MHz copper half-cells obtained through EHF or through spinning followed by machining [26].

As shown in Figure 16, the thickness distribution found for EHF is homogeneous on the whole half-cavity profile with values between 3 mm and 2.3 mm as required by CERN. Only the thickness at the iris was slightly lower than 2.3 mm, but this was accepted by CERN after structural verification of the loads encountered by the SRF cavity during the operation. The thickness distribution obtained from spinning and successive machining shows a higher variability. Moreover, at the equator, the thickness is slightly below 2.3 mm, and in most of the profiles, the thickness is higher than 3 mm, which can result in difficulties when the half-cell needs shape deformation to obtain proper RF response (tuning of the cavity [30]).

Finally, EHF is the process chosen by CERN to supply spare parts for the LHC. In addition to the manufacturing benefits, the RF results obtained on a 400 MHz full cavity formed by EHF were best, as shown in Figure 17 disclosed by CERN [31]. This graph showing the quality factor (Q_0) versus the accelerating electric field (E_{acc}) is typically used to measure RF performance, relating the conductance of the inner surface of a given cavity geometry to the maximal electric field before quench where the cavity loses its superconducting quality (here, above 10 MV/m for the EHF cavity).

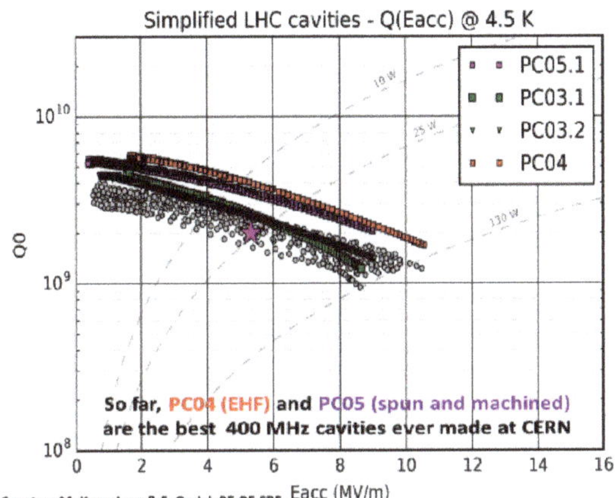

Figure 17. Quality factor (Q_0) versus accelerating electric field (Eacc) to measure RF performances of 4 different 400 MHz RF cavities, including cavities formed by spinning and EHF [31].

Therefore, from an industrial point of view, EHF is very attractive to reduce fabrication time, decrease the cost of half-cells of RF cavities and improve their performance due to higher precision and better inner surface quality.

7. Oil Deflector for Helicopters

A study was led concerning the forming of an oil deflector using the EHF process, and D. Allehaux presented the results [32]. This work is the result of a close collaboration between the industrial end-user (Airbus Helicopters) and the technology supplier (Bmax). The objective was to verify the ability of the EHF process to form the part starting from a plate having the final expected temper state. The motivation was to perform global manufacturing lead time optimization and cost reduction compared to the applied conventional manufacturing method, but also to check the EHF process robustness and to use a greener technology.

The blank is made of Al 6061 blank, 1 mm thick, and the part has a final diameter of approximately 400 mm.

The traditional method to manufacture this part is to start with a blank in a T0 state so annealed. It employs two dies (one male and one female), three stamping steps on a rubber press that includes a final restriking operation on the female die, two intermediate aging thermal treatments, a full thermal treatment made of solution heat treatment, quench, and natural aging and a final manual rework after the restriking operation. The EHF die and the part with details are shown in Figure 18. The die has been machined directly from the theoretical geometry of the part. Details show how EHF makes it possible to form undercuts and complex shapes without time-consuming die adjustment.

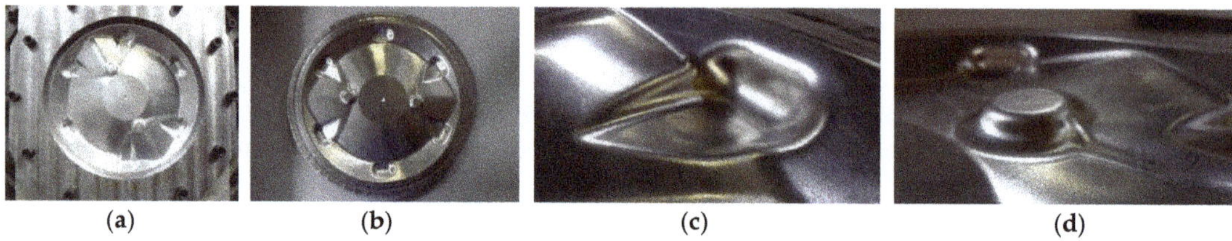

(a)   (b)   (c)   (d)

Figure 18. Oil deflector female die (**a**); formed part (**b**); formed details (**c**,**d**).

The part formed by EHF was initially a flat blank in the T4 state, so hardened by heat treatment, which is the final expected temper for the part in the operation. The geometry was fully formed in three shots, using a three electrodes system due to the size of the part.

Simulations were performed to define the process parameters of the three different shots. Figure 19a–e presents one of the simulations of the first shot showing the progressive forming process and how dynamic it is. Instead of the bulging that would occur in conventional hydroforming, the part is pushed onto the die by radial ironings from the center to the periphery and from the periphery to the center. Both then make some extra material appear in the (c) picture, flattened later without wrinkles in the (d) picture, and helping to form the grooves without tearing. The two additional shots not shown here finalize the forming of the grooves.

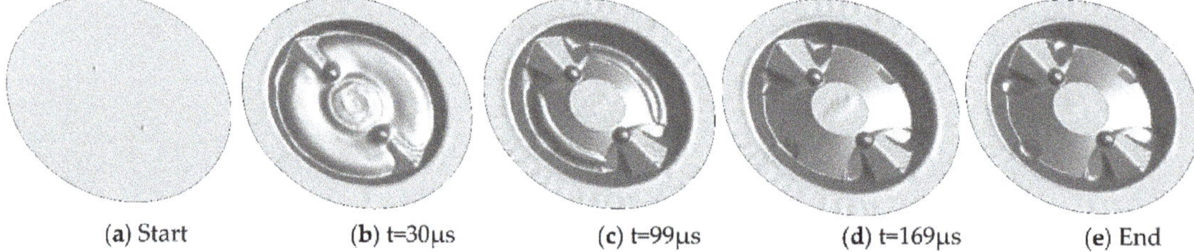

(a) Start   (b) t=30μs   (c) t=99μs   (d) t=169μs   (e) End

Figure 19. Oil deflector simulation results at different times (**a**–**e**) during the first EHF step.

Several analyses have been performed on the part after EH forming by Airbus Helicopters, such as geometrical measurement including thickness, crack detection using dye penetrant analysis, local conductivity, and hardness. No tearing occurs, although a maximum of 35% thinning was observed in the grooves located on the periphery (see C and D points in Figure 20c). With the conventional slow forming method, only 15% maximum thinning was possible before tearing at these locations. As expected, local hardenings have been observed in the highly deformed areas (zones B, C, and D in Figure 20a), and the conductivity measured was close to the T6 temper in the D areas (see D1 and D2 points in Figure 20a), so harder than T4.

The process has been shown to be reproducible from one plate to another. Getting the part without tearing was possible only by controlling the drawing of the part below the blank holder to avoid excessive deformation and the location and the energy level of the discharges. The drawing of the material under the blank holder was obtained by successive EHF discharges without combining them with hydrostatic pressure or stamping. As for the forming examples presented before, high strain rates combined with ironing and bending have been observed in simulations and explain the gain in formability that was noticed.

With these results, cost reduction by Airbus on the product was estimated to be up to 60%, including recurring and non-recurring costs.

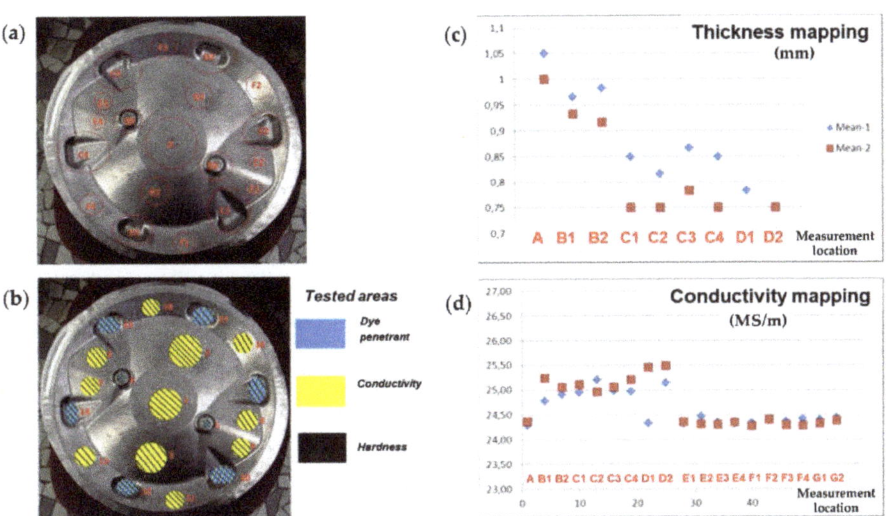

Figure 20. Part analysis—(**a**) Tested localization labels on the part; (**b**) Tested area for dye penetrant, conductivity, and hardness; (**c**) Mean thickness in different areas; (**d**) Mean conductivity in different areas [32].

8. Some Limitations When Forming by High Strain Rates

Psyk et al. [1] gave a global overview of limitations concerning the EMF process. Even if some issues of this technology have been solved since then, it is interesting to refer to this article for a deeper view.

Among the main limitations of EMF and EHF, even if higher strain rates can bring potentially better formability, the impact speed often has to be limited to avoid too high stresses on the die and surface defects on the formed part due to bounce-back or jetting effects. For the latter, similar to impact welding processes, jetting can appear at high velocity between the part and the die due to hydrodynamic phenomena when the angle between the two parts is favorable (typically between 5° and 15°), tending to remove micrometers of materials from the surfaces of both the part and the die.

A limitation also comes from the thickness of the parts to be formed. Indeed, energy requirements increase directly with the thickness, which tends to increase the cost of the generators and decrease the lifetime of the tools (coils in EMF processes, electrodes, and insulators in EHF processes) as higher pressures are required to deform thicker parts.

For high strain rate forming processes, the lifetime of these tools is generally lower than the one of metallic punches in traditional stamping processes. This is often related to areas where significant dynamic forces and/or temperatures are applied. Therefore, the mechanical strength and lifetimes of high voltage insulators, as well as conductors driving high energy densities, are a point of attention for high volume pulsed forming applications. Their lifetime depends on the design, energy, peak pressure, current density, or electric fields involved. Therefore, to optimize the design of these components and to choose the proper materials, strongly coupled simulations are of interest. Currently, high-volume industrial solutions are running using either long-life or cheap and easy-to-replace materials in critical areas. Their lifetime can be in a range from several thousand to more than a hundred thousand pulses.

9. Conclusions

Firstly, a synthesis was proposed giving an overview to explain why high strain rates can increase the formability of metals, as this ability is one of the important competitive advantages compared to quasi-static processes. The phenomenon of bending downstream the impact on the die has been proposed as one of the possible reasons for this increase in formability.

Several recent industrial applications of EMF and EHF processes were then described for high and low volume production. A particular focus was made on simulation results in order to deal with the formability of these parts and to better visualize the type of deformations specific to high-speed forming. An application in deep forming has been presented, combining stamping and EHF to take the advantages of both processes. An example of postforming by MPF showed how to create sharp radii in the automotive industry, possibly for high-volume applications. A high-volume embossing application using EHF in luxury packaging demonstrated how highly dynamic forming can make it possible to form complex shapes rather than cast them. Two other examples for the aeronautic and particle accelerator industries disclosed how the improvement in forming limits can reduce manufacturing costs, in particular by avoiding annealing and/or machining steps.

Nonetheless, to help have a more global picture from an engineering point of view, some limitations of the EMF and EHF processes have been pointed out.

As a whole, the presented applications brought specific insight into the possibilities to use these high strain rates processes for different purposes, including lightweight designs, manufacturing cost reductions, new shapes, improved perceived quality, and higher precision.

Author Contributions: Simulations and resources, G.M., E.C., F.B., J.-P.C.-L., J.D.; writing—original draft preparation, G.A., E.C.; writing—review and editing, G.A., G.M, E.C.; All authors have read and agreed to the published version of the manuscript.

Funding: Parts of the results presented come from contracts with Renault, Audi, Dior, CERN and Airbus Helicopters. This research received no external funding.

Data Availability Statement: The data presented in this study are available in the cited references.

Acknowledgments: The authors would like to thank forming groups from Renault, Audi, Dior, CERN, Airbus Helicopters, as well as Jeanson, Held, Tiercelin, Croteau, Golovashchenko, and Bmax teams for their involvement in the material used for this article.

Conflicts of Interest: The authors declare no conflict of interest.

References

1. Psyk, V.; Risch, D.; Kinsey, B.L.; Tekkaya, A.E.; Kleiner, M. Electromagnetic forming—A review. *J. Mater. Process. Technol.* **2011**, *211*, 787–829. [CrossRef]
2. Daehn, G.; Vohnout, V.J.; Datta, S. Hyperplastic forming: Process potential and factors affecting formability. *MRS Online Proc. Libr.* **1999**, *601*. [CrossRef]
3. Jenab, A.; Green, D.E.; Alpas, A.T.; Golovashchenko, S.F. Experimental and numerical analyses of formability improvement of AA5182-O sheet during electro-hydraulic forming. *J. Mater. Process. Technol.* **2018**, *V.255*, 914–926. [CrossRef]
4. Demir, K.; Goyal, S.; Hahn, M.; Tekkaya, E. Novel approach and interpretation for the determination of electromagnetic forming limits. *Materials* **2020**, *13*, 4175. [CrossRef] [PubMed]
5. Golovashchenko, S.F. Material Formability and Coil Design in Electromagnetic Forming. *J. Mater. Eng. Perform.* **2007**, *16*, 314–320. [CrossRef]
6. Renaud, J.Y.; Leroy, M. Formabilité à Grande Vitesse de Déformation: Application Aux Techniques de Formages Electro-magnétique et Electrohydraulique. Ph.D. Thesis, Université de Nantes, Nantes, France, 1980. (In French).
7. Kim, S.; Huh, H.; Bok, H.; Moon, M. Forming limit diagram of auto-body steel sheets for high-speed sheet metal forming. *J. Mater. Process. Technol.* **2011**, *211*, 851–862. [CrossRef]
8. Allwood, J.M.; Shouler, D.R. Generalised forming limit diagrams showing increased forming limits with non-planar stress states. *Int. J. Plast.* **2009**, *25*, 1207–1230. [CrossRef]
9. Emmens, W.C.; Van den Boogarard, A.H. Extended tensile testing with simultaneous vending. In Proceedings of the IDDRG 2008 International Conference, Olofström, Sweden, 16 June 2008.
10. Duroux, P.; Bellut, X.; Canivenc, R. Alternative forming criterion to FLC when sheet metal has been drawn over the die entry radius. In Proceedings of the IDDRG Conference, Paris, France, 1–4 June 2014.
11. Yamada, T.; Kani, K.; Sakuma, K.; Yubisui, A. Experimental study on the mechanics of springback in high speed sheet metal forming. In Proceedings of the 7th International Conference on High Energy Rate Fabrication, Leeds, UK, 14–18 September 1981.

12. Gillard, A.J.; Golovashchenko, S.F. Effect of quasi-static prestrain on the formability of dual phase steels in electrohydraulic forming. *J. Manuf. Process.* **2013**, *15*, 201–218. [CrossRef]
13. Golovashchenko, S.F. *Electrohydraulic Forming of Near Net Shape Automotive Panels*; Technical Report DE-FG36-08-GO18128; Ford Motor Company: Detroit, MI, USA, 2013; 224p.
14. Golovashchenko, S. Hydromechanical Drawing Process and Machine, Ford Global Technologies. Patent US 9,375,775, 28 June 2016.
15. Avrillaud, G.; Mercier, R. Method for Electrohydraulic Forming and Associated Device. Patent WO2018091481, 24 May 2018.
16. Avrillaud, G.; Beguet, F. Hybrid Forming Method and Corresponding Forming Device. Patent WO2020165538, 20 August 2020.
17. Deroy, J. Modélisation et Méthodologie de Caractérisation d'une Décharge Electrique Impulsionnelle Dans L'eau. Ph.D. Thesis, Ecole Polytechnique, Palaiseau, France, 2014. (In French).
18. Jeanson, A.C.; Avrillaud, B.; Mazars, G.; Taber, G. Identification of Material Constitutive Parameters for Dynamic Applications: Magnetic Pulse Forming (MPF) and Electrohydraulic Forming (EHF). In Proceedings of the 5th International Conference on High-Speed Forming, Daejeon, Korea, 29 May 2014.
19. Jeanson, A.C.; Bay, F.; Jacques, N.; Avrillaud, G.; Arrigoni, M.; Mazars, G. A coupled experimental/numerical approach for the characterization of material behavior at high strain-rate using electromagnetic tube expansion testing. *Int. J. Impact Eng.* **2016**, *98*, 75–87. [CrossRef]
20. Felts, R. Method and Apparatus for Forming Metal, Cincinnati Shaper. Patent US3273365, 20 September 1966.
21. Brejcha, R.J.; Bazell, S.; Radnik, J.L. Diaphragm Member of Elastomeric Material. Patent US3358487, 19 December 1967.
22. Held, C.; Plaut, R. Development of an electromagnetic tool for postforming of Aluminum hood, Doors & closures in car body eng. In Proceedings of the Automotive Circle International Conference, Bad Nauheim, Germany, 1–3 April 2014.
23. Cappelaere, M.; Boutin, J.C. Magnetic Pulse Forming process integration on a Renault press line to sharpen radii on outer panels, Forming in Car Body Engineering. In Proceedings of the Automotive Circle International, Bad Nauheim, Germany, 28–29 September 2021.
24. Atieh, S.; Carvalho, A.A.; Santillana, I.A.; Bertinelli, F.; Calaga, R.; Capatina, O.; Favre, G.; Garlasché, M.; Gerigk, F.; Langeslag, S.A.E.; et al. First Results of SRF Cavity Fabrication by Electro-Hydraulic Forming at CERN. In Proceedings of the SRF2015, Whistler, BC, Canada, 13–18 September 2015; pp. 1–7.
25. Cantergiani, E.; Atieh, S.; Léaux, F.; Fontenla, A.T.P.; Prunet, S.; Dufay-Chanat, L.; Koettig, T.; Bertinelli, F.; Capatina, O.; Favre, G.; et al. Niobium superconducting RF cavity fabrication by electrohydraulic forming. *Phys. Rev. Accel. Beams* **2016**, *19*, 114703. [CrossRef]
26. Cantergiani, E.; Avrillaud, G.; Clemente, C.A.; Atieh, S.; Favre, G.; Deroy, J.; Raveleau, F. First Results of Superconducting RF (SRF) cavity fabrication by Electro-Hydraulic Forming. In Proceedings of the 7th International Conference on High-Speed Forming, Columbus, OH, USA, 13–16 May 2018.
27. Singer, W.; Singer, X.; Jelezov, I.; Kneisel, P. Hydroforming of elliptical cavities. *Phys. Rev. ST Accel. Beams* **2015**, *18*, 022001. [CrossRef]
28. Marhauser, F. JLAB SRF Cavity Fabrication Errors, Consequences and Lessons Learned. In Proceedings of the IPAC11 Conference, San Sebastian, Spain, 4–9 September 2011.
29. Wong, C.C.; Dean, T.A.; Lin, J. A review of spinning, shear forming and flow forming processes. *Int. J. Mach. Tools Manuf.* **2003**, *43*, 1419–1435. [CrossRef]
30. Cantergiani, E.; Avrillaud, G.; Clemente, C.A.; Atieh, S.; Favre, G.; Deroy, J.; Raveleau, F. *First Results of Large Size SRF Cavity Fabrication by Electrohydraulic Forming*; Universitätsbibliothek Dortmund: Berlin, Germany, 2017.
31. Atieh, S. Novel Technologies Applied to SRF Fabrication at CERN, TTC 2020, CERN. Available online: https://indico.cern.ch/event/817780/contributions/3716472/ (accessed on 5 February 2020).
32. Allehaux, D.; Re, R. Electro-Hydro-Forming on aeronautical parts: Expectations, preliminary investigations and further improvement. In Proceedings of the 2nd International Conference for Indus. Magnetic Pulse Welding and Forming, Munich, Germany, 24 January 2013.

Journal of
*Manufacturing and
Materials Processing*

Article

Electrohydraulic Forming of Low Volume and Prototype Parts: Process Design and Practical Examples

Alexander V. Mamutov [1,*], Sergey F. Golovashchenko [1], Nicolas M. Bessonov [2] and Viacheslav S. Mamutov [3]

[1] Mechanical Engineering Department, School of Engineering and Computer Science, Oakland University, Rochester, MI 48309, USA; golovash@oakland.edu

[2] Institute for Problems in Mechanical Engineering of the Russian Academy of Sciences, 199178 St. Petersburg, Russia; nickbessonov1@gmail.com

[3] Peter the Great St. Petersburg Polytechnic University, Institute of Machinery, Materials, and Transport, 195251 St. Petersburg, Russia; mamutov_vs@spbstu.ru

* Correspondence: a.mamutov@yahoo.com

Abstract: Electro-Hydraulic Forming (EHF) is a high rate sheet metal forming process based on the electrical discharge of high voltage capacitors in a water-filled chamber. During the discharge, the pulsed pressure wave propagates from the electrodes and forms a sheet metal blank into a die. The performed literature review shows that this technology is suitable for forming parts of a broad range of dimensions and complex shapes. One of the barriers for broader implementation of this technology is the complexity of a full-scale simulation of EHF which includes the simulation of an expanding plasma channel, the propagation of waves in a fluid filled chamber, and the high-rate forming of a blank in contact with a rigid die. The objective of the presented paper is to establish methods of designing the EHF processes using simplified methods. The paper describes a numerical approach on how to define the shape of preforming pockets. The concept includes imposing principal strains from the formed blank into the initial mesh of the flat blank. The principal strains are applied with the opposite sign creating compression in the flat blank. The corresponding principal stresses in the blank are calculated based upon Hooke's law. The blank is then virtually placed between two rigid plates. One of the plates has windows into which the material is getting bulged driven by the in-plane compressive stresses. The prediction of the shape of the bulged sheet provides the information on the shape of the preforming pockets. It is experimentally demonstrated that using these approaches, EHF forming is feasible for forming of a fragment of a decklid panel and a deep panel with complex curvature.

Keywords: electro-hydraulic; pulsed forming; numerical simulation; preforming

Citation: Mamutov, A.V.; Golovashchenko, S.F.; Bessonov, N.M.; Mamutov, V.S. Electrohydraulic Forming of Low Volume and Prototype Parts: Process Design and Practical Examples. *J. Manuf. Mater. Process.* **2021**, 5, 47. https://doi.org/10.3390/jmmp5020047

Academic Editor: Verena Psyk and Steven Y. Liang

Received: 30 March 2021
Accepted: 10 May 2021
Published: 13 May 2021

Publisher's Note: MDPI stays neutral with regard to jurisdictional claims in published maps and institutional affiliations.

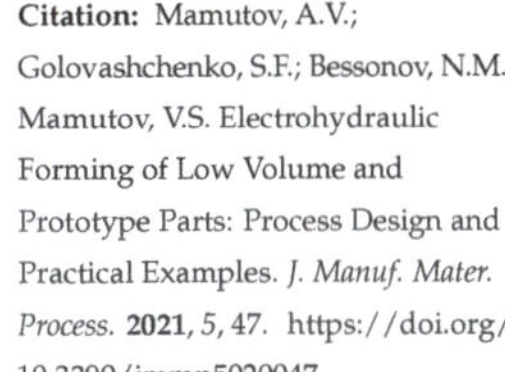

1. Introduction

The trend of creating global vehicle architecture in automotive industry described by Ferreira and Kaminski [1] is broadly spreading among automotive manufacturers. Sharing components between different platforms and vehicles leads to overall increase in production volumes and generates a need for high productivity manufacturing processes. In stamping of sheet metal components, servopress equipment becomes more popular due to increased productivity compared to the mechanical presses broadly used in industry. However, the automotive industry has a strong demand for low volume technologies, which would enable low cost production of prototype parts. Sheet metal forming processes where one side of the stamping die (typically a punch) is replaced by pressure were described in a number of publications and reference books on metal forming [2]. These processes have the strong advantages of lower tool cost due to elimination of the need to cast and machine one side of the die as well as due to no further need for accurate alignment of matching portions of the die. These simplifications also shorten the tool development cycle. This group of metal forming processes are typically employed for deep drawing and stretch drawing of sheets and forming of tubular blanks by applying forming pressure from inside the tubes. Among them are: (a) quasistatic processes, such as superplastic forming (SPF), hydroforming (HF), rubber/polyurethane forming (RF), and (b) high rate forming processes, such as electromagnetic forming (EMF), electrohydraulic forming (EHF) and explosive forming (EF). EMF, EHF, and EF belong, according to Wilson [3], to the group of high velocity forming technologies.

Quasistatic processes are simpler for analysis and experimental tryout, since they are similar to traditional sheet metal forming operations where stamping presses are utilized. A major drawback of these processes is usually in the significant force required to form the blank: pressure that is needed to form the smallest radius of the die cavity defines the pressure that has to be applied to the overall surface of the blank inside the die cavity. This condition leads to significant investments for forming equipment required to implement these processes. The need to manufacture the prototype parts motivated broad usage of hydroforming processes employing a polyurethane sheet as a membrane to separate the sheet metal blank from the fluid. This technology is known as the flexform process [4]. The term of crush forming is often utilized for these processes because no binder is being used to control sheet metal flow into the forming die.

The high speed forming technologies were mentioned for industrial applications starting from 1960s [5], and more extensively in early 1970s [6], mainly due to several advantages which these processes have compared to quasistatic processes. The most impressive advantage was in the capability to form parts with dimensions of several meters. Forming such parts was impossible during that stage of development of metal forming technology and equipment. Callender [7] reported the results of EHF with a net of wire bridges to form a dish component 3 m in diameter. Epoxy lined dies were used in these experiments which started showing degradation after forming twelve blanks. Felts [8] demonstrated EHF forming of dish components with flanged holes from Aluminum 6016-T0 as well as deep drawing box type shapes from deep drawing quality steel. A variety of die materials starting from steel-reinforced concrete, chopped glass-fiber-reinforced epoxy resin and, in some configurations, tool steel were used.

Hanley [9] carried out a very detailed review of applications of high speed forming technologies at General Dynamics Corporation: EHF was considered the most efficient method for forming of tubular blanks with potential punching of holes in the same operation as forming the shape of the part. A typical diameter of tubular blanks was 30-0 mm. The electrode system employed in studies reviewed in [9] was based upon wire bridges, which had to be replaced after each high-voltage discharge. The dies were designed consisting of two halves based upon the clam shell approach. Implementation of EHF allowed a reduction in the number of manufacturing steps and eliminated the need to subdivide one part into several independently stamped and further joined smaller parts. The production cycle for the EHF processes described in [9] was reported as several minutes. Feddersen [10] illustrated a number of applications of tube expansion by EHF including one case where a window was pierced in the tubular component.

Callender [7] also described experiments on EHF of a conical shape welded blank into an ogival shape. Schrom [11] demonstrated a laboratory EHF process for bulging sheet metals and expanding thin cylindrical shells indicating that Aluminum foil of 12.7 mm width can substantially increase the efficiency of the process compared to other bridge materials and no-bridge tooling configuration. The attempt to study the efficiency of energy transfer in EHF was made by Duncan and Johnson [12] for both sheet forming and tube forming. Duncan and Johnson [12] demonstrated forming of a dish with a lot of fine features which certainly increased the complexity of the EHF operation compared to previous cases.

Bruno [5] illustrated several examples of duct parts from special alloys for aerospace applications: various shapes were formed by electric discharges inside cylindrical shells. This type of part is difficult to make by other methods. Similarly, Davies, and Austin [6] illustrated an application of the EHF process for the piercing and flanging of holes in the tubular thin walled extrusion made of pure aluminum. However, very limited details of the studied EHF process were disclosed. Most of the applications belonged to the defense industry which limited publishing of technical details. More importantly, the methodologies of formability analysis were in their infancy: the initial ideas on Forming Limit Diagram development were published several years later for biaxial stretching by Keeler [13] and for stretching–compression by Goodwin [14].

During the last twenty-five years, a number of experimental studies illustrated that the very significant improvement in sheet metal formability can be attributed to high strain rate, low friction, and the coining effect. This effect is based upon through thickness compression during high velocity impact between the blank and the die surface. In [15] a significant improvement of formability was observed for AA6061-T4. Imbert et al. reported visible improvement in formability for two aluminum alloys: 6111-T4 and 5754 [16]. Authors [17] reported formability improvements for AA5182. Dariani et al. [18] indicated an extension of formability for 1045 steel and 6061-T6 aluminum alloy. Golovashchenko et al. [19] quantified formability improvements for four dual phase steels: DP500, DP590, DP780, and DP980. Analysis of microstructure and porosity development in quasistatic and EHF processes [20] indicated that both high velocity impact and high strain rate create favorable conditions for DP780 and DP500. Jenab et al. [21] arrived to similar conclusions for AA5182. In addition, calibration of the formed part is provided in the same tool, minimizing springback. In a traditional stamping process, either a restrike die would be needed to minimize springback, adjustment of the die surface to compensate for springback, or additional stretching of the blank using a lockbead at the very end of the forming process to minimize springback.

Even though explosive forming processes provide nearly unlimited capabilities from the perspective of achievable pressure and impact speeds, the safety implications often limit its application for industrial processes. A review of more recent results on EF processes was published by Mynors and Zhang [22].

The intention to achieve similar benefits without safety issues motivated studies of EHF and EMF forming technologies. Psyk et al. [23] performed a very detailed analysis on various aspects of EMF processes. These aspects included coil designs, details on forming and joining processes, and recently developed numerical and analytical models. The major limitation of EMF technology is in a significant reduction in forming pressure applied to the sheet when it moves further away from the inductor generating electromagnetic field as well as in the requirement of good electrical conductivity for the material of the blank and for the material of the coil. A typical EMF coil includes insulation material which, in most cases, limits the structural strength of the coil. A reduction in pulsed pressure as a result of the blank getting further away from the coil limits the application of EMF processes to either rather shallow shapes or the shapes where the final forming is possible without backing pressure.

Recent studies on EHF mostly concentrate on the improved formability demonstrated in free bulging, for example by Maris et al. [24] and the forming of a sheet into very simple shapes such as conical or V-shape, for example, by Cheng et al. [25]. The general trend is that substantial improvement can be achieved for the forming of Aluminum alloys, dual phase steels and a variety of other materials.

Based upon experimental study performed by Golovashchenko et al. [26], EHF technology has the strongest potential among pulsed forming processes to be applied to forming deep cavities with sharp corners and shapes that are more complex. Authors [26] demonstrated that a sheet metal blank can be formed by sequential discharges of several pairs of electrodes with no specific requirement to the parameters of grain structure (as for SPF), or high electrical conductivity (as in EMF processes). Employment of several pairs of electrodes enables the pressure distribution applied to the sheet metal blank to be tailored in a broad range: several discharges generated by multiple electrodes may create a more favorable mechanism of sheet metal flow into the cavity of the die.

Another opportunity for formability improvement was discussed in [26] where the blank was preformed into a shape which utilized formability of the adjacent portions of the blank to fill the corners of the die shape. Authors [27] demonstrated that EHF has a better capability to form sharp corners than quasistatic hydroforming described in [28] and also requires a lower clamping force employing the inertia of the tool and the clamping press, which requires a much smaller capital investment. A reconfigurable EHF tool comprising of several small size chambers can be assembled together for the forming of various shapes, as suggested by Golovashchenko [29]. The downside of the EHF process is in the necessity to evacuate and refill at least some portion of the EHF chamber with water for every formed part.

A very important advancement which started changing the sheet metal forming industry in the late 1980s was the developing capability to simulate sheet metal forming processes. The explicit integration procedures initially developed for simulation of explosions in defense applications very naturally fit into high velocity forming processes. The models considering EHF process as dynamic hydroforming with uniformly applied pressure were developed in the late 1980s, for example by Vagin et al. [30]. The dynamics of pressure distribution in the EHF chamber were studied in [30] experimentally by: (a) inserting piezo sensors capable of surviving such a high level of dynamic pressure in the chamber and (b) by using a membrane method as proposed by Cole [31] accounting for the pressure level on the surface of the membrane by the local displacement of the membrane in a small diameter hole. Knyazev and Zhovnovatuk [32] performed similar measurements for a multi-electrode chamber. Such experimental measurements provided the justification for applying the dynamic pressure to the surface of the blank.

Melander et al. [33] analyzed formation of a conical shape and also free bulging of the blank into an open die cavity. Simplifying assumptions regarding pressure distribution in the chamber and percentage of reflection of pressure from the walls of the chamber were employed. Rohatgi et al. [17] approximated the pressure distribution on the surface of the round blank with a linear function along the radius of the blank assuming the exponential decay of pressure as a function of time. Hassannejadasl et al. [34] assumed hydrodynamic pressure in the model as an acceleration of fluid particles on a spherical surface inside the chamber. The model was calibrated using the experimental data on free bulging of the blank. The time function for the acceleration was selected as a half of a sinusoid. In this case, the pressure on the sheet metal blank was not considered a uniform or constant distribution: it was calculated from the hydrodynamic analysis. The simplifying assumption regarding the spherical impact as a method of energy deposition worked well for the analysis of sheet metal forming into a conical die. However, in order to design an EHF chamber of a more complex shape, more realistic energy deposition accounting for the shape of the chamber is needed.

Vohnout et al. [35] analyzed pressure heterogeneity in a cylindrical EHF chamber by using a membrane penetration method. The numerical simulation was performed using CTH code simulating explosions. The equivalence between the explosion and EHF processes was not very clearly established; however, the dynamic pressure distribution could be similar with the exception that EHF provides substantially slower energy deposition compared to a chemical explosion. Based on the performed analysis, it was concluded that pressure distribution is very sensitive to changes in the location of the energy source (explosion in the case of the performed theoretical analysis).

Mamutov et al. [36] described a more general algorithm based upon LS-DYNA commercial code capable of predicting pressure wave formation as a result of a high voltage discharge in a water filled chamber. This algorithm is capable of analyzing the blank formation for a general configuration of the electrode system. The drawback of this approach is that, at current computational capabilities, many hours of computations on a multiprocessor supercomputer are required to predict multi discharge formation of the blank. However, without pre-existing chamber design, these calculations might show that (a) a different configuration of the EHF chamber is needed, (b) the blank has insufficient formability to fill the required shape, and (c) the available equipment cannot provide a sufficient discharge current. If any of these situations occur, there may be a need to run multiple other configurations of the EHF chamber and sheet metal forming process. In this study, a simplified approach is proposed which is expected to assist in a much faster design of the EHF chamber configuration capable of forming the desired shape from the targeted sheet material.

Woo et al. [37] studied formation of pressure pulses in the EHF chamber by employing an ALE numerical approach in LS-DYNA, illustrating that this powerful approach enables a detailed study of wave reflection and propagation in EHF chamber. Woo et al. [38] simulated the EHF free forming process into an open round window and used the model described in [37] to identify the material parameters of Aluminum sheet 6061-T6 which would have the best correlation with experimental data in free forming conditions.

Jenab et al. [39] described a simplified methodology of EHF process analysis by assuming that the EHF load is applied as a pressure pulse similar to an explosive load known from the literature instead of taking into account the history of pressure wave propagation through the EHF chamber. Authors [39] concentrated on studying the details of AA5182 sheet forming into an open round window or into a conical die analyzing different material models and neural algorithms accounting for material high rate behavior.

A tryout study was performed experimentally in [40] to demonstrate the overall feasibility to form various sheets with EHF using a pulse generator assembled by researchers from high voltage capacitors. This study confirmed that the EHF technology can be employed as a very simple process for forming tryout.

Xiong et al. [41] developed a simplified engineering model predicting pulse pressure parameters as a function of distance from the discharge location in an electrohydraulic chamber and applied it to the analysis of impact cracking test conducted on cement samples.

As it can be seen in the presented literature review, the overall advantages of the EHF processes is in extended material formability due to high strain rates, coining effects, and no friction on one side of the die. However, an additional advantage of redistributing strains in a formed blank based upon flexible loading in media forming processes compared to traditional stamping in two-sided dies has not been explored in significant detail.

The idea of preforming the sheet in a traditional stamping die and then applying the pulsed pressure to form the material into a sharp radius was discussed by Daehn et al. [42] as electromagnetic forming of a door inner part where sheet metal blank was originally preformed in two-sided dies, and then a restrike electromagnetic forming operation was applied. An idea of preforming a sheet by gaining the metal in the pockets adjacent to the areas of excessive strains was introduced by Golovashchenko et al. [43]. The initial concept was to form the preform shape similarly to the final shape of the part with the exception of the areas with sharp corners, which would be supplemented with donor pockets minimizing stretching of sheet in these areas. The cost of such a preforming die would be similar to the cost of the die with the final shape of the part. Overall, this technology would require an incremental increase in the cost of the dies. In this paper a further advancement of this concept in the direction of significantly simplifying the shape and minimizing the cost of the preform die [44] will be discussed.

The objective of this paper is to introduce a simplified methodology of design for EHF processes and to illustrate a low-cost preform concept for the case where extended formability of EHF process is not sufficient to make the targeted shape. Therefore, the proposed concept can be viewed as a further step to enable usage of higher strength, lighter and less formable materials in the automotive industry.

2. Proposed Method of Developing EHF Processes

This paper represents an attempt to formulate the necessary and optional steps to develop the EHF prototype process for a new application where benefits of this pulsed forming technology can be substantial either due to a simpler tooling design, due to the capability of the EHF method produce more deformation from a deformed sheet without fracture, or take advantage of the soft application of the forming load compared to stamping in traditional rigid dies. The authors admit that the proposed method is still in an early phase of development and will further mature with broader implementation of EHF processes in production.

There are following important questions which need to be answered at the development stage of the prototype EHF process:

(a)   whether the part can be formed using the EHF process from the sheet material proposed by the product designer;

(b)   if the sheet material candidate does not provide sufficient formability even with extra formability offered by EHF, decide whether the preforming process lowering the maximum strains can help to fill the shape without fracture;

(c)   whether the available pulsed equipment and the EHF chamber provide a sufficient amount of pulsed pressure to fill all the details of the formed shape;

(d)   if a new EHF chamber is being developed, decide where the electrodes should be positioned to achieve required pressures in the most difficult to form areas of the part, which are typically in the areas of sharper corners at the bottom of the formed cavity.

Numerical simulation or experimental tryout can be selected to execute these steps, and the specific actions strongly depend on accumulated experience and necessary efforts to successfully achieve the goal taking into account the existing experience in EHF technology at a particular organization, the timeline and available resources. An important point of this paper is the selection of quick and efficient simplified approaches not requiring lengthy development and very significant computational efforts. The following sequence of steps addresses the questions listed above.

1.   Identify the areas of the part that are the most difficult to fill and require the largest pressure. A simplified numerical approach assuming a uniform pressure forming load in quasi static formulation is capable to achieve this goal. The areas of the cavity which are filled last require the largest pressure. Usually it occurs at the sharp corners located at the bottom of the die cavity. The deformation of the blank in these areas is usually the largest, since the rest of the cavity is already filled, and filling of the corners is achieved by local stretching of the material. The locations of these difficult to fill cavities will indicate where to place the electrodes and, potentially, how many electrodes are needed. This step can use various commercial software suitable for media forming processes. Based on these simulation results and available data on sheet metal formability in EHF processes reviewed in the Introduction, it is possible to define whether the part can be formed using the EHF process without any preforming.

2.   Develop the preforming process to enable the redistribution of the peak strains in a formed part. This step is optional and is needed only in case when direct application of EHF will not be enough to manufacture the part without fracture. It can be also applied for further weight reduction of the component and using higher strength and less formable material. The concept of this step will be explained later in this chapter.

3.   Analyzing the pressure distribution in the chamber and configuring the chamber design and the electrode system to provide sufficient pressure to fill the most difficult areas. This step is optional and can be very useful if it is anticipated to build a new chamber, or if the available energy is not sufficient for the initial tryout. This step would be beneficial to further improve the process, lower the discharge energy and extend the life of the electrodes. However, for the initial tryout or low volume production, using an existing chamber might be more economical. This analysis can be done based on the Lagrangian model described by Mamutov et al. [36] where only a hydrodynamic model is used without accounting for blank deformation. However, if this technology is not available, it is possible to use simplified methods for relative comparison estimating pressure based upon the distance from the discharge channel. This approach was employed in early publications analyzing pressure distribution during an explosion in water or early EHF analysis reviewed by Mamutov et al. [36].

4. Validate the developed process and chamber design using the full methodology for EHF analysis, accounting for pressure pulse propagation through the water filled chamber. This step is certainly optional but can be very helpful if this technology is readily available to the user. LS-DYNA software has all the necessary capabilities to perform this step. The details of such a detailed simulation are described in Mamutov et al. [36]. However, it is very computationally intensive and requires lengthy simulation on a supercomputer.

The step 2 described above is a unique step not previously described in the technical literature on EHF. In this case, a solution to the split issue was to introduce a preforming operation in which the material is bulged in the areas of low strain to provide additional metal to the areas of high strain, as suggested by Golovashchenko [43]. The overall concept of preforming of gaining pockets is known in literature, typically in the area of stretch flanging [44]. In this case, the preforming pockets are attempting to increase the length of the trim line before the trimming operation while stretch flanging occurs after trimming; therefore, minimizing the amount of stretching along sheared edges.

For corner filling processes, defining the geometry of the preforming pockets is much more challenging because the pocket needs to provide sufficient, but not extra, material to spread along the surface rather than along the trim line. The overall idea is based upon a non-uniform distribution of strains in the parts with local features. Sharp corners are the areas of high concentration of strains while the rest of the blank has significantly smaller strains. Designing the process in which the strains are spread through larger areas and the peak strains are significantly reduced is the major step towards achieving this goal. The problem is how to identify the location and the shape of the preforming pocket. The following steps were made to achieve this goal:

2.1. The forming process was simulated based upon uniformly distributed pressure, as described in step 1 of the overall methodology. From this model, the principal strain tensor components from the mid-surface of the sheet were extracted from each element and assigned to the corresponding elements of the initial sheet metal blank with an opposite sign meaning that stretching is replaced by compression as shown in Figure 1a,b.

2.2. The initial sheet metal blank with the same Lagrangian mesh as at the beginning of the forming process in step 2.1 was positioned between two flat rigid plates as shown in Figure 1c. One of the plates had windows in which the strained sheet could bulge out. These windows were positioned above the elements which were insufficiently stretched in the initial forming process and could tolerate more deformation safely. Selection of the size and position of the windows was defined by the iterations, but the model was providing the depth of bulging in each selected window opening.

2.3. The deformation process of sheet metal bulging into specified windows (as shown in Figure 1d) was simulated in an elastic membrane formulation previously described by Golovashchenko et al. [26]. This process was simulated in explicit formulation with linear viscosity, so the bulging stopped after few cycles of vibration. The shape of the bulged blank was then considered to be the die surface for the preforming operation. The elastic formulation of this model allowed to have a proper amount of material bulged into the windows. The yield stress was considered nonexistent for this step of the analysis.

2.4. The preforming process of the defined pockets was simulated in elastoplastic formulation applying uniform pressure to bulge the flat sheet into the shape defined in step 2.3. The blank was then moved to the die, identical to step 2.1, and the forming process continued. As a result, the maximum strains are expected to be lowered approximately by half.

The software used for simulation in steps 1, 3, and 4 in this Chapter was LS-DYNA 971 DP.

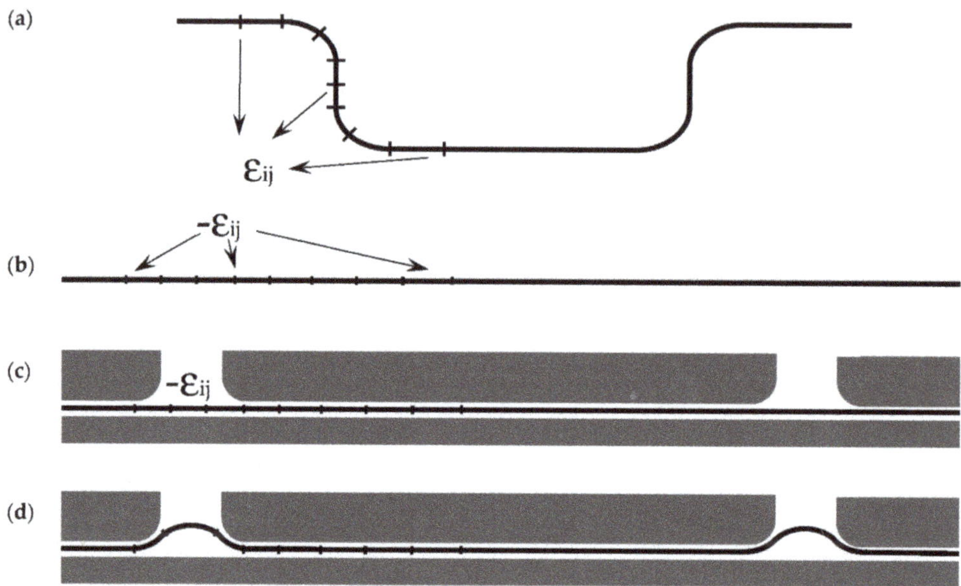

Figure 1. Schematic illustrating the steps of building a preforming shape: (**a**)—simulation of forming under uniform pressure, (**b**)—applying the obtained opposite sign field of strains to the initial flat blank, (**c**)—placing the prestressed blank within the restraining shape, (**d**)—identifying the preforming geometry by simulating the bulging process.

3. Case 1: Decklid Panel

3.1. Materials and Methods

The decklid panel is shown in Figure 2. The material used for this panel was an AA6111-T4 0.93 mm thick sheet.

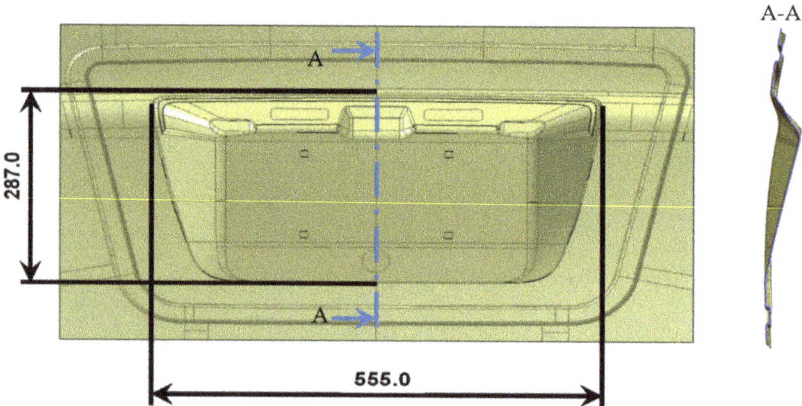

Figure 2. Design of the part formed by EHF.

The panel is formed using the EHF process. The perimeter of the part was initially locked by the lockbead during the binder closure. Therefore, no material inflow from outside the chamber occurred during the EHF process.

A full-scale simulation of the EHF process is a very computationally expensive tool. The most complicated and CPU-consuming part of such a model is simulating the expanding plasma bubble and simulating a water stream that moves the deformable metal sheet. Therefore, the major factor in time-saving will be eliminating one or more of these objects from consideration.

A feasibility study for the part would include the following tasks:

1. *Identify the areas of the part that are most difficult to fill and require the largest pressure.* Since the goal of the study is to develop a simplified, less computationally intensive approach, the simulation is performed assuming uniform pressure distribution. This approach allows to eliminate plasma and water from the model, which, combined, account for over 90% of the computational time. The resulting simplified model does not account for effects related to water–blank interaction, but the required result mostly depends on the blank–die interaction, so the simplification is justifiable and can be further refined during the final simulation step, where a more accurate model [36] can be employed.

The areas of the cavity which are filled last require the largest pressure. Usually it occurs at the sharp corners located at the bottom of the die cavity. The deformation of the blank in these areas is usually the largest, since the rest of the cavity is already filled, and filling of the corners is achieved by local stretching of the material. The locations of these difficult to fill cavities will indicate where to place the electrodes and, potentially, how many electrodes are needed.

2. *Analyzing the pressure distribution in the chamber and configuring the chamber design and the electrode system to provide sufficient pressure to fill the most difficult areas.* This analysis can be done based on the Lagrangian model described by Mamutov et al. [36] where only a hydrodynamic model is used without accounting for blank deformation. It should be admitted that the duration of the process at the final discharge where almost all the cavity is filled is much shorter compared to the initial forming step where large deflections of the blank take place. It also allows the usage of Lagrangian approach vs the more computationally expensive Arbitrary Lagrangian–Eulerian (ALE) approach.

This simplified model allows for virtual movement of the electrodes at the simulation stage and modification of the shape of the chamber to provide sufficient pressure to fill the sharp corners in all necessary areas of the part.

3. *Conduct a formability study based on numerical or experimental verification of the decisions made in the previous two steps.* During this step, the available data on pulse forming formability should be reviewed as well as expected loads on the die and on the electrode system.
4. *Perform the analysis to clarify whether a preforming step enabling the redistribution of the peak strains in a formed part is needed.*

The simulation software used for all other numerical models was LS-DYNA 971 DP.

3.2. Results and Discussion

The strain distribution in the formed panel at the end of the simplified uniform pressure analysis is shown in Figure 3. The symmetry along the cross-section A-A was taken into account, so only the half of the panel was simulated. The results of numerical simulation indicated that the styling lines with sharper radii do not represent a problem, since the material can be pulled from the adjacent areas of the blank. The area of deep drawing and sharp corners shown in red and yellow in Figure 3 represent a problem from both the required pressure and formability perspectives. Similar results were received from a more detailed numerical model [36] where all the details of pulsed pressure propagation through the water filled chamber as well as dynamic deformation of the blank were taken into account: the corners shown in red were filled last and required higher voltage discharges to be completed. The quantitative comparison of simulation results using the simplified approach and the model [36] in the area of maximum strains in Figure 3 indicated that the simplified approach gives slightly higher strains in the area of corners: 0.25 major principal strain vs. 0.22 in the simulation using method [36]. It should be admitted, however, that in [36], the ALE Method was employed while in this study, the traditional Lagrangian mesh was used. It should be indicated that in both numerical models, friction was accounted for based upon Coulomb's friction law using a constant coefficient of friction. In real forming conditions, the coefficient of friction may vary significantly depending on contact pressure and the possibility to squeeze the lubricant from the contact between the sheet and the die. This factor alone might lead to more significant uncertainty in major strains than the difference between the results obtained by both methods.

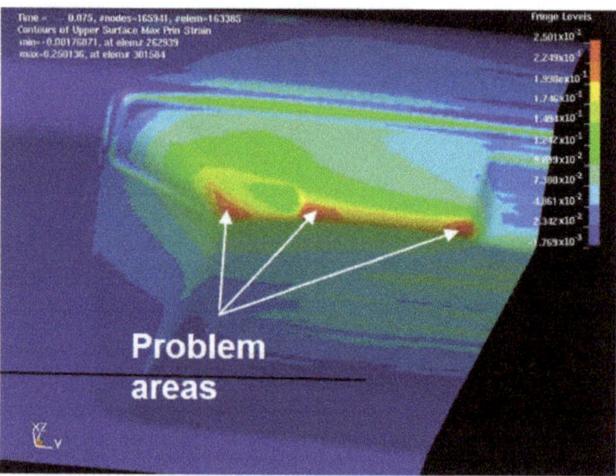

Figure 3. Results of quasistatic numerical simulation of forming a sheet metal blank with uniformly distributed pressure.

One of the major drawbacks of using the full-scale simulation approach described in [36] is significant hardware and CPU-time demand. This issue exists on multiple levels. First of all, the approach [36] requires simulation of the discharge channel, pressure transmitting water, and the sheet being deformed in contact with the die. The computational effort to simulate pressure propagation through the fluid takes over 90% of the computational time.

Second, the contact between the water and the sheet metal requires correlation between the element sizes of water and blank meshes: the Fluid Structure Interaction algorithm in LS-DYNA dictates that they must be about the same size. This leads to a cubic-power growth of the number of fluid elements with the refinement of sheet metal elements. As a result, even when using powerful cluster computers, the user has to select element size based not on the desired mesh size, but based on the available hardware resources.

Third, a typical EHF process is a multidischarge process, which means that three to four such simulations must be performed correlating to the actual number of performed EHF discharges in order to simulate a full forming process of a panel. One additional complication related to this is the need to prepare the models and transfer simulation data while simulating sequential discharges.

Comparison of required computations could be done based upon the following example: to simulate the EHF process in [36], 8–10 h of computation on a powerful cluster machine with 32 processors was required for each discharge. It required multiple days of simulation with just a small improvement in mesh quality.

Even though the growth of CPU power and available memory as well as software development will make the full scale approach eventually available for an engineer working on the design of sheet metal forming processes, it is still difficult to use the approach [36] outside of research facilities. Alternatively, using the simplified approach presented in this paper only requires a desktop PC and much shorter simulation time of under one hour, which opens the possibility to analyze multiple configurations of the EHF forming process and make practical design decisions.

Figure 4 shows the blank formed by quasistatic hydroforming with 2.1 MPa pressure of the fluid. The clamping force was provided by a 100 ton hydraulic Dake Dura press. It can be seen that the overall cavity is filled except the corners. This simplified analysis indicated that approximately 30 MPa of pressure is needed to fill the sharp corners. Taking into account the full area of the panel where the quasistatic pressure is applied, the maximum affordable pressure for a 100-ton clamping force is approximately 3 MPa. To form this panel quasistatically, a factor-of-10 larger clamping press would be needed. The next step is to design a chamber which can deliver this level of pulsed pressure using available energy from an existing pulse generator.

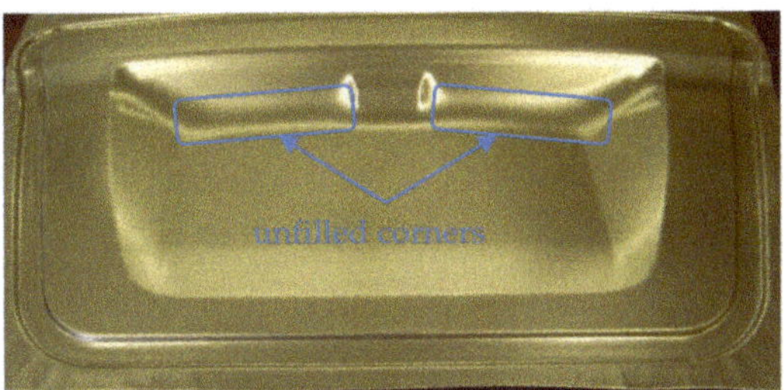

Figure 4. Aluminum 6111-T4 panel formed by quasistatic hydroforming using 2.1 MPa of pressure.

An example of such analysis for the part shown in Figure 2 is illustrated in Figure 5. Due to the limitations of the Lagrangian approach used in this simulation, the shape of the chamber is slightly simplified, and the shape of the electrodes is not taken into account. Since the volume of the electrodes is significantly smaller than the overall volume of the chamber, the introduced numerical error is negligible.

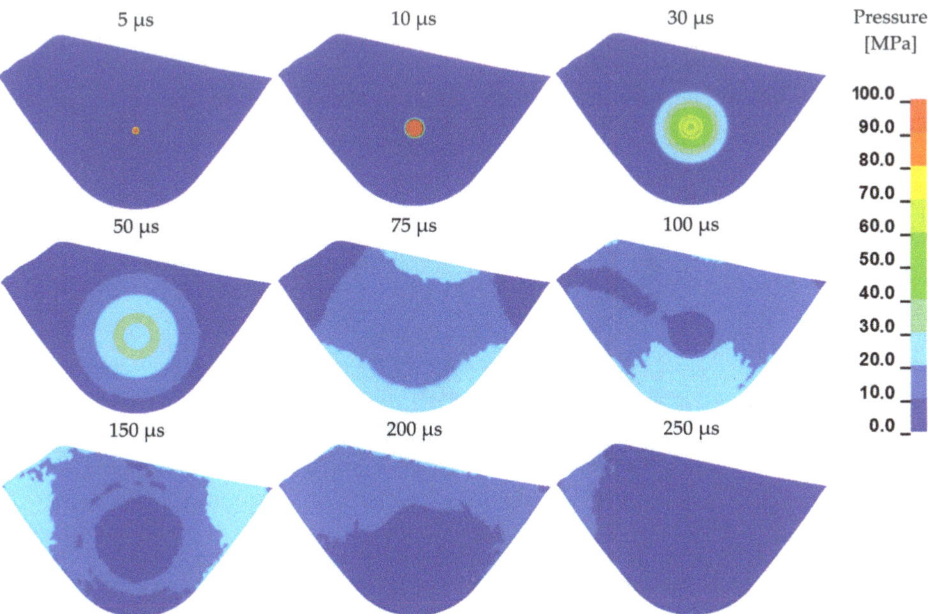

Figure 5. Distribution of pressure in the EHF chamber along the cross-section A-A for the last discharge when the die cavity is completely filled with the blank.

Based on this analysis, it was concluded that a single electrode chamber design is capable of providing the required pressure. The next step is to optimize the position of the electrodes and the size of the chamber using a purely hydrodynamic model. By minimizing the volume of the chamber, higher pressure can be accomplished from a very simple and approximate consideration that the energy of water compression can be understood as an integral of pressure by volume: a smaller volume of the chamber with the same level of discharged energy will result in a higher average pressure in the chamber.

At the initial stage of formulating the chamber design, the general considerations that the propagation of a shock wave initially has spherical symmetry, as it was admitted in early publications on the hydrodynamic analysis of explosions, for example by Cole [31], until the wave contacts the walls of the chamber, and reflections from the walls start influencing the pressure distribution. Therefore, positioning the electrodes closer to the most difficult to form locations might help to improve the efficiency of the chamber by utilizing the energy of the initial shock wave as well as minimizing the overall volume of the chamber. However, one important limitation needs to be taken into consideration: the initial position of the sheet metal blank in the chamber should be at a distance sufficient enough to avoid arcing on the blank. For the prototype conditions, in addition to just increasing the distance from the electrodes to the sheet metal blank, the conductive wire can be placed between the electrodes for the first discharge.

Alternatively, a moveable electrode head proposed by Golovashchenko [29] can be used to keep the volume of the chamber small and have a capability to adjust the distance between the discharge channel and the blank at each following discharge. The resulting die and the chamber design for making the part is shown in Figure 6a,b.

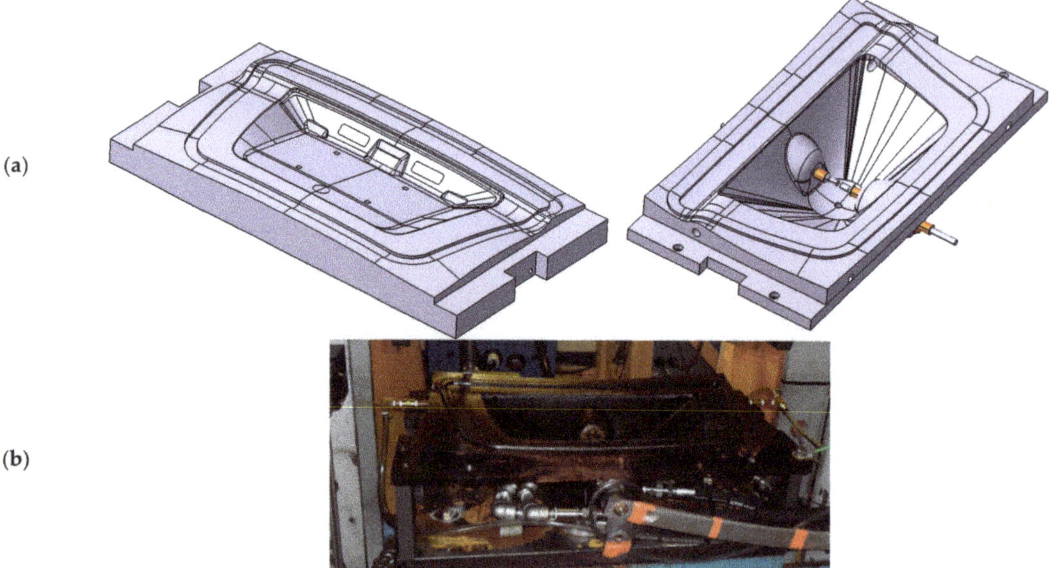

Figure 6. Design of the die-chamber tool set (**a**), and experimental chamber (**b**).

The next step in the proposed algorithm is formability analysis. Making a decision based upon a traditional Forming Limit Diagram (FLD) and simulation results is safe but might be too conservative. As indicated in Figure 3, the major strain in the corner filling operation was 0.25 with approximately the same level of minor strain. Based upon formability studies performed by Graf and Hosford [45] and Chow et al. [46], the equibiaxial stretching of this level is marginal to fracture.

However, EHF brings visible improvement in formability due to high strain rate and coining effects, especially for corner filling operations, as it was discussed in multiple publications quantifying formability in pulsed forming. The existing data on formability improvement should be taken with caution, especially when applied to parts of complex geometry when multiple discharges are necessary for successful filling of the die cavity. Based upon studies of formability for AA6111-T4 and AA5754 by Imbert et al. [16], AA5182 by Rohatghi et al. [17], and steels DP500, DP600, DP780, and DP980 by Golovashchenko et al. [19], a relative formability improvement of 50% is achievable. A higher percentage of improvement could be achieved with higher energies and higher forming velocities in exchange for a risk of damaging the die, especially if a low cost die material were used for prototype applications. The experimental results published by Golovashchenko et al. [47] indicated that this panel is safe to form with EHF. Therefore, no preforming step was necessary.

The tryout experiments for the part illustrated in Figure 2 were performed successfully employing an EHF chamber with 11 liters of volume. The air from the described chamber between the blank and the water as well as from the area between the sheet and the surface of the die was removed before starting the EHF process. In order to minimize impact loading on the surface of the die and extend the life of the electrodes, the EHF forming process was done in three discharges of 8 kV, 9 kV, and 13 kV using a Magnepress pulse generator which had 200 μF capacitance and 200 nH internal inductance. The process started from closing the binder followed by air evacuation from the chamber and from the die. The next step was EHF forming itself, which included three discharges. In order to estimate forming results without opening the die, the amount of water added to the chamber was measured experimentally. After the last discharge, the amount of added water was approximately 20 mL to compensate for the displacement of the blank towards the die cavity. This indicated that the process was successfully completed. At the tryout stage, an extra discharge of 13 kV was performed with nearly no water added to the chamber. The formed parts from AA6111-T4 material of 0.93 mm thickness are shown in Figure 7. For this application, no preforming step is needed.

Figure 7. EHF formed prototype part from Aluminum Alloy 6111-T4.

4. Case 2: Complex Automotive Panel

4.1. Materials and Methods

In order to demonstrate another possible scenario to form a prototype part, a part with deep local cavities but relatively dull radii is illustrated in Figure 8.

The die and the chamber design for making the part is shown in Figure 9a,b.

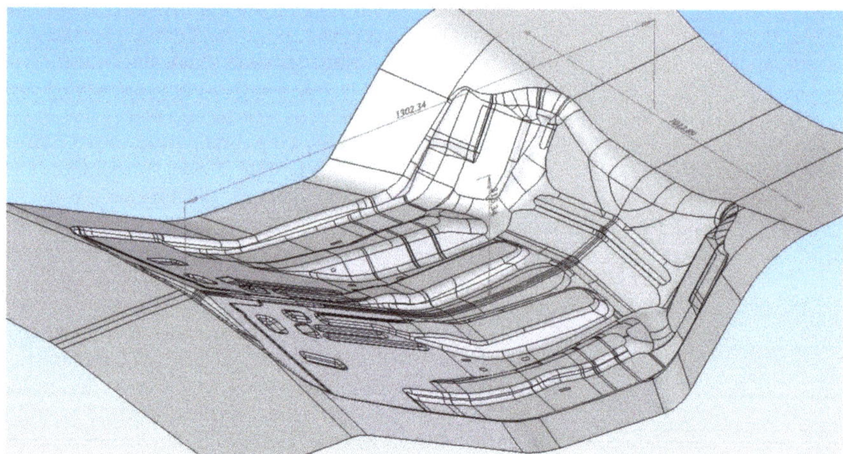

Figure 8. Design of the part for the prototype process.

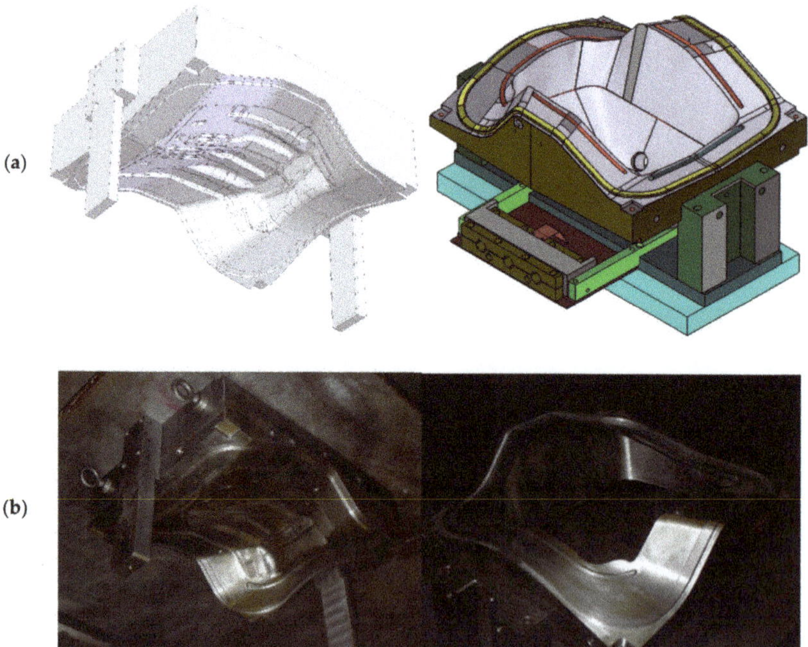

Figure 9. Die and chamber design (**a**), and fabricated die and chamber (**b**).

This part has multiple channels and a deep cavity with curvatures of the flange in two directions. This part was formed as a prototype in 40% scale: every dimension was reduced by a factor of 2.5. The in-plane dimensions of the EHF formed part were 521 mm (instead of 1302) × 448 mm (instead of 1012 mm), and the maximum depth of the cavity was 126 mm (instead of 315 mm). The sheet material used for this prototype process was 0.55 mm DP500 steel. The experimental results on sheet metal behavior at high strain rates for numerical simulation were taken from the study by Baumer et al. [48], and the experimental results on formability of DP500 were used from the experimental work by Golovashchenko et al. [19].

4.2. Results and Discussion

To reduce the simulation computational costs and speed-up the design process, the simulation was performed using a simplified approach. Instead of using a costly full-scale approach, which included simulating the discharge channel, water, and blank deforming at high velocity, a simplified approach was used such that the blank was deformed by uniform pressure. The result of such a simulation is shown in Figure 10.

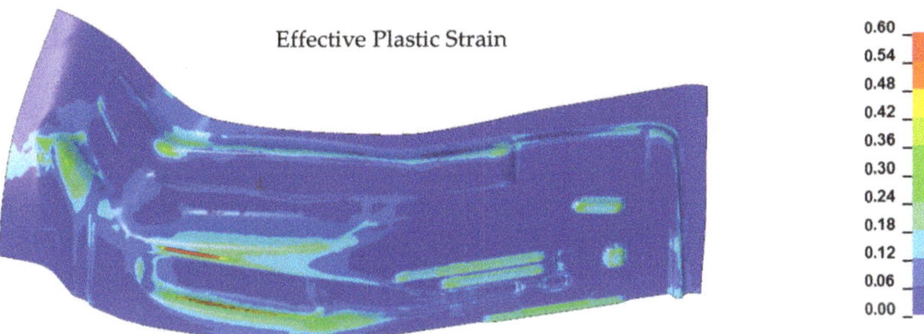

Figure 10. Results of numerical simulation based upon the assumption of uniformly distributed pressure. Maximum strain is 1.21.

Some material inflow was observed during the forming process which helped to form the peripheral channels. The maximum strain in the simulation was observed in the central elongated channel with a true strain of 1.21. In this case, the benefits of extended formability would not be sufficient. A split in the blank was predicted in the form of a strain localization in simulation results in Figure 10. This split was observed in experimental tryout shown in Figure 11, which confirmed the original expectations. Please note that even though the simulation was performed in simplified form, it was able to predict this split and its location.

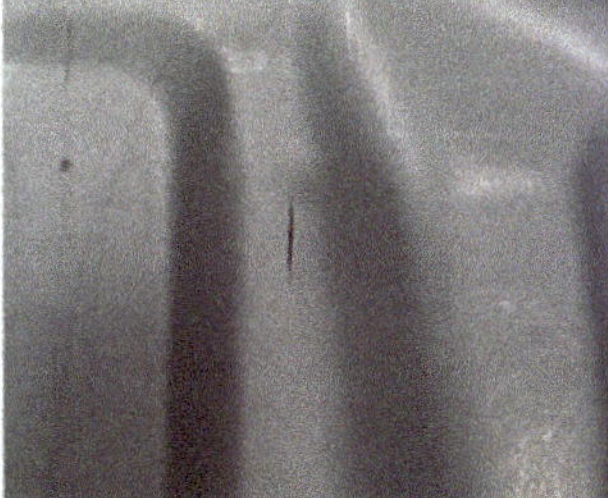

Figure 11. EHF formed part with splits in the areas of maximum strain.

In this case a solution to the split issue was to introduce the preforming operation in which the material is bulged in the areas of low strain to provide additional metal inflow to the areas of high strain, as suggested by Golovashchenko [43]. In order to define where these pockets should be, the principal strains of the mid-surface of the sheet obtained from the initial numerical simulation were numerically imposed as the initial condition in the initial mesh of the flat blank with opposite signs. In other words, since the strains of the mid surface of the formed part were always tensile, they were imposed into the initial flat blank as compressive strains, which led to compressive elastic stresses in the flat sheet restrained between the two rigid plates.

At this stage the sheet was assumed to be purely elastic, similar to the approach earlier suggested by Golovashchenko et al. [26]. The blank was fully clamped at the edges between the two flat rigid plates. In order to identify the necessary pockets and their depth, the cavities on the upper flat plate were open in specified areas of the flat blank, allowing the material to bulge. These specified areas were selected in the groups of elements which would experience low strain in the formed final shape, but being adjacent to the areas of the flat blank which would experience high strain during forming of the targeted shape. These boundary conditions permitted only in-plane displacement of the areas corresponding to high strain and free flow into the open windows where the strain was low.

Driven by elastic compressive internal stresses, the sheet was bulged into these open cavities. The depth of bulging was defined by the level of internal compressive stresses assigned to each element of the mesh in accordance with the amount of strain the material received during the forming step. Overall, this process had multiple iterations and multiple bulged cavities were observed and analyzed.

At the next step, the initial flat blank was formed into the designed pockets in a traditional elasto-plastic formulation which led to stretching in the areas of otherwise low strain. This preformed blank then was virtually placed in a simplified model of the EHF process where uniform pressure was applied to the preformed blank, and the deformation process of EHF forming into the shape shown in Figure 9 was further continued. During the forming of prebulged blanks, the material would flow from the preformed pockets to the areas of higher strain lowering the maximum level of strain. The strain distribution was analyzed from the perspective of peak strain. The best configuration, which had the minimum peak strain, was selected for the experimental validation.

The most successful pockets configuration is shown in Figure 12. The pockets were formed with flexform process using a one sided tool. They also could be formed through an EHF process.

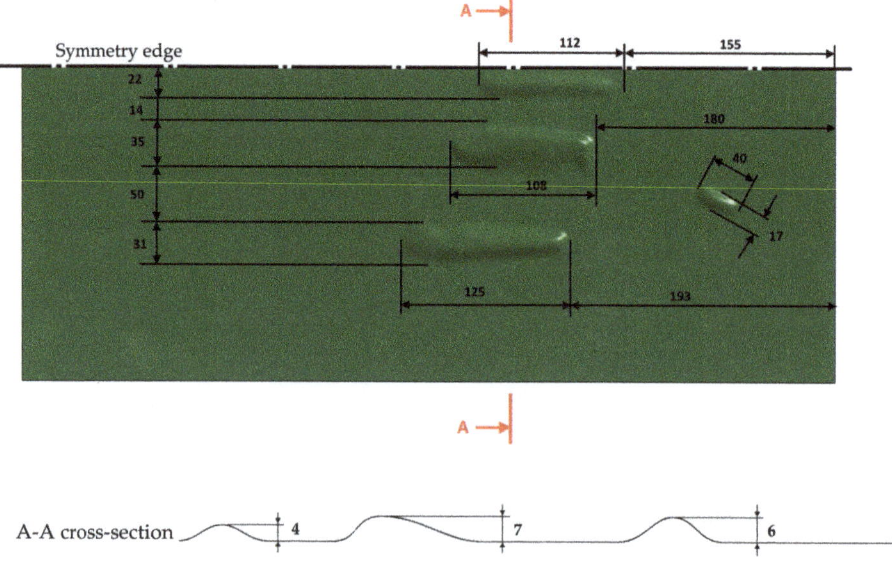

Figure 12. Preform shape.

One can note that the maximum strain of 0.7 shown by the simplified simulation approach is significantly less than in the configuration without preforming pockets, but is still larger than that realistically achievable in a typical forming process. The purpose of this simulation was not to achieve formable conditions in the simulation result but rather to find the way to reduce the final strain. The ultimate feasibility in this case was proven experimentally. As a result of introducing the preforming step, the part was formed without splits using the same die and chamber design shown in Figure 9a,b. The successful panel was formed using a Magnepress pulse generator with the following sequence of nine discharges: 9 kV, 9 kV, 10 kV, 11 kV, 12 kV, 13 kV, 14 kV, 14.5 kV, and 14.5 kV. The last discharge was to confirm that the blank is fully formed, and no more water is added to the EHF chamber. Forming blanks with multiple discharges permits the forming of more complex parts and avoid fracture due to reduced friction: when the blank moves in several steps with unloading, the local areas of metal-to-metal contact typically leading to increases in friction are getting separated. This effect needs more thorough study in future work.

The results of simulation using the same simplified approach are shown in Figure 13.

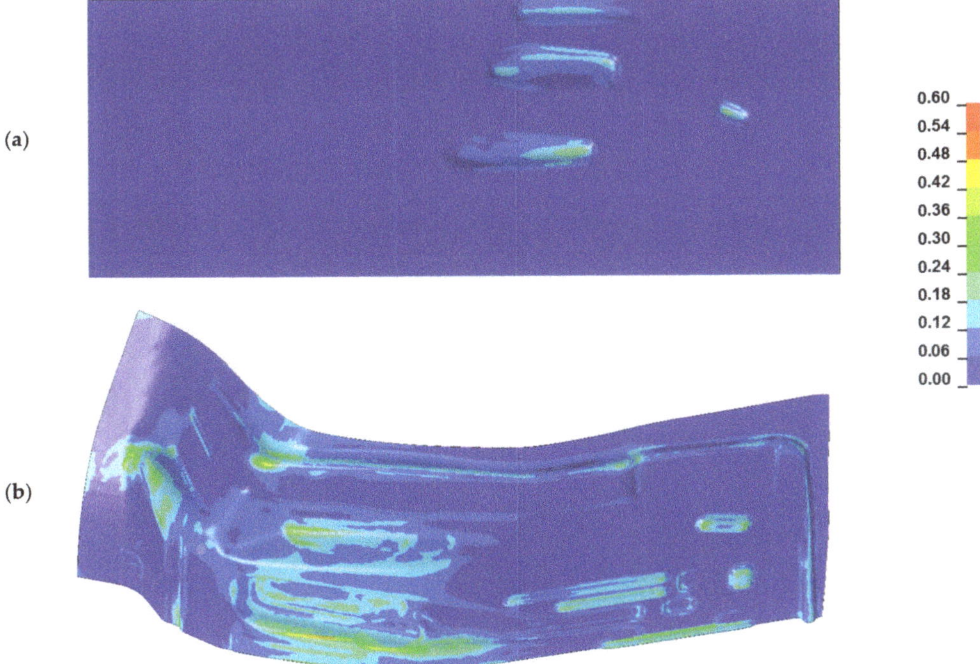

Figure 13. Simulation of two-step forming process: (**a**) preforming step, maximum strain is 0.213; (**b**) fully formed part, maximum strain is 0.7.

This study validated another advantage of EHF technology: due to "soft" applications of pressure, the blank can have pockets of material which enable redistribution of strain in sheet metal by allowing an easier flow of material from the pockets to heavily strained areas. By tracking the heavily strained elements and low strained elements, the areas of high strain and low strain can be projected on the initial blank. In this case, the pockets can be formed on the flat blank which make the preforming die simple and inexpensive. A preforming step with a very shallow preform can be added to the process if a traditional EHF approach does not provide sufficient formability to successfully form the part. A similar effect is difficult to achieve in a conventional die set, because the rigid upper die tends to crush pockets before they can spread.

5. Conclusions

1. Based upon performed demonstrations, the electrohydraulic forming process is a viable technology for prototype and low volume production of sheet metal components formed from flat sheets.
2. A simplified simulation based on quasistatic forming of the blank under uniformly distributed pressure helps to identify critical areas of the part. Such a simulation is significantly less computationally-demanding than a full-scale model, yet is able to provide information about potential splits and difficult to form areas. This simplified approach helps to understand which areas of the die are filled last and require the highest pressure, which enables designing the EHF chamber in a way that sufficient pressure would be provided, to make decisions on electrodes configuration, and also makes preforming analysis possible.
3. A simplified simulation based on a purely hydrodynamic model, which does not include the blank, helps to analyze pressure distribution at the last discharge and enables optimization of the chamber's shape and volume.
4. Preforming sheet metal blanks in order to redistribute the strains involving areas of low strain is a low cost additional process, which can help if insufficient formability occurs in local areas of the formed blank.

Author Contributions: Conceptualization, methodology, validation, investigation, A.V.M., S.F.G., N.M.B. and V.S.M.; software, formal analysis, A.V.M.; resources, data curation, S.F.G.; writing—original draft preparation, writing—review and editing, A.V.M. and S.F.G.; visualization, A.V.M.; supervision, project administration, and funding acquisition, S.F.G. All authors have read and agreed to the published version of the manuscript.

Funding: This material is based upon work supported by the Advanced Manufacturing Office of the United States Department of Energy under Award Number DE-FG36-08GO18128 and Ford Research and Advanced Engineering.

Institutional Review Board Statement: Not applicable.

Informed Consent Statement: Not applicable.

Data Availability Statement: Data is contained within the article.

Acknowledgments: Authors would like to acknowledge Andrey Ilinich and Georgji Demirasi for their contribution at the initial stage of the project.

Conflicts of Interest: The authors declare no conflict of interest. The funders had no role in the design of the study; in the collection, analyses or interpretation of data; in the writing of the manuscript or in the decision to publish the results.

References

1. Ferreira, C.d.S.; Kaminski, P.C. Global Vehicle Architectures Development in the Automotive Industry. *SAE Tech. Pap. Ser.* **2007**, *1*, 2575. [CrossRef]
2. Lange, K. *Handbook of Metal Forming*; McGraw-Hill Book company: New York, NY, USA, 1985; pp. 27.1–30.11.
3. Wilson, F.W. *High-Velocity Forming of Metals*; Prentice-Hall Inc.: Englewood Cliffs, NJ, USA, 1964.
4. Altan, T.; Tekkaya, A.E. *Sheet Metal Forming: Processes and Applications*; ASM International: Materials Park, OH, USA, 2012; pp. 160–167.
5. Bruno, E.J. *High-Velocity Forming of Metals. American Society of Tool and Manufacturing Engineers*; Prentice Hall, Inc.: Dearborn, MI, USA, 1964.
6. Davies, R.; Austin, E.R. *Development in High Speed Metal Forming*; Industrial Press Inc.: New York, NY, USA, 1970.
7. Callender, E.M. Electric Hydroforming by Wire Vaporization. In *Proceedings of "Advanced High Energy Rate Forming", Creative Manufacturing Seminars*; ASME: Detroit, MI, USA, 1961–1962; paper SP62-81; pp. 1–21.
8. Felts, R. Application of Electrohydraulic Effect to Metal Forming. In *Proceedings of "Advanced High Energy Rate Forming", Creative Manufacturing Seminars*; ASME: Detroit, MI, USA, 1961–1962; paper SP62-08; pp. 1–10.

9.   Hanley, F. High Energy Rate Forming at General Dynamics/Fort Worth. In *Proceedings of "Advanced High Energy Rate Forming"*, *Creative Manufacturing Seminars*; ASME: Detroit, MI, USA, 1961–1962; paper SP62–15; pp. 1–22.

10.   Feddersen, E.W. Electrohydraulic Process of High Energy Rate Forming. In *Proceedings of "Advanced High Energy Rate Forming"*, *Creative Manufacturing Seminars*; ASME: Detroit, MI, USA, 1961–1962; paper SP60–164; pp. 1–9.

11.   Schrom, E.C. Metal Forming with Capacitor Discharge Electro-Spark. In *Proceedings of "Advanced High Energy Rate Forming"*, *Creative Manufacturing Seminars*; ASME: Detroit, MI, USA, 1961–1962; paper SP62–80; pp. 1–9.

12.   Duncan, J.L.; Johnson, W. Electrohydraulic Forming. In Proceedings of the NATO Seminar "High Rate Working of Metals", Sandefjord, Norway, 24–25 September 1964; pp. 107–118.

13.   Keeler, S.P. Determination of Forming Limits in Automotive Stampings. *SAE Tech. Pap. Ser.* **1965**, 1–9. [CrossRef]

14.   Goodwin, G.M. Application of Strain Analysis to Sheet Metal Forming Problems in the Press Shop. *SAE Tech. Pap. Ser.* **1968**, *1968*, 1–8. [CrossRef]

15.   Balanethiram, V.; Daehn, G.S. Hyperplasticity: Increased forming limits at high workpiece velocity. *Scr. Met. Mater.* **1994**, *30*, 515–520. [CrossRef]

16.   Imbert, J.M.; Winkler, S.L.; Worswick, M.J.; Oliveira, D.A.; Golovashchenko, S.F. The effect of tool-sheet interac-tion on damage evolution in electromagnetic forming of aluminum alloy sheet. *J. Eng. Mater. Technol.* **2005**, *127*, 145–153. [CrossRef]

17.   Rohatgi, A.; Soulami, A.; Stephens, E.V.; Davies, R.W.; Smith, M.T. An investigation of enhanced formability in AA5182-O Al during high-rate free-forming at room-temperature: Quantification of deformation history. *J. Mater. Process. Technol.* **2014**, *214*, 722–732. [CrossRef]

18.   Dariani, B.M.; Liaghat, G.H.; Gerdooei, M. Experimental investigation of sheet metal formability under various strain rates. Proceedings of the Institution of Mechanical Engineers. *J. Eng. Manuf.* **2009**, *223*, 703–712. [CrossRef]

19.   Golovashchenko, S.F.; Gillard, A.J.; Mamutov, A.V. Formability of dual phase steels in electrohydraulic forming. *J. Mater. Process. Technol.* **2013**, *213*, 1191–1212. [CrossRef]

20.   Samei, J.; Green, D.E.; Golovashchenko, S.; Hassannejadasl, A. Quantitative Analysis of Microstructure Defor-mation Improve-ment in Dual Phase Steels Subject to Electrohydraulic Forming. *J. Mater. Eng. Perform.* **2012**, *22*, 2080–2088. [CrossRef]

21.   Jenab, A.; Green, D.; Alpas, A. Microscopic investigation of failure mechanisms in AA5182-O sheets subjected to elec-tro-hydraulic forming. *Mater. Sci. Eng. A* **2017**, *691*, 31–41. [CrossRef]

22.   Mynors, D.; Zhang, B. Applications and capabilities of explosive forming. *J. Mater. Process. Technol.* **2002**, *125–126*, 1–25. [CrossRef]

23.   Psyk, V.; Risch, D.; Kinsey, B.L.; Tekkaya, A.E.; Kleiner, M. Electromagnetic forming–A review. *J. Mater. Process. Technol.* **2011**, *211*, 787–829. [CrossRef]

24.   Maris, C.; Hassannejadasl, A.; Green, D.E.; Cheng, J.; Golovashchenko, S.F.; Gillard, A.J.; Liang, Y. Comparison of quasi-static and electrohydraulic free forming limits for DP600 and AA5182 sheets. *J. Mater. Process. Technol.* **2016**, *235*, 206–219. [CrossRef]

25.   Cheng, J.; Green, D.E.; Golovashchenko, S.F. Formability enhancement of DP600 steel sheets in electro-hydraulic die forming. *J. Mater. Process. Technol.* **2017**, *244*, 178–189. [CrossRef]

26.   Golovashchenko, S.F.; Bessonov, N.M.; Ilinich, A.M. Two-step method of forming complex shapes from sheet metal. *J. Mater. Process. Technol.* **2011**, *211*, 875–885. [CrossRef]

27.   Homberg, W.; Beerwald, C.; Probsting, A. Investigation of the Electrohydraulic Forming Process with Respect to the Design of Sharp Edged Contours. In Proceedings of the 4th International Conference on High Speed Forming, Columbus, OH, USA, 9–10 March 2010; pp. 58–64.

28.   Singh, H. *Fundamentals of Hydroforming*; Society of Manufacturing Engineers: Dearborn, MI, USA, 2003; pp. 29–35.

29.   Golovashchenko, S.F. Electrohydraulic Forming Process with Electrodes that Advance within a Fluid Chamber towards a Workpiece. U.S. Patent 8,844,331B2, 30 September 2014.

30.   Vagin, V.A.; Zdor, G.N.; Mamutov, V.S. *Methods of Research of High-Speed Deformation of Metals*; Nauka I Technika: Minsk, Belarus, 1990. (In Russian)

31.   Cole, R.H.; Weller, R. Underwater Explosions. *Phys. Today* **1948**, *1*, 35. [CrossRef]

32.   Knyazev, M.K.; Zhovnovatuk, Y.S. Measurements of Pressure Fields with Multi-Point Membrane Gauges at Elec-Trohydraulic Forming. In Proceedings of the 4th International Conference on High Speed Forming, Columbus, OH, USA, 9–10 March 2010; pp. 75–82.

33.   Melander, A.; Delic, A.; Bjorkblad, A.; Juntunen, P.; Samek, L.; Vadillo, L. Modelling of electrohydraulic free and die forming of sheet steels. *Int. J. Mater. Form.* **2013**, *6*, 223–231. [CrossRef]

34.   Hassannejadasl, A.; Green, D.E.; Golovashchenko, S.F.; Samei, J.; Maris, C. Numerical Modeling of Electrohydrau-lic Free-Forming and Die-Forming of DP590 steel. *J. Manuf. Process.* **2014**, *16*, 391–404. [CrossRef]

35.   Vohnout, V.J.; Fenton, G.; Daehn, G.S. Pressure heterogeneity in small displacement electrohydraulic forming process. In Proceedings of the 4th International Conference on High Speed Forming, Columbus, OH, USA, 9–10 March 2010; pp. 65–74.

36.   Mamutov, A.V.; Golovashchenko, S.F.; Mamutov, V.S.; Bonnen, J.J. Modeling of electrohydraulic forming of sheet metal parts. *J. Mater. Process. Technol.* **2015**, *219*, 84–100. [CrossRef]

37.   Woo, M.; Lee, K.; Song, W.; Kang, B.; Kim, J. Numerical Estimation of Material Properties in the Electrohydraulic Forming Process Based on a Kriging Surrogate Model. *Math. Probl. Eng.* **2020**, *2020*, 1–12. [CrossRef]

38.   Jenab, A.; Green, D.E.; Alpas, A.T.; Golovashchenko, S.F. Experimental and numerical analyses of formability im-provement of AA5182-O sheet during electro-hydraulic forming. *J. Mater. Process. Technol.* **2018**, *255*, 914–926. [CrossRef]

39. Woo, M.; Song, W.-J.; Kang, B.-S.; Kim, J. Evaluation of formability enhancement of aluminum alloy sheet in elec-trohydraulic forming process with free-bulge die. *Int. J. Adv. Manuf. Technol.* **2019**, *101*, 1085–1093. [CrossRef]

40. Zhang, F.; Zhang, J.; He, K.; Hang, Z. Application of Electro-hydraulic Forming (EHF) Process with Simple Dies in Sheet Metal Forming. In Proceeding of the International Conference on Information and Automation, Wuyi Mountain, China, 11–13 August 2018; pp. 401–405.

41. Xiong, L.; Liu, Y.; Yuan, W.; Huang, S.; Li, H.; Lin, F.; Pan, Y.; Ren, Y. Cyclic shock damage characteristics of electrohydraulic discharge shockwaves. *J. Phys. D Appl. Phys.* **2020**, *53*, 185502. [CrossRef]

42. Daehn, G.S.; Vohnout, V.J.; Dubois, L. Improved Formability with Electro-Magnetic Forming: Fundamentals and Practical Example. In Proceedings of the TMS Symposium "Sheet Metal Forming Technology", San-Diego, CA, USA, 28 February–4 March 1999; Demeri, M.Y., Ed.; pp. 105–115.

43. Golovashchenko, S.F. Method and Apparatus for Making a Part by First Forming an Intermediate Part that Has Donor Pockets in Predicted Low Strain Areas Adjacent to Predicted High Strain Areas. U.S. Patent 9,522,419B2, 20 December 2016.

44. Chen, X.; Hsiung, C.-K.; Schmid, K.; Du, C.; Zhou, D.; Roman, C. Evaluation of Metal Gainers for Advanced High Strength Steel Flanging. *SAE Tech. Pap. Ser.* **2014**. [CrossRef]

45. Graf, A.; Hosford, W. The influence of strain-path changes on forming limit diagrams of AL 6111 T4. *Int. J. Mech. Sci.* **1994**, *36*, 897–910. [CrossRef]

46. Chow, C.; Yu, L.; Tai, W.; Demeri, M. Prediction of forming limit diagrams for AL6111-T4 under non-proportional loading. *Int. J. Mech. Sci.* **2001**, *43*, 471–486. [CrossRef]

47. Golovashchenko, S.F.; Mamutov, V.S.; Dmitriev, V.V.; Sherman, A.M. Formability of Sheet Metal with Pulsed Electromagnetic and Electrohydraulic Technologies. In Proceedings of the TMS symposium "Aluminum-2003", San-Diego, CA, USA, 2–6 March 2003; pp. 99–110.

48. Bäumer, A.; Jiménez, J.A.; Cugy, P. *Investigation of the Strain-Hardening Behavior of Modern Lightweight Steels Considering the Forming Temperature and Forming Rate*; Directorate-General for Research and Innovation, European Commission: Brussel, Belgium, 2004; pp. 78, 90.

Journal of
*Manufacturing and
Materials Processing*

Article

Adiabatic Blanking: Influence of Clearance, Impact Energy, and Velocity on the Blanked Surface

Sven Winter *, Matthias Nestler , Elmar Galiev, Felix Hartmann, Verena Psyk , Verena Kräusel and Martin Dix

Fraunhofer Institute for Machine Tools and Forming Technology IWU, 09126 Chemnitz, Germany;
matthias.nestler@iwu.fraunhofer.de (M.N.); elmar.galiev@iwu.fraunhofer.de (E.G.);
felix.hartmann@iwu.fraunhofer.de (F.H.); verena.psyk@iwu.fraunhofer.de (V.P.);
verena.kraeusel@iwu.fraunhofer.de (V.K.); martin.dix@iwu.fraunhofer.de (M.D.)
* Correspondence: sven.winter@iwu.fraunhofer.de

Abstract: In contrast to other cutting processes, adiabatic blanking typically features high blanking velocities (>3 m/s), which can lead to the formation of adiabatic shear bands in the blanking surface. The produced surfaces have excellent properties, such as high hardness, low roll-over, and low roughness. However, details about the qualitative and quantitative influence of significant process parameters on the quality of the blanked surface are still lacking. In the presented study, a variable tool is used for a systematic investigation of different process parameters and their influences on the blanked surface of a hardened 22MnB5 steel. Different relative clearances (1.67% to 16.67%), velocities (7 to 12.5 m/s), and impact energies (250 J to 1000 J) were studied in detail. It is demonstrated that a relative clearance of $\leq 6.67\%$ and an impact velocity of ≥ 7 m/s lead to adiabatic shear band formation, regardless of the impact energy. Further, an initiated shear band results in the formation of an S-shaped surface. Unexpectedly, a low impact energy results in the highest geometric accuracy. The influence of the clearance, the velocity, and the impact energy on the evolution of adiabatic shear band formation is shown for the first time. The gained knowledge can enable a functionalization of the blanked surfaces in the future.

Keywords: adiabatic blanking; adiabatic shear band; high velocity; clearance; blanked surface; stress triaxiality; FE simulation

Citation: Winter, S.; Nestler, M.; Galiev, E.; Hartmann, F.; Psyk, V.; Kräusel, V.; Dix, M. Adiabatic Blanking: Influence of Clearance, Impact Energy, and Velocity on the Blanked Surface. *J. Manuf. Mater. Process.* **2021**, *5*, 35. https://doi.org/10.3390/jmmp5020035

Academic Editor: Steven Liang

Received: 18 March 2021
Accepted: 10 April 2021
Published: 13 April 2021

Publisher's Note: MDPI stays neutral with regard to jurisdictional claims in published maps and institutional affiliations.

1. Introduction

The conservation of resources and the saving of energy-intensive process steps is a constant challenge in the field of materials and production technology. Here, adiabatic blanking offers high potential, because it has several technological and economic advantages when compared to other cutting processes, such as laser cutting and conventional or fine blanking [1,2]. The process is lubricant free with short cycle times, very small web widths, and low component deformation [3]. Due to the high blanking speed, the kinetic energy of the tool is strongly localized in the shear zone and therefore almost completely converted into cutting energy. The resulting blanking surfaces feature excellent properties, such as high hardness, low roll-over, and low roughness [4,5]. Due to these properties, time-consuming mechanical reworking and subsequent hardening is avoided, which significantly shortens the process chain.

The adiabatic blanking process typically features tool velocities of >3 m/s [6], local strain rates of >10^2 s^{-1}, and process times in the magnitude of a millisecond. These conditions cause a local temperature increase in the shear zone, because the generated heat cannot dissipate that quickly especially if the thermal conductivity of the material is only moderate. As a consequence, quasi-adiabatic or, in simplified terms, adiabatic behavior is forced. The local heating leads to thermal softening and a locally reduced strength followed by an enhanced strain localization in these areas. This self-reinforcing process of thermal

179

softening and strain localization leads to the formation of an adiabatic shear band (ASB) in the shear zone [7,8].

The microstructure and properties of ASBs directly depend on the deformation and strain rate [9] and the bands are classified into deformation and transformation shear bands [10]. Deformation shear bands exhibit a pronounced strain localization without phase transformation. They represent a precursor to transformation shear bands, which are narrower than deformation shear bands and clearly separated from the surrounding microstructure [11]. The microstructure in transformation shear bands is nano-crystalline [12], thus influencing the local material properties in the shear band. It appears as a bright band when observed under an optical microscope, due to a higher chemical resistance to the etchant [13]. According to Nesterenko et al. [14], this microstructural transformation is caused by dynamic recrystallization under compressive or shear stress. High elastic energy in the material (typically coupled to high strain hardening) leads to a much easier initiation of the ASB [15,16] and furthermore to a blanked surface of higher quality. Therefore, the technology is particularly suitable for high-strength and ultra-high-strength steels, such as the manganese-boron press-hardening steels used primarily in automotive applications.

Schmitz et al. [4] identified dynamically recrystallized microstructures on a 20MnB5 and a C75S steel after adiabatic blanking and showed that the ASB has a significantly higher hardness than the surrounding microstructure and a brittle character. This results in material failure almost always occurring directly in the shear band [17,18]. The investigations by Schmitz et al. [4] demonstrated that adiabatic blanking can lead to the formation of an S-shaped blanking surface, especially if materials with high rate sensitivities m are regarded.

Due to the extreme stress and temperature conditions in the shear zone during adiabatic blanking [19], the numerical prediction of the formation of the blanking surface is very complex. Most approaches use the Johnson–Cook damage model [4,20], which considers material specifications at very high strain rates and temperatures for sufficiently accurate modeling. The required material data can be determined, e.g., via Split–Hopkinson pressure bar tests for strain rates of up to 10^3 s^{-1} and temperatures of up to 1000 °C [21]. An alternative approach allowing strain rates of up to 10^4 s^{-1} uses a test setup with an electromagnetically accelerated punch (up to 50 m/s) and subsequent inverse numerical simulation [22]. However, experimental tests with varying process parameters are indispensable for validating the numerical simulation.

In summary, adiabatic blanking has high potential to influence the local microstructure of the blanked surfaces and thus the blanked surface properties of ultra-high-strength steel sheets in a target-oriented manner. The sophisticated exploitation of the adiabatic effect can enable the blanking of sheet metal, which currently must be laser cut, resulting in long process duration and high processing costs. However, for a target-oriented design of the adiabatic blanking process, detailed knowledge about the qualitative and quantitative influences of adjustable process parameters on the process and the resulting quality properties of the blanked part is necessary. It is well known that the initiation and propagation of ASB requires a compressive or shear stress state or a combination of both [23] and that in adiabatic blanking, the stress state in the shear zone can be varied by adjusting the clearance. Furthermore, it is obvious that the tool speed and impact energy mainly determine the strain rate during the process. However, there is still a lack of systematic studies investigating the influence of these parameters on the resulting blanked surface. Therefore, the presented study focuses on identifying and quantifying these correlations. Thus, it contributes to understanding ASB formation specifically as a function of different process parameters (clearance, impact energy, and impact velocity) on shear band initiation and blanking surface geometry. The focus is on analyzing the qualitative and quantitative influence of the process parameters on the properties of the produced blanked surface and the utilization of this surface as a potential functional area.

2. Materials and Methods

For the investigation, the press hardening steel 22MnB5 (chemical composition in Table 1) in sheet form was used. The sheet thickness was 3 mm, and the width of the sheet strips was 50 mm. The material was hardened to 430 ± 8 HV10. The heat treatment was carried out under an active carbon gas atmosphere with a carbon content of 0.2%, so that decarburization could be reduced to a minimum.

Table 1. Chemical composition of the used 22MnB5 steel in wt.-%.

in wt.-%	C	Mn	Si	Cr	B (ppm)	Fe
22MnB5	0.23	1.21	0.15	0.18	21.00	balance

The blanking tests were performed using an ADIAflex® adiabatic blanking machine from MPM France (formerly ADIAPRESS). A 15-kg striking unit is accelerated to various speeds via high-speed hydraulics. This allows for variation of the impact energy introduced into the tool. The energy was varied from 250 J to 1000 J in 250-J steps. The principal setup of the tool specially designed for the investigations and used in the experimental tests is illustrated in Figure 1. The process starts when the striker unit (not depicted in Figure 1) impacts the so-called pusher (mushroom-shaped, Toolox44 steel from *SSAB*, m = 9.95 kg, shown in red in Figure 1). A powder metallurgical steel CPM Rex76 punch (blue, Figure 1) with a diameter of 20 mm is screwed and therefore firmly integrated into the pusher and stands directly on the sheet to be blanked. The maximum punch penetration depth was 2 mm (defined in Figure 2). The die (light blue, Figure 1) is located under the sheet. Different dies were used (Ø 20.1 to 21.0 mm) to vary the blanking clearance from 0.05 mm to 0.5 mm. The used dies and the resulting clearance (absolute value and relative value, referring the clearance to the sheet thickness) are shown in Table 2. The blank holder force was 100 kN. The displacement of the pusher was recorded during the process by an LK-H157 Ultra High Speed High Precision Laser Displacement Sensor (brown, Figure 1) from *KEYENCE*. The corresponding pusher velocity was calculated by differentiating the measured displacement over time curve.

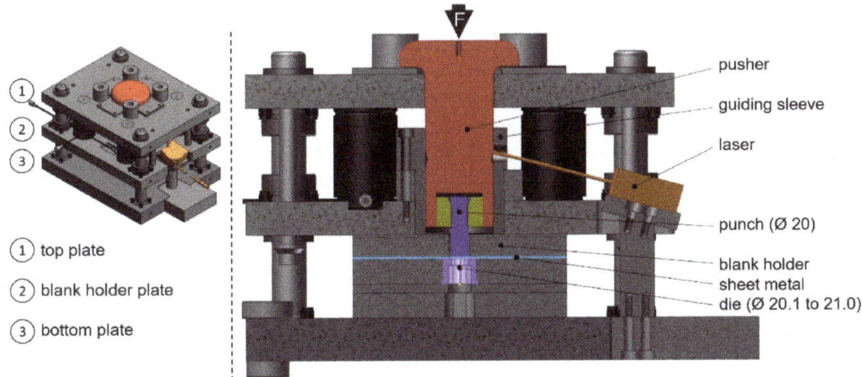

Figure 1. Setup of the tool used for adiabatic blanking. The force is applied to the pusher via a striker unit of the ADIAflex® machine.

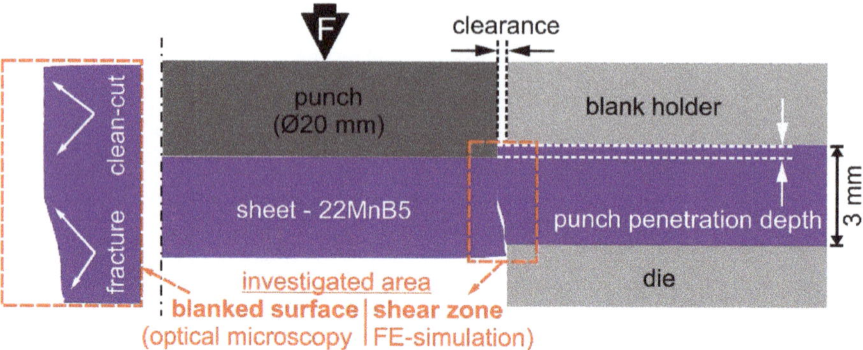

Figure 2. Two-dimensional model for the numerical calculation of the stress triaxiality in the shear zone in LS-DYNA. Additionally, the investigated area (blanked surface) for the optical microscopy is schematically shown.

Table 2. Investigated dies (sample designation) and the resulting blanking and relative clearances.

Ø Die in mm (Sample Designation)	20.1	20.2	20.4	20.6	20.8	21.0
blanking clearance in mm	0.05	0.10	0.20	0.30	0.40	0.50
relative clearance in%	1.67	3.34	6.67	10.00	13.34	16.67

In addition to the tests with the described steel pusher, further experiments were carried out using a pusher made of the aluminum alloy EN AW-7075 with identical dimensions and a mass of m = 3.44 kg. This allowed for investigating the influence of a tool velocity increase in the system, with identical impact energy of the striker unit. Due to the lower mass and constant impact energy, the velocity must increase during blanking according to the conservation of momentum. This effect was investigated on an exemplary relative clearance of 10% (die Ø 20.6 mm). After the blanking tests with different parameter sets, the blanked surfaces of the sheets were examined and evaluated by optical microscopy. The used optical microscope was the BX53 from Olympus.

According to [9,23], the local stress state in the shear zone is important for the formation of adiabatic shear bands and significantly influences the blanked surface topography. Higher local compressive stresses inhibit cracking and enable the material to plastically flow for a longer period of time. In order to determine the influence of the clearance on the stress triaxiality, numerical simulations of the adiabatic blanking process were performed. The finite element simulation was carried out with the explicit solver of LS-DYNA (*DYNAmore GmbH*). The 2-D axisymmetric model illustrated in Figure 2 was used. The punch, blank holder, and die were modelled as rigid bodies, while the sheet was considered to be deformable. The punch velocity was assumed to be 10 m/s, and the blank holder force and the coefficient of friction were set to 10 kN and $\mu = 0.15$, respectively. The mesh size was 0.01 mm in the shear zone. The damage in the shear zone was calculated using the Johnson–Cook model Equation (1). The required material constants A, B, C, n, k, the heat conductivity λ (at room temperature), and the strain rate sensitivity m were taken from Schmitz et al. [4] and are listed in Table 3. The other parameters in the Johnson–Cook model are the effective plastic strain $\bar{\varepsilon}$, the effective strain rate $\dot{\bar{\varepsilon}}$, the reference strain rate $\dot{\bar{\varepsilon}}_0$, the room temperature T_{room}, and the melting point T_{melt}. The stress triaxiality in the shear zone (Figure 2, red marking) was investigated for two exemplary blanking clearances (Ø 20.2 and 20.6) and different punch penetration depths:

$$\bar{\sigma} = [A + B\bar{\varepsilon}^n] \cdot \left(1 + C \cdot ln\left(\frac{\dot{\bar{\varepsilon}}}{\dot{\bar{\varepsilon}}_0}\right)\right) \cdot \left(1 - \left(\frac{T - T_{room}}{T_{melt} - T_{room}}\right)^k\right) \tag{1}$$

Table 3. Parameters used for the Johnson–Cook model from Schmitz et al. [4].

A in MPa	B in MPa	C	n	k	m	λ in W/(mK)
1380	502	0.0011	0.15	0.41	0.0040	48.7

3. Results and Discussions

3.1. Influence of the Clearance on the Adiabatic Shear Band Initiation and the Blanked Surface Geometry

Figure 3 illustrates the influence of the blanking clearance on the microstructure of the blanked surface by comparing micrographs of the specimens, which were adiabatically blanked using an impact energy of 1000 J and varying the die diameters of 20.1 mm to 20.6 mm, which corresponds to a variation of the relative clearance between 1.67% and 10% (see Table 2). For larger clearances, no adiabatic shear bands were detected and therefore the corresponding specimens are not shown.

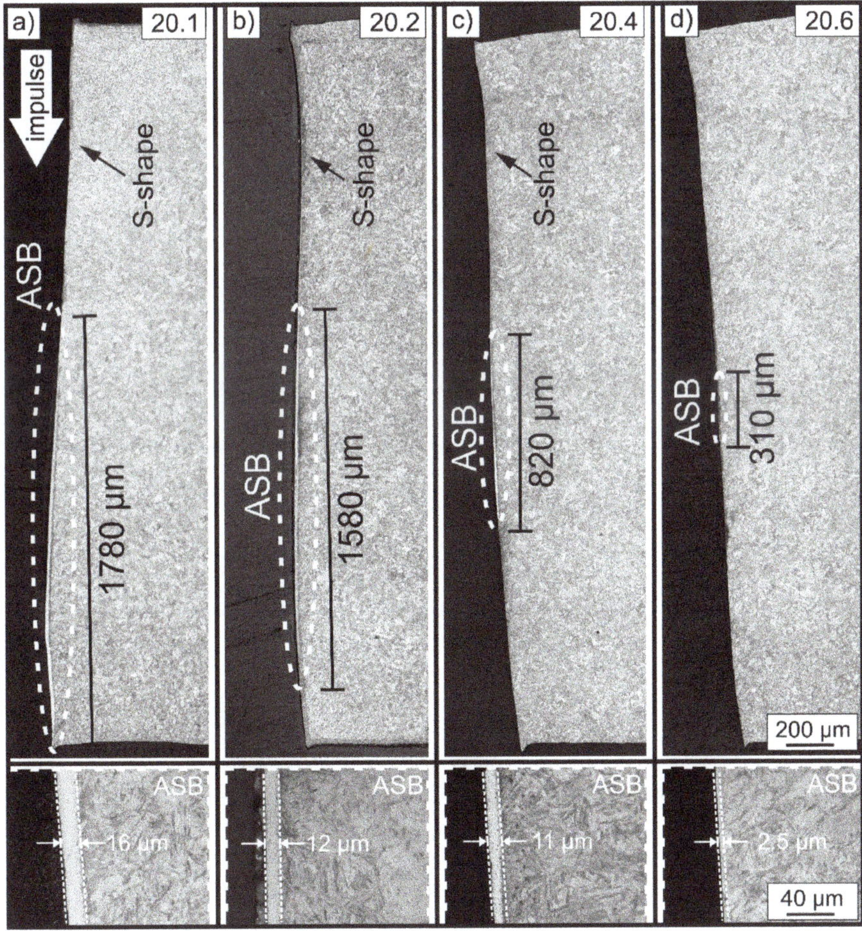

Figure 3. Blanked surface of the sheet for 1000 J impact energy and a relative clearance of (**a**) 1.67%, (**b**) 3.34 %, (**c**) 6.67% and (**d**) 10.00%. With decreasing blanking clearance, the adiabatic shear band (ASB) becomes longer and wider.

The bright lines on the surface indicate the typical nano-crystalline microstructure of a transformation shear band as explained in Section 1. It is obvious that the smallest blanking clearance resulted in the longest and widest adiabatic shear band (Figure 3a) and that the length and width of the ASB decreases continuously with increasing clearance (Figure 3b–d). It should be noted that the ASB, which was actually formed in the experimental process, was cut lengthwise in the area of most significant thermal softening, i.e., in the middle of the shear band and the blanked surfaces of both the blanked part and punched sheet feature identical halves of the ASB. Thus, the actual with of the ASB is twice as high as the values given in Figure 3, i.e., 32 μm in case of the smallest blanking clearance.

The microstructure near the shear band is almost undeformed in all cases. This is a well-known effect [24] and an advantage of adiabatic blanking. Due to the high local energy input, very limited thermal deformation occurs. The deformation is concentrated directly in the shear zone and thus the deformation in the surrounding microstructure is minimal. Another effect of adiabatic blanking is the change in the shape of the blanked surface. Specimen 20.6 still exhibits a classic fracture surface with minimal clean-cut. In contrast, specimen 20.1, 20.2, and 20.4 show an S-shaped blanked surface without commonly observed roll-over and burr. This phenomenon in correlation with the formation of ASBs was also observed in [4] and explained by the influence of the strain rate sensitivity m of the different investigated materials. However, Figure 3 clearly shows the direct relation of the initiation of the ASB to the resulting blanked surface. The larger the ASB, the more pronounced the formation of an S-shaped blanked edge. Consequently, the initiation of an ASB causes a significant change in the blanked surface.

The micrographs shown in Figure 3 also suggest that the shear band formation is initiated in the center of the respective fracture surfaces. This effect becomes evident, when the blanked surfaces from the largest clearance (10%, sample 20.6, Figure 3d) to the smallest clearance (1.67%, sample 20.1, Figure 3a) are compared. The barely detectable ASB in specimen 20.6 is nearly centered in the blanked surface. If the clearance is reduced further, the ASBs become longer and spread further to the outer edge of the blanked surface. In sample 20.1, the ASB extends almost to the outer edge. Presumably, this effect is a result of the predominantly compressive stresses in the middle region of the shear zone. This is particularly evident in Figure 5a–c, where the stress triaxiality at a relative clearance of 3.34% is depicted and discussed in detail. The outer regions are more dominated by shear stresses, while compressive stresses are more dominant in the center of the blanking surface. According to [14,23], compressive stresses significantly favor shear band formation and are considered as a reason for shear band initiation in the center, while shear stresses are more likely to promote failure.

Figure 4 summarizes the results of the experimental tests concerning the ASB formation in dependence of the clearance and the impact energy. Lower impact energies do not promote the formation of an ASB in the tests with a relative clearance of 10% or more. However, at lower clearances (sample 20.1–20.4), ASB formation already occurs at relatively low impact energies of 250 J. In general, the length and width of the ASB increases with increasing energy. Consequently, the longest ASB can be observed on the specimen produced with the smallest blanking clearance (1.67%) and the highest impact energy (1000 J). In summary, reducing the blanking clearance and increasing the impact energy raise the length and width of the ASB. Thus, blanking clearance and impact energy are dominant factors influencing the formation of the ASB.

To achieve a deeper understanding of the influence of the blanking clearance on the shear band formation, numerical simulations were performed to determine stress triaxiality in the shear zone. According to [14,25], an adiabatic shear band can only evolve under a dominant compressive or shear stress condition. Dynamic recrystallization and the transformation of the initial microstructure to a nanocrystalline structure can only occur under compressive or shear stresses [14]. Contrary, tensile stresses promote crack initiation and consequently failure of the material, thus preventing transformation of the microstructure. Figure 5 shows the stress triaxiality for two exemplary relative clearances (3.34% and 10%) as a function of the punch penetration depth. Even at a low penetration depth, the smaller clearance (Figure 5a) leads to a significantly higher and more concentrated proportion of compressive stresses in the shear zone than the higher clearance (Figure 5e), which consequently promotes shear band formation.

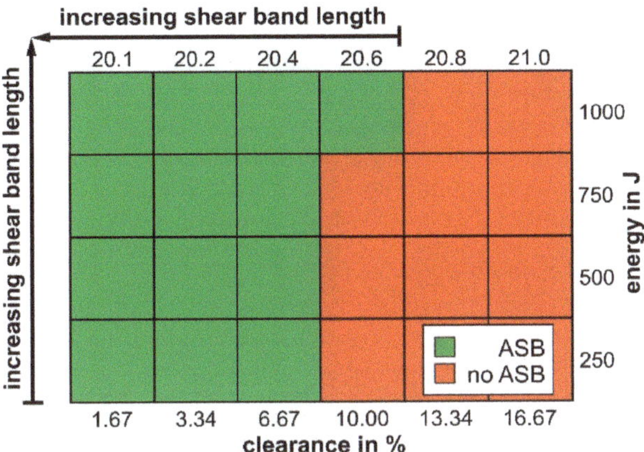

Figure 4. Process windows of the blanking tests performed with varying clearance and impact energies and their influence on the formation of an ASB.

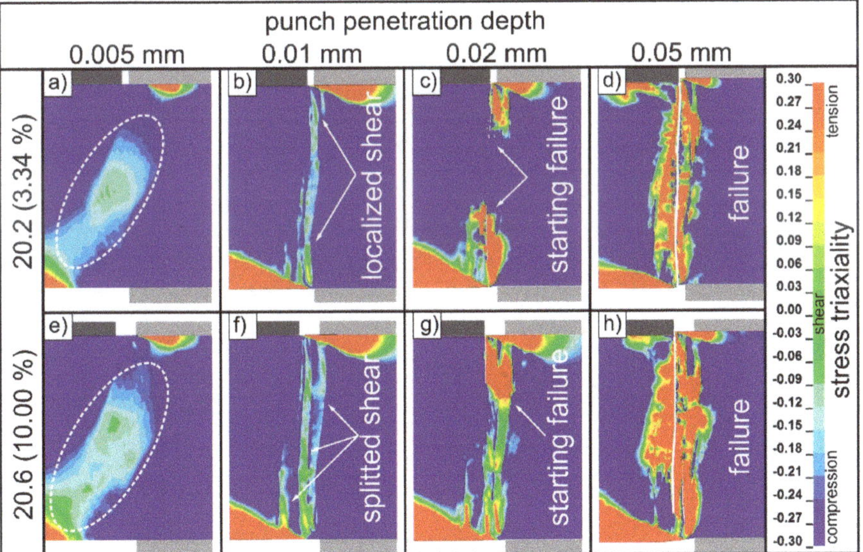

Figure 5. Results of the finite element simulation of the stress triaxiality in the shear zone with two blanking clearances ((**a–d**) 20.2 and (**e–h**) 20.6) and different punch penetration depths.

As the penetration depth of the punch progresses, the smaller clearance results in a much higher concentration of compressive and shear stresses in the shear zone. As already mentioned in Figure 3, both specimens, but especially specimen 20.2, exhibit significantly higher compressive stresses in the center of the shear zone than at the edge. This favors the initiation of the ASB in the center of the blanking surfaces observed in Figure 3. The S-shaped formation in the shear zone at the smaller clearance also becomes visible (Figure 5b). The larger clearance leads to a significant distribution and splitting, respectively, of the occurring stresses and to dominating shear stresses in the shear zone. When compared to Figure 5c,g depicts that smaller clearances result in an earlier failure due to the larger concentration of stress. The initiated crack has already grown through more than 50% of the sheet thickness here, while the initiation of the crack has just begun for the larger clearance. Finally, it is noticeable in Figure 5d,h that the tensile stresses occurring in the shear zone are significantly lower when the clearance is reduced. A stronger localization of stresses and, consequently, of deformation and impact energy is more significant at a clearance of 3.34% (20.2) than at a clearance of 10% (20.6). Higher local energy input possibly results in higher local temperature in the shear zone during adiabatic blanking [4,20]. Due to the dynamic recrystallization, the formation of adiabatic shear bands is enabled. Moreover, the necessary proportion of compressive stresses is significantly higher with a reduced clearance and thus supports the formation of the ASB further.

3.2. Influence of Impact Energy and Velocity on the Blanked Surface

In order to vary the velocity independently from the impact energy, an aluminum pusher (EN AW-7075) was used in addition to the steel pusher. Due to the conservation of momentum, the impact velocity increases during blanking, if the pusher mass is lower and the same impact energy is constant. The tests presented in the following were performed with a relative clearance of 10% and are referred to as samples 20.6 v↑. Due to the identical blanking clearance, these tests were compared with the 20.6 tests with a steel pusher. Figure 6 shows the results of the laser velocity measurement directly at the pusher. Obviously, the velocity increases with increasing impact energy. The specimens 20.6 with the steel pusher showed velocities from 7.2 to 9.7 m/s. As expected, the tests carried out with the aluminum pusher showed a higher impact velocity. Already at an impact energy of 250 J, the measured velocity of 9.8 m/s is higher than the fastest velocity measured during the test series with the steel pusher. With an increase in impact energy, the velocity increases continuously to a maximum of 12.5 m/s. This demonstrates that the speed in adiabatic blanking can be significantly increased by varying the mass of the pusher. The 20.6 v↑ samples reach a maximum velocity that is about 28% higher than the maximum of the samples with the steel pusher.

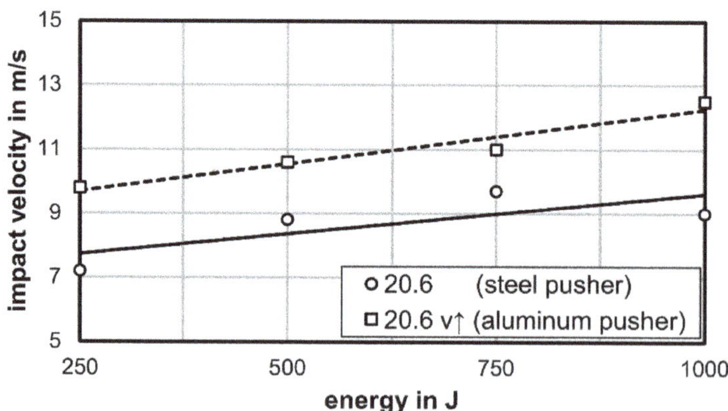

Figure 6. Evaluation of the impact velocity for a relative clearance of 10% with varying impact energies and different pushers.

Regardless of the formation of adiabatic shear bands, the high velocity and especially the impact energy in adiabatic blanking affects the blanked surface in several ways. Figure 7 shows the influence of the impact energy exemplarily on the specimen cut with a die diameter of 20.6 mm and with a relative clearance of 10%, respectively. In this test series, the energy varied between 250 J and 1000 J. All samples show similar blanked surfaces, which exhibit a predominantly fractured surface. The clean-cut fraction seen in the upper part of the blanked surface is less than 100 μm for all specimens. In the 750 J impact energy sample, it is almost non-existent. The S-shaped surface, as shown by the specimens with smaller clearances (see Figure 3a–c), cannot be observed with a 10% relative clearance, regardless of the impact energy. On the contrary, the fracture surface of the specimens is very straight but inclined. It is clearly visible that the angle α between the fracture surface and the normal to the sheet surfaces becomes smaller with increasing impact energy. In case of the specimen cut with an impact energy of 250 J, the angle α amounts to 8°, while the specimen cut with an impact energy of 1000 J features an α of 4.5°. Consequently, the angle is reduced by more than 40% for the largest impact energy due to a significant increase of impact velocity and energy. The displacement measurements in the tool by laser triangulation showed that the velocity in the process increases from about 7.2 m/s for the 250 J sample to over 9.7 m/s for the 1000 J sample (Figure 6). The higher velocity leads to a stronger localization of the deformation in the shear zone. The heat generated during blanking cannot dissipate quickly enough due to the high velocity and the correspondingly short process time, resulting in a very strong local increase in temperature followed by thermal softening of the material. Further deformation takes place in the thermally softened area due to the locally lower flow stress. Thus, strain localization occurs. Consequently, as the velocity increases, the deformed material volume is reduced, and the fracture angle must become smaller. This effect was also observed in [5,26]. For smaller clearances (Figure 3), the higher stress concentration (see Figure 5) and the increased proportion of compressive stresses can lead to shear band initiation, which influences the fracture surface as a result. It must be taken into account that the influence of velocity, energy, and clearance always depends on the material to be blanked and its mechanical, thermal, and microstructural properties.

The effect of the increased velocity is clearly evident in the optical micrographs of the 20.6 v↑ samples in Figure 8. For impact energies of 250 J to 750 J (Figure 8a–c), the combination of speed and energy leads to the formation of an almost vertical blanked surface despite the large relative clearance of 10%. These blanked surfaces have the highest geometric accuracy in the entire parameter space investigated, especially in direct comparison with the specimens of the same clearance in Figure 7a–c. The sample cut with an impact energy of 750 J (Figure 8c) shows a larger deviation from the vertical dotted line in the upper part of the blanked surface compared to those cut with lower impact energies. This seems to be a starting formation of the S-shaped blanked surface, indicating the formation of a shear band, which is only detectable to a limited extent by optical microscopy methods, however (see details in Figure 9b). Figure 8d shows a stronger offset and the occurrence of a fracture angle. This principal shape is similar to the specimens cut with the steel pusher at a correspondingly lower impact velocity (Figure 7d), but the angle is smaller here due to the increased velocity. Although, the blanked surface also has a much wavier (slightly more S-shaped) character than the 20.6 specimen at 1000 J. However, the clear formation of an ASB in the center area of the blanked surface is evident (see Figure 3). This demonstrates again that the initiation of shear band formation changes the geometry and further the quality of the fracture surface. In this case, the ASB forms despite the large blanking clearance due to the increased velocity and the resulting reduced time for heat conduction from the shear zone. When compared to lower velocities, the local temperature increases more significantly and the dynamic recrystallization processes for the transformation of the microstructure can start. However, the impact velocity cannot be discussed separately without considering the impact energy. Above a critical velocity, the energy is the driving force for the microstructural transformation and the resulting surface geometry. Therefore, specimens with identical impact velocities and blanking clearances (e.g., 20.6: 750 J, $v = 9.7$ m/s and 20.6 v↑: 250 J, $v = 9.8$ m/s) show different blanking surfaces due to varying energies.

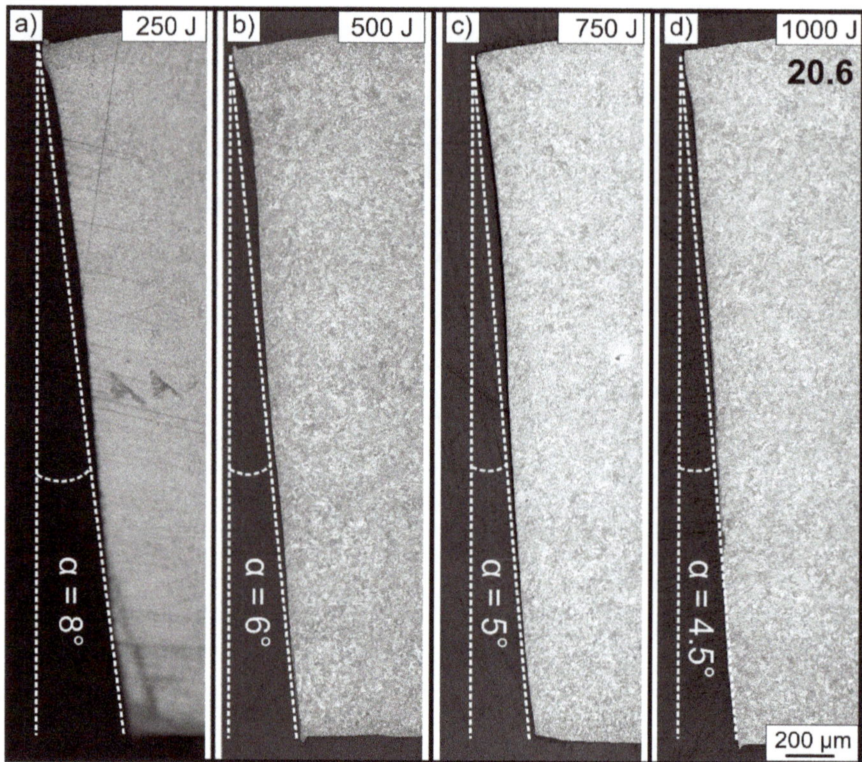

Figure 7. The blanked surface of the sheet is shown for a relative clearance of 10% (specimen 20.6) and varying impact energies from (**a**) 250 J, (**b**) 500 J, (**c**) 750 J to (**d**) 1000 J. The fracture angle α becomes smaller with increasing impact energy.

Figure 9 shows that due to the increased velocity, shear band formation is detectable from an impact energy of 500 J (Figure 9a). However, the length of the ASB cannot be analyzed meaningfully at either 500 J or 750 J (Figure 9b), although qualitatively the ASB is slightly longer for the 750 J sample. Both specimens show a jagged ASB, because the thermal softening in the ASB is too low to cause plain failure here. This shows that the parameter set of a blanking clearance of 10%, an impact energy of 500 J, and an impact velocity of 10.6 m/s describes a lower limit for the formation of an adiabatic shear band for this material. Very high-quality and perpendicular blanked surfaces are achieved if the process is carried out just below this limit. The best blanked surfaces were reached for the samples cut with an impact energy of 250 J and 500 J. This clearly shows the different evolutions of the blanking surface as a function of the blanking parameters. The predominant shear band type here is presumably a deformation shear band and not the classical transformation shear band. However, for accurate identification, the blanked surfaces need to be further examined using scanning electron microscopy techniques.

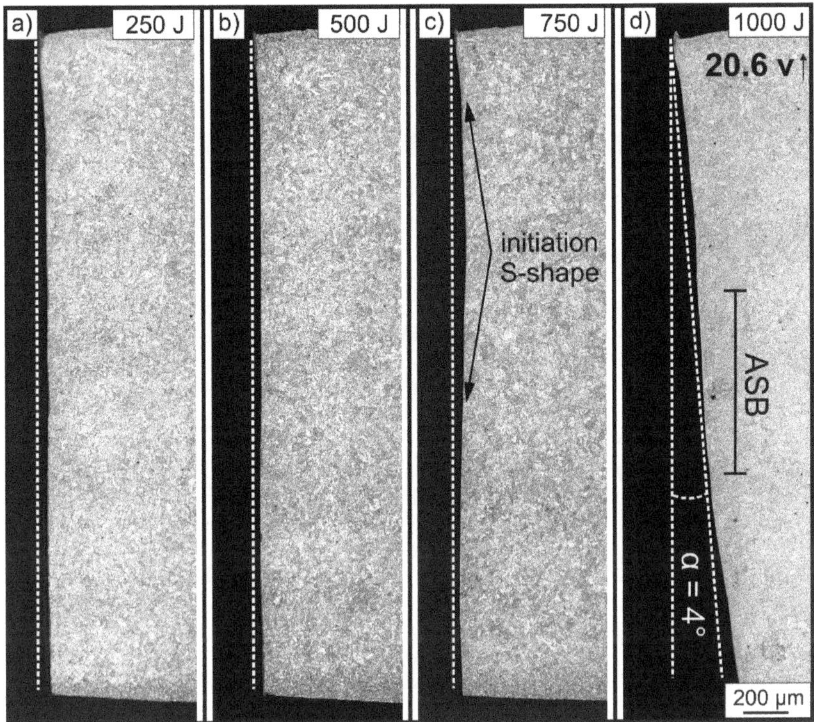

Figure 8. Blanked surfaces of the sheet cut with a relative clearance of 10% (sample 20.6 v↑) and varying impact energies from (**a**) 250 J, (**b**) 500 J, (**c**) 750 J to (**d**) 1000 J and velocities, respectively.

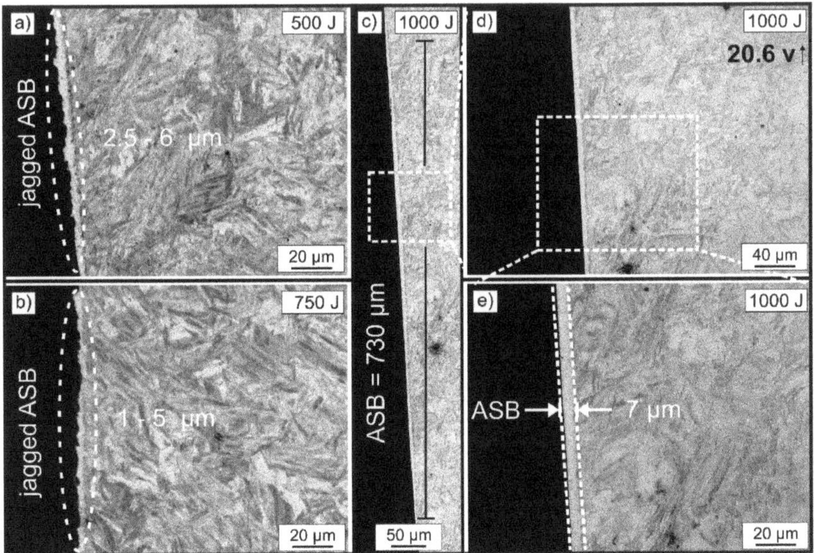

Figure 9. Optical microscopy images with increased resolution of the ASB of the 20.6 v↑ samples from (**a**) 500 J, (**b**) 750 J and (**c–e**) 1000 J.

With increasing impact energy and correspondingly increasing impact velocity, the amount of converted microstructure and thus ASB becomes larger. Figure 9c–e show the ASB of the 1000 J sample blanked with the aluminum pusher at a velocity of 12.5 m/s. The length of the ASB was determined to be 730 µm, which is more than 100% longer than the 310-µm-long ASB detected for the specimen cut with the steel pusher using the same impact energy but significantly lower speed (Figure 3d). Furthermore, the increased velocity and temperature in the shear zone lead to a three times thicker ASB due to the facilitated dynamic recrystallization and the resulting transformation of the microstructure [27]. Precisely, the width is 7 µm in case of the specimen cut with the aluminum pusher compared to 2.5 µm in case of the specimen cut with the steel pusher. However, the large blanking clearance and especially the initiated ASB change the quality of the fracture surface to negative due to a fracture angle of $\alpha = 4°$. If higher velocities and higher energies are used in adiabatic blanking, the formation of an ASB over the whole cross-section of the blanked surface could probably significantly improve the quality further and enhance the cutting surface with unique properties, such as a very high hardness [4].

4. Summary and Conclusions

The presented study systematically investigated the influence of the blanking clearance (relative clearance: 1.67–16.67%), the impact energy (250 J–1000 J), and the impact velocity (7–12.5 m/s) on the adiabatic blanking process of a hardened 22MnB5 steel sheet (3 mm thickness) and on the quality of the resulting blanking surface. Numerical simulation served for determining the stress state during the process and corresponding experiments were carried out. The blanked surfaces were analyzed by optical microscopy.

The investigated blanking parameters were proven to have a significant effect on the geometry of the blanking surface. Certain parameter combinations promote the initiation of ASB. Especially, the blanking clearance, which highly influences the stress triaxiality in the shear zone, is an important factor. Surprisingly, not the blanked surfaces with ASB, but those that are close to forming an ASB showed the highest quality in terms of geometric accuracy. The following key results were determined:

- If the relative clearance is ≤6.67% and the impact velocity is ≥7 m/s, an ASB is always formed regardless of the investigated impact energies (250 J to 1000 J). A smaller blanking clearance increases the amount of compressive stress in the shear zone, which is necessary for dynamic recrystallization and for the formation of an ASB. For larger clearances, the energy and velocity must be increased to form an ASB. Thereby, the impact velocity cannot be considered separately from the impact energy.
- The initiation of the ASB is centered in the shear zone in the area of the highest compressive stresses. A smaller blanking clearance, a higher impact energy, and a higher speed result in longer and wider ASB.
- Three different blanking surface types were identified depending on the process parameters. The initiation of an ASB promotes the formation of an S-shaped blanking surface. If the relative clearance is ≥10% and the velocity is less than 10 m/s, an angled fracture surface occurs, whereby an increase in velocity results in smaller angles. High velocity (10–12.5 m/s) combined with a blanking clearance of 10% and a maximum impact energy of 500 J results in a very straight blanking surface with high quality. This surface exhibits selectively jagged and very small ASB. This demonstrates that lower impact energies combined with increased velocity can lead to blanked surfaces of very high quality.

As ASB provide special properties, such as very high hardness [4], an aim for further research is the formation of an ASB over the entire blanked surface. For this purpose, the impact energy and the velocity must be increased while maintaining a small relative blanking clearance (≤3.34%). Under these conditions, the benefits of the ASB can be further utilized for the blanked surfaces.

Author Contributions: Conceptualization, S.W. and M.N.; methodology, S.W.; investigation, S.W., M.N., E.G. and F.H.; resources, M.D.; writing—original draft preparation, S.W.; writing—review and editing, S.W., M.N., V.P., V.K.; visualization, S.W.; supervision, M.D., V.K., V.P.; project administration S.W.; funding acquisition, S.W. All authors have read and agreed to the published version of the manuscript.

Funding: The authors would like to thank the Free State of Saxony for their financial support of the study, which was funded by the European Regional Development Fund (EFRE) together with the state funds made available by the Free State of Saxony.

Institutional Review Board Statement: Not applicable.

Informed Consent Statement: Not applicable.

Data Availability Statement: Not applicable.

Conflicts of Interest: The authors declare no conflict of interest.

References

1. Sachnik, P.; Hoque, S.E.; Volk, W. Burr-free cutting edges by notch-shear cutting. *J. Mater. Process. Technol.* **2017**, *249*, 229–245. [CrossRef]
2. Sahli, M.; Roizard, X.; Assoul, M.; Colas, G.; Giampiccolo, S.; Barbe, J.P. Finite element simulation and experimental investigation of the effect of clearance on the forming quality in the fine blanking process. *Microsyst. Technol.* **2020**, *7*. [CrossRef]
3. Neugebauer, R.; Weigel, P.; Westkämpfer, E.; Verl, A.; Eicher, F. *Investigation of Application Potentials and Limits of Adiabatic Cutting and Punching Operations*; Germany Report P774; Verlag und Vertriebsgesellschaft mbH: Düsseldorf, Germany, 2010.
4. Schmitz, F.; Winter, S.; Clausmeyer, T.; Wagner, M.F.X.; Tekkaya, A.E. Adiabatic blanking of advanced high-strength steels. *CIRP Ann.* **2020**, *69*, 269–272. [CrossRef]
5. Neugebauer, R.; Kräusel, V.; Weigel, P. Hochgeschwindigkeitsscherschneiden hält Einzug in die Blechbearbeitung. *Wt. Werkstattstech. Online* **2008**, *98*, 813–814.
6. Rosenthal, S.; Maaß, F.; Kamaliev, M.; Hahn, M.; Gies, S.; Tekkaya, A.E. Lightweight in Automotive Components by Forming Technology. *Automot. Innov.* **2020**. [CrossRef]
7. Zener, C.; Hollomon, J.H. Effect of Strain Rate Upon Plastic Flow of Steel. *J. Appl. Phys.* **1944**, *15*, 22–32. [CrossRef]
8. Meyers, M.A.; Xu, Y.B.; Xue, Q.; Pérez-Prado, M.T.; McNelley, T.R. Microstructural evolution in adiabatic shear localization in stainless steel. *Acta Mater.* **2003**, *51*, 1307–1325. [CrossRef]
9. Bai, Y.; Dodd, B. *Adiabatic Shear Localization*; Pergamon Press: Oxford, UK, 2012; ISBN 978-0-08-097781-2.
10. Rogers, H. Adiabatic plastic deformation. *Annu. Rev. Mater. Sci.* **1979**, *9*, 283–311. [CrossRef]
11. Wang, B.; Sun, J.; Wang, X.; Fu, A. Adiabatic shear localization in a near beta Ti-5Al-5Mo-5 V-1Cr-1Fe alloy. *Mater. Sci. Eng. A* **2015**. [CrossRef]
12. Xu, Y.; Bai, Y.; Meyers, M.A. Deformation, phase transformation and recrystallization in the shear bands induced by high-strain rate loading in titanium and its alloys. *J. Mater. Sci. Technol.* **2006**, *22*, 737–746.
13. Meyers, M.A.; Wittman, C.L. Effect of metallurgical parameters on shear band formation in low-carbon (approx. 0.20 wt pct) steels. *Metall. Trans. A, Phys. Metall. Mater. Sci.* **1990**, *21A*, 3153–3164. [CrossRef]
14. Nesterenko, V.F.; Meyers, M.A.; LaSalvia, J.C.; Bondar, M.P.; Chen, Y.J.; Lukyanov, Y.L. Shear localization and recrystallization in high-strain, high-strain-rate deformation of tantalum. *Mater. Sci. Eng. A* **1997**, *229*, 23–41. [CrossRef]
15. Osovski, S.; Rittel, D.; Landau, P.; Venkert, A. Microstructural effects on adiabatic shear band formation. *Scr. Mater.* **2012**, *66*, 9–12. [CrossRef]
16. Rittel, D.; Wang, Z.G.; Merzer, M. Adiabatic shear failure and dynamic stored energy of cold work. *Phys. Rev. Lett.* **2006**, *96*, 1–4. [CrossRef] [PubMed]
17. Mendoza, I.; Villalobos, D.; Alexandrov, B.T. Crack propagation of Ti alloy via adiabatic shear bands. *Mater. Sci. Eng. A* **2015**, *645*, 306–310. [CrossRef]
18. Gaudillière, C.; Ranc, N.; Larue, A.; Lorong, P. Investigations in high speed blanking: Cutting forces and microscopic observations. *EPJ Web Conf.* **2010**, *6*, 19003. [CrossRef]

19. Winter, S.; Pfeiffer, S.; Bergelt, T.; Wagner, M.F.-X. Finite element simulations on the relation of microstructural characteristics and the formation of different types of adiabatic shear bands in a β-titanium alloy. *IOP Conf. Ser. Mater. Sci. Eng.* **2019**, *480*, 012022. [CrossRef]
20. Winter, S.; Schmitz, F.; Clausmeyer, T.; Tekkaya, A.E.; Wagner, M.F.-X. High temperature and dynamic testing of AHSS for an analytical description of the adiabatic cutting process. *IOP Conf. Ser. Mater. Sci. Eng.* **2017**, *181*, 012026. [CrossRef]
21. Pouya, M.; Winter, S.; Fritsch, S.; Wagner, M.F.-X. A numerical and experimental study of temperature effects on deformation behavior of carbon steels at high strain rates. *IOP Conf. Ser. Mater. Sci. Eng.* **2017**, *181*, 012022. [CrossRef]
22. Psyk, V.; Scheffler, C.; Tulke, M.; Winter, S.; Guilleaume, C.; Brosius, A. Determination of Material and Failure Characteristics for High-Speed Forming via High-Speed Testing and Inverse Numerical Simulation. *J. Manuf. Mater. Process.* **2020**, *4*, 31. [CrossRef]
23. Dodd, B.; Bai, Y. Width of adiabatic shear bands formed under combined stresses. *Mater. Sci. Technol.* **1989**, *5*, 557–559. [CrossRef]
24. Xue, Q.; Meyers, M.A.; Nesterenko, V.F. Self-organization of shear bands in titanium and Ti-6Al-4V alloy. *Acta Mater.* **2002**, *50*, 575–596. [CrossRef]
25. Nesterenko, V.F.; Meyers, M.A.; Wright, T.W. Self-organization in the initiation of adiabatic shear bands. *Acta Mater.* **1998**, *46*, 327–340. [CrossRef]
26. Landgrebe, D.; Müller, R.; Sterzing, A.; Mauermann, R.; Rennau, A. Hochfeste Stähle—Chance für Leichtbau und für Effizienzsteigerung in der Produktion. In *Proceedings of the 18 Werkstofftechnischen Kolloquium in Chemnitz 2016*; Lampke, T., Ed.; Schriftenreihe Werkstoffe und Werkstofftechnische Anwendungen 59; Technische Universität Chemnitz: Chemnitz, Germany, 2016; pp. 34–44.
27. Rodríguez-Martínez, J.A.; Vadillo, G.; Rittel, D.; Zaera, R.; Fernández-Sáez, J. Dynamic recrystallization and adiabatic shear localization. *Mech. Mater.* **2015**, *81*, 41–55. [CrossRef]

Journal of
*Manufacturing and
Materials Processing*

Article

An Efficient Methodology towards Mechanical Characterization and Modelling of 18Ni300 AMed Steel in Extreme Loading and Temperature Conditions for Metal Cutting Applications

Tiago E. F. Silva [1,2], Afonso V. L. Gregório [3], Abílio M. P. de Jesus [1,2] and Pedro A. R. Rosa [3,*]

[1] INEGI, Faculdade de Engenharia, Universidade do Porto, Rua Dr. Roberto Frias, 4200-465 Porto, Portugal; tesilva@inegi.up.pt (T.E.F.S.); ajesus@fe.up.pt (A.M.P.d.J.)
[2] DEMec, Faculdade de Engenharia, Universidade do Porto, Rua Dr. Roberto Frias, 4200-465 Porto, Portugal
[3] IDMEC, Instituto Superior Técnico, Universidade de Lisboa, Av. Rovisco Pais 1, 1049-001 Lisboa, Portugal; afonsogregorio@tecnico.ulisboa.pt
* Correspondence: pedro.rosa@tecnico.ulisboa.pt

Abstract: A thorough control of the machining operations is essential to ensure the successful post-processing of additively manufactured components, which can be assessed through machinability tests endowed with numerical simulation of the metal cutting process. However, to accurately depict the complex metal cutting mechanism, it is not only necessary to develop robust numerical models but also to properly characterize the material behavior, which can be a long-winded process, especially for state-of-stress sensitive materials. In this paper, an efficient mechanical characterization methodology has been developed through the usage of both direct and inverse calibration procedures. Apart from the typical axisymmetric specimens (such as those used in compression and tensile tests), plane strain specimens have been applied in the constitutive law calibration accounting for plastic and damage behaviors. Orthogonal cutting experiments allowed the validation of the implemented numerical model for simulation of the metal cutting processes. Moreover, the numerical simulation of an industrial machining operation (longitudinal cylindrical turning) revealed a very reasonably prediction of cutting forces and chip morphology, which is crucial for the identification of favorable cutting scenarios for difficult-to-cut materials.

Keywords: mechanical characterization; tribological characterization; high strain rate; elevated temperature; stress triaxiality; additive manufacturing; 18Ni300 maraging steel; constitutive modelling; damage modelling; machining simulation

Citation: Silva, T.E.F.; Gregório, A.V.L.; de Jesus, A.M.P.; Rosa, P.A.R. An Efficient Methodology towards Mechanical Characterization and Modelling of 18Ni300 AMed Steel in Extreme Loading and Temperature Conditions for Metal Cutting Applications. *J. Manuf. Mater. Process.* **2021**, *5*, 83. https://doi.org/10.3390/jmmp5030083

Academic Editor: Steven Y. Liang

Received: 13 June 2021
Accepted: 21 July 2021
Published: 28 July 2021

Publisher's Note: MDPI stays neutral with regard to jurisdictional claims in published maps and institutional affiliations.

1. Introduction

The numerical simulation of metal cutting operations has become an everyday practice to assist in the design of better cutting tools and in the optimization of cutting conditions aiming to improve the performance of the machining processes. This is because, in a general sense, numerical simulation of manufacturing processes has scientifically matured and become consolidated as engineering commercial software packages. The quality of the numerical predictions in metal cutting simulation depends mainly on how accurate the constitutive models are to describe the tribo-thermomechanical response of the workpiece materials in a given application. This is of tremendous importance in a context of metal cutting simulation where plastic deformation, fracture, and friction occur under extreme conditions of strain, strain rate, temperature, and complex state-of-stress which make mechanical and tribological testing challenging and leading frequently to intricate the calibration of constitutive parameters by inverse analysis. Similar constitutive model calibration challenges are also being experienced in other manufacturing technologies, such as electromagnetic forming and friction stir welding.

The development of constitutive models and experimental methodologies for the calibration of input data towards the numerical simulation of metal cutting has made significant progress over the last years, especially those considering the state-of-stress influence (i.e., stress invariants) on flow stress and damage [1,2]. Numerical estimates are now closer to laboratory measurements, or even to industrial machining operations measurements. Despite the increasingly good quality of these numerical estimates, the steady progress in materials science and cutting tools promotes the complexity of constitutive modelling, leading frequently to intricate calibration procedures. Typically, a significant number of tests is required to allow the identification of a vast number of model coefficients to better describe the mechanical and tribological response in a given application [3]. In parallel to material improvements, the increasing performances of the manufacturing processes (e.g., high-speed machining) makes input data calibration even more challenging and a time-consuming process each year.

Input data calibration is complex and involves a large number of experimental and theoretical procedures. Some of them are based on long-established laboratory tests, while others use "trial and error" inverse methods to calibrate a constitutive model. Fracture and tribological contributions can significantly expand the already time-consuming calibration procedure of a simple thermo-viscoplastic constitutive model. To put it in another way: calibration should be able for accompanying changes and requirements of the industrial sector, not hindering its development. Therefore, in order to properly handle the model complexity during calibration, the tuning process of the model coefficients requires a deep understanding of the manufacturing process and what assumptions on the mechanical and tribological characterization can be made. Indeed, each manufacturing process has its own specificities, being advantageous to simplify the numerical problem aiming to reduce the computational costs, without penalizing the quality of the theoretical estimates [4]. Likewise, there should be a concern to simplify constitutive model data calibration, which limits the usage of unnecessary experimental and theoretical procedures. This is not an obvious exercise since mechanical and tribological response under typical metal cutting conditions are highly nonlinear, influenced by temperature, state-of-stress invariants (e.g., hydrostatic pressure), and with strong rate dependencies [5].

The uniqueness of the model coefficients values is also of concern, since different coefficients sets can be generated for the same material and constitutive model, depending on the used calibration methodology and experimental techniques. For example, experimental techniques such as the split-Hopkinson pressure bars are commonly employed, despite having a loading signature distinct from the chip formation mechanics [6]. Similarly, pin-on-disc tribometers are regularly employed despite the low contact pressure and existence of metallic oxides, different from the tribological condition in the tool-chip contact interface [7,8]. Even though these traditional mechanical and tribological tests are not able to guarantee the desired accuracy, they are commonly used in the direct calibration of the model coefficients for metal cutting simulation. This is due to scarcity of specific apparatus able to reproduce realistic operative conditions of metal cutting processes under laboratory-controlled conditions and time-consuming experiments. Maybe due to this, most of the published research in metal cutting modelling is based on inverse calibration methodologies [9].

The inverse calibration methodologies are based on the reverse simulation of previously performed experiments in order to determine the best set of coefficients that allows for a better correlation between the numerical estimates and the experimental measurements. Typically, the model coefficients are adjusted in order to replicate the experimental load-time histories and the stress and strain fields. Most of these experiments are based on simulative tests, such as the orthogonal cutting test and even practical metal cutting operations. The accuracy of the constitutive models increases with the diversity of experiments to be replicated by the numerical methods. Although some phenomena are omitted in order to simplify the numerical problem and some coefficients reportedly differ from what should be physically expected, it appears that inverse calibration methodologies are able to collect some of these statistics and can usefully be directed to support numerical modelling of metal cutting processes [10,11].

The majority of constitutive models used in the numerical simulations of metal cutting considers only the plastic deformation contribution to represent the material response during the chip formation, neglecting the damage and consistently omitting the contribution of the ductile fracture. However, the model coefficients are very sensitive to small changes in the metallurgical conditions of the workpiece material. This is particularly relevant for difficult-to-machine materials such as super-alloys, stainless steels, maraging steels, and titanium alloys, among others. The same applies to the process by which the workpiece material is obtained (e.g., additively manufactured materials). The combination of these two factors requires special attention from the point of view of the constitutive modelling. For example, machining post-processing stages of additively manufactured metallic parts bring to focus the importance of machinability assessment. The difficulty in machining such materials with standard tools and cutting parameters has recently been the focus of research as reported in the literature. For example, Marbury et al. [12] reported tool breakage when applying standard cutting parameters thought for conventionally manufactured (CMed) 316 L stainless steel, on the AMed counterpart. On the same alloy, Leça et al. [13] have shown that porosity directly influences the cutting loads and reported a relationship between chip segmentation and relative density. Focusing on the effect of cutting velocity, Bai et al. [14] have compared the machinability of the AMed A131 steel with its conventional counterpart, focusing on the effect of cutting speed. The authors state that the microstructure differences between AMed and CMed materials (existence of melt pools, layer construction, and porosity) impact directly on chip morphology (continuous for the CMed), cutting forces and tool wear (higher for the AMed). Such relative decrease in the machinability of AMed parts has motivated several authors in finding the most appropriate cutting parameters for their processing.

Additively manufactured 18Ni300 maraging steel was selected for this research to better illustrate the need to move in two directions at once: on the one hand, develop an accurate constitutive model calibration for complex thermo-viscoplastic material respons and, on the other hand, optimize the cutting conditions on a new and emerging difficult-to-machine material. In what follows, a combination of experimental tests and adequate numerical methodologies are presented, to help the effective and practical calibration of constitutive models for metal cutting simulation. Therefore, the present study focuses on the mechanical and tribological characterization, accounting for the influence of the strain rate and temperature, as well as the state-of-stress. The choice of the characterization tests and specimen geometries were based on a logical flow to support the input data calibration with physically representative values and its sensitivity to the operative conditions. With the goal of numerically simulate the metal cutting processes, an extensive mechanical characterization of the AMed 18Ni300 maraging steel has been performed. A coupled plastic-damaged model based on Johnson–Cook model was employed and model coefficients determined using a combined direct characterization and inverse calibration procedure. The constitutive model validation was done by comparison of calculated and experimental load and chip curling for orthogonal metal cutting tests. Finally, the validated numerical model has been successfully used in simulations of industrial turning operations performed on AMed 18Ni300 maraging steel.

2. Efficient Methodology for Constitutive Modelling Calibration

This section proposes a progressive and efficient calibration procedure for constitutive modelling and numerical simulation in the metal cutting domain (refer to Figure 1). This procedure seeks a compromise between the direct characterization of physical parameters and the inverse calibration of theoretical models for a quick and effective determination of the constitutive model coefficients. A minimum set of complementary mechanical and tribological tests were selected. These experimental tests allow to reproduce the typical operative conditions similar to those of real metal cutting processes, some allowing to obtain absolute reference values (e.g., yield stress, flow stress under uniaxial compressive loading and friction coefficient at room temperature), whereas others relate to assessing the sensitivity of the mechanical and tribological response to the extreme operative conditions of metal cutting (e.g., strain rate, temperature, triaxiality). Regarding the material modelling, the Johnson–Cook model was selected as the basis since it is one of the mostly used material laws, as well it is available in most of the commercial FE codes. It is commonly believed that the Johnson–Cook model is able to reproduce flow stress behavior of materials under uniaxial impact loading and high temperatures [15].

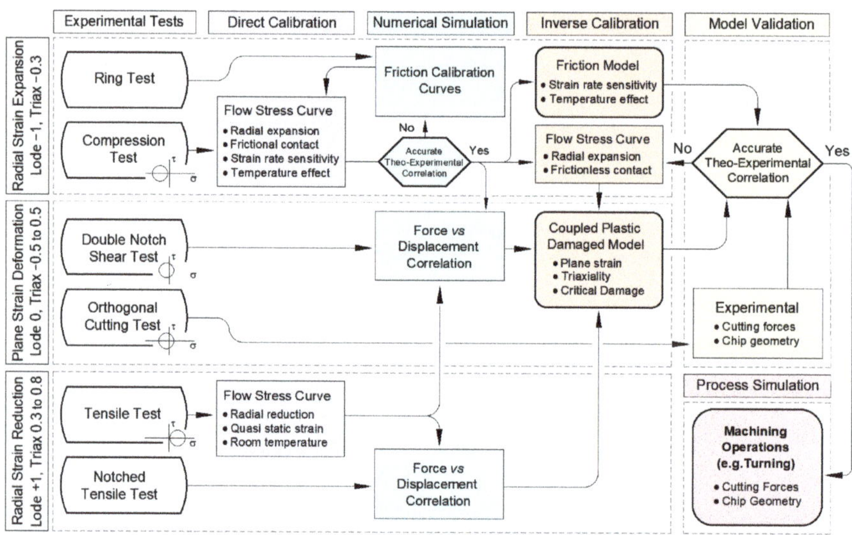

Figure 1. Flowchart of the optimization-based methodology for the identification of the plasticity and damage model coefficients for metal cutting simulation.

Tensile testing is one of the most common mechanical tests performed on materials. It provides a simple, accurate, and direct characterization of the material yield stress and flow stress–strain behavior until fracture, which occurs under a positive stress triaxiality (e.g., triaxiality equal to 0.3 or higher) and Lode angle parameter equal to unity. Under these stress conditions, fracture strain is a function of stress triaxiality, being common reduced plastic strains at failure for high stress triaxialities, e.g., effective strains of 0.1, even for ductile metallic alloys. The tensile test is typically performed using universal testing machines capable of being used for a maximum strain rate of 10 s^{-1} and the occurrence of necking hinders the direct characterization after the necking onset. Non-uniform stress–strain fields during necking impact on stress triaxiality and therefore on the effective strain at failure. After necking, inverse characterization techniques are the most common approaches.

Notwithstanding the foregoing, given the loading signature of metal cutting, it appears logical to complete the mechanical characterization with the uniaxial compression test to undergo high plastic strains (e.g., effective strains up to 1 or above) without cracking due the negative state of stress (e.g., triaxiality of −0.3 or lower and Lode angle of −1). This mechanical test can be carried out using impact testing machines (e.g., strain rates of 10^4 s^{-1}) inside furnaces (e.g., 750 °C). Contrarily to tensile tests, compression tests allow a direct characterization of the material flow behavior under severe plastic deformation. In addition, they allow for sensitivity analysis concerning the operative conditions influencing the material response rate- and temperature-dependent mechanical characteristics, which is hardly achieved by means of tensile tests under such severe conditions [16].

For reasons of simplification, it is considered that the rate- and temperature-dependent characteristics are independent of the state of stress and can be adequately determined by the uniaxial compression test. These uncoupled effects hypothesis is the basis of the Johnson–Cook constitutive model for the flow stress [17]. This rate- and temperature-dependent response under uniaxial compression can subsequently be generalized to the entire range of triaxialities and Lode angle parameters combinations (axisymmetric tensile and plane strain tests), where the mechanical testing under high strain rates and elevated temperatures are extremely difficult or even impossible to accomplish. As a clear example of this complementarity, the material flow stress which can be determined by the double-notch plane strain test introduced by Abushawashi et al. [18] (quasi-static and room temperature) to better representing the metal cutting conditions, can be further enhanced by the rate- and temperature-dependent sensitivity determined by uniaxial compression testing under high strain rates and elevated temperatures.

Although the compression test simulates the operative conditions (i.e., strain rate and temperature) of the metal cutting processes, the friction between the cylindrical specimen and the compression platen promotes an overestimation for the flow stress. The flow stress should be corrected by subtracting the contribution of the frictional energy. This can be done by performing ring compression tests using the same compression apparatus and thus maintaining constant the tribological interface for all mechanical tests. The ring-shaped specimen is compressed between two flat platens with (similar to the compression test) and without lubrication (similar to dry friction of metal cutting), and the experimental evolution of the inner diameter is measured as a function of the specimen height. The friction coefficient is determined by correlating the experimental measurement of the inner diameter and the friction calibration curves. These calibration curves are inversely determined by numerical simulation, changing the friction coefficient in order to reproduce the experiments. However, an optimization-based procedure is needed for an inverse determination of the flow stress curve under frictionless conditions, and new friction calibration curves and a friction coefficient should be determined. After, the achieved input data for the numerical simulation of the compression test (frictionless flow stress and friction coefficient value) should provide an accurate loading-displacement curve. The method is confirmed to provide flow stress curves within an error of 5%. It is worth noting that the ring test can be carried out in the impact testing machine under the extreme conditions of pressure, strain rate and temperature of the practical machining processes.

The above mentioned procedure has traditionally been considered acceptable for calibrating input data parameters for the numerical simulation of metal cutting processes. However, the radial expansion (compression) and the radial reduction (tensile) of the commonly used mechanical tests may differ from each other, and from the desired plane strain conditions of the chip formation mechanics (refer to Figure 2). This is of relevance when the mechanical response of engineering materials (e.g., flow stress curve, critical damage), which includes the recent AMed metallic alloys, exhibits high sensitivity to the state of stress (e.g., stress triaxiality, Lode angle). The literature shows appropriate procedures to account for the influence of the state-of-stress in the constitutive modelling [19] based on a combination of tensile, compressive and shear tests and inverse calibration methodologies. The geometry of the test specimens plays a determinant role in the resulting state-of-stress of the plastic zone to better replicate the micro cracks initiation and damage propagation occurring in manufacturing processes [20]. The coupled plastic-damage model should be able to capture and reproduce that influence of the triaxiality and Lode angle on the mechanical response of the workpiece material. An adequate selection of the constitutive model and related calibration methodologies can significantly influence the quality of the numerical estimates. The accuracy of the numerical models increases with the range and combination of the triaxiality and Lode angle considered in the inverse calibration procedures [21].

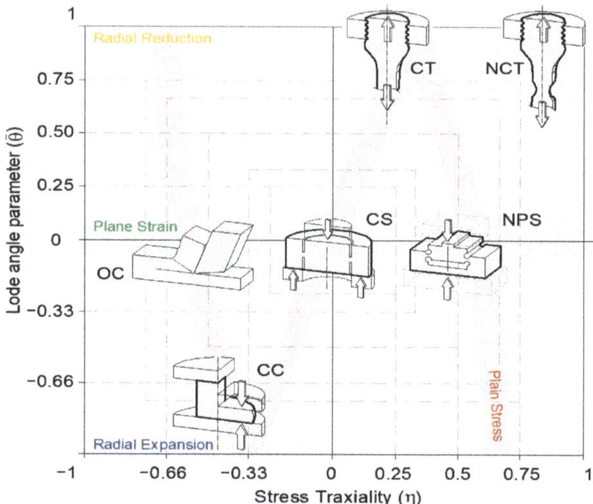

Figure 2. Schematic representation of the stress triaxiality and Lode angle parameter space for different specimen geometries (C_ stands for cylinder, NC_ for notched cylinder and NP for notched plane strain specimen) and loading conditions (_C stands for compression, _S for shear and _T for tensile). CC refers to uniaxial compression of cylinders and OC to orthogonal cutting test.

Nevertheless, the cost of implementing a full-range constitutive model valid for any triaxiality and Lode angle parameter is very high, both in terms of the number of tests required as well as in terms of model parameters to be identified thru intricate direct/indirect optimization procedures. In the present research, recognizing the plane strain nature of the metal cutting operations, double notched plane strain specimens introduced by Abushawashi et al. [18] will assume a central role for the constitutive model calibration regarding the reference conditions of temperature and strain rate (e.g., room temperature and quasi-static deformation) based on standard yield theories (e.g., J2 plasticity). The triaxiality of the chip formation mechanism can be well reproduced by the inclined double-notch plane strain tests under controlled laboratory conditions. The triaxiality value is influenced by the tilted notches angle and can be determined based on an optimization-based procedure including physical constraints and initial input values. Thus, given the plane strain deformation of the double-notch plane strain test (aprox. triaxiality ranging from −0.5 to 0.5 and null Lode angle), it appears logical that the optimal solution should be framed between the tensile test (triaxiality 0.3 and Lode angle 1) and the compression test (triaxiality −0.3 and Lode angle −1) for meaningful values. As initial input values the frictionless flow curve obtained from the uniaxial compression test should be used due to its capability to reproduce the extreme conditions of metal cutting, namely the large plastic strain. Using inverse simulation techniques, this kind of test will be used to validate a stress–strain relation. Further, the developed reference flow stress relation can be updated with the thermal and rate dependent influences, from compression testing assuming independent contributions, as underlined by the Johnson–Cook model. Regarding the damage behavior evaluation, the tensile tests covering various triaxialities and the double notched plane strain specimens may provide enough insight covering both positive and negative triaxialities.

The small number of parameters and relatively simple calibration contribute to extensive usage of the Johnson–Cook plasticity model (refer to Equation (1)) in metal cutting simulation. The model describes the equivalent plastic stress (σ) in function of the uncoupled effects of strain hardening, through material parameters of the equation's first term A, B, and n, as well as the equivalent plastic strain (ε_p). The viscoplastic response is modelled by the second term through the material parameter C as well as strain rate

($\dot{\varepsilon}$) and reference strain rate ($\dot{\varepsilon}_0$). The temperature softening is modelled by the third term through the definition of the material parameter m, room temperature T_0, and melting temperature, T_m. In addition to the flow stress evolution, the Johnson–Cook damage threshold model (Equation (2)) depicts that the damage initiation strain (ε_f) as a function of stress triaxiality (η), strain rate, and temperature as well as material parameters d_1 to d_5. In order to model damage evolution (stiffness degradation), the Hillerborg model (Equation (3)) can be applied due to the idealized combination between damage mechanics and finite element degradation and may be regarded as a material parameter capable of expressing the energy dissipation relative to crack opening. The critical damage dissipation energy, G_f, is depicted in function of the characteristic length (L), which is associated with an integration point.

$$\sigma = \left(A + B\varepsilon_p{}^n\right)\left[1 + Cln\left(\frac{\dot{\varepsilon}}{\dot{\varepsilon}_0}\right)\right]\left[1 - \left(\frac{T - T_0}{T_m - T_0}\right)^m\right] \tag{1}$$

$$\varepsilon_f = \left(d_1 + d_2 e^{d_3\eta}\right)\left[1 + d_4 ln\left(\frac{\dot{\varepsilon}}{\dot{\varepsilon}_0}\right)\right]\left[1 + d_5\left(\frac{T - T_0}{T_m - T_0}\right)\right] \tag{2}$$

$$G_f = \int_{\varepsilon_0}^{\varepsilon_f} L\sigma \, d\varepsilon_p \tag{3}$$

The above mentioned direct characterization and inverse calibration methodologies are expected to provide accurate input data parameters for the numerical simulation of the metal cutting processes [22]. Nevertheless, it must be pointed out that assumptions and simplifications have been done in order to allow the development of a progressive and efficient calibration procedure. Thus, an experimental validation concerning metal cutting simulation is also desirable. This can be achieved by comparing the numerical simulation estimates to the experimental measurements on the simple orthogonal cutting test carried out under controlled laboratorial conditions. It should be noted that the tribological condition of metal cutting differs from metal forming due the absence of metallic oxides and thus slight variations in the friction coefficient may occur influencing the chip curling.

3. Additively Manufactured 18Ni300 Maraging Steel

In order to address the characterization requirements, an AMed 18Ni300 material batch was built, through laser powder bed fusion. An ytterbium laser with a spot size of 70 µm was used for processing the material samples, according to the manufacturers' recommendation for optimal printing conditions and overall quality. This corresponds to a laser power of 400 W, scan speed of 0.86 m/s, hatch spacing of 95 µm and a layer thickness of 40 µm. Additively manufactured parts may present a considerably different mechanical performance, depending on a multitude of aspects in their manufacture. In order to fully understand the materials' as-built condition, a comprehensive analysis regarding physical, metallurgical and mechanical properties was carried.

The micrograph cross-section method for porosity measurement was adopted in this study. In this technique, AMed samples were cut in two distinct directions (perpendicular and parallel to the build direction), embedded in resin (through compression mount). Manual and semi-automatic grinding using sandpaper, followed by polishing of the cut cross-sections (Struers Pedemax-2, INEGI), allowed for its optical microscope observation (Olympus PMG3 and Zeiss Axiophot). Etchant is not used in these samples, given that the goal is to simply reveal the absence of material (pores) in a polished surface. In order to estimate relative density, metallographic samples were post-processed in ImageJ software for the generation of binary maps that more easily allow for porosity assessment. Figure 3 presents two representative metallographic image samples (one for each direction of interest) for the additively manufactured maraging steel. A computed relative porosity of 99.7% was found, which is identical to the manufacturer's specification (99.8%). No pattern- or direction-related porosity was found in the analyzed samples.

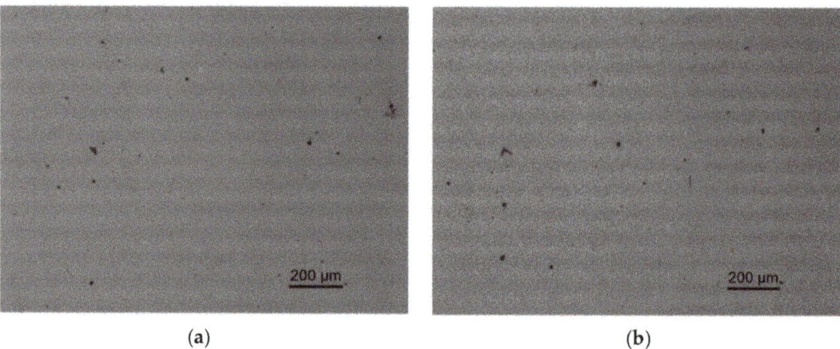

(a)   (b)

Figure 3. Unetched metallographic samples showing voids occurrence in both: (**a**) perpendicular, and (**b**) parallel-to-build directions.

Microstructural analysis was carried by the same optical equipment used in the porosity analysis. For such task, material samples were etched in order to reveal the melt pool baths as well as the grain morphology. Samples were chemically etched with nital solution (2%) whilst oxalic acid was used for electrolytical etching (6 V DC for 50 s). Grain morphology can easily be observed with chemical etching, whereas melt pool geometry and AMed-related microscopic features, such as laser trace, require electrolytical etching. Figure 4a shows a chemically etched sample of AMed 18Ni300 maraging steel in perpendicular-to-build direction, characterized by an uneven distribution of lath and plate martensite among austenite. Figure 4b shows an electrolytical etched sample in the parallel-to-build direction, where melt pool geometry is evidenced. The melt pool approximate geometry allows for the calculation of an energy density of 122 J/mm^3, which is coherent with the literature values for steel alloys processing [23].

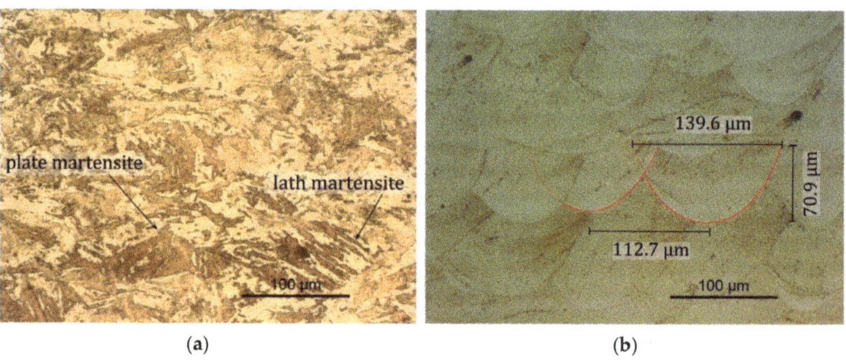

(a)   (b)

Figure 4. (**a**) Chemically etched metallographic samples in perpendicular-to-build direction, and (**b**) Electrolytically etched samples in parallel-to-build direction.

Chemical composition analysis was performed using spark emission spectroscopy (SPECTROMAXx). The comparison/verification compliance with material standards, for each alloying element is depicted in Table 1, where the low carbon content is noticed. Rather than the typical high carbon content, the martensitic microstructure of AMed 18Ni300 maraging steel is achieved by nickel addition, which is the main element. The lack of carbon enhances weldability, which in turn makes such alloys a good choice for additive processing [24,25].

Table 1. Chemical composition (wt%) of 18Ni300 AMed steel with respective reference values [26].

	Ni	Co	Mo	Ti	Si	Mn	C	P	S
Standard min	18.0	8.5	4.6	0.5	-	-	-	-	-
Standard Max	19.0	9.6	5.2	0.8	0.10	0.10	0.03	0.01	0.01
Current	18.80	8.84	5.15	0.65	0.05	0.03	0.02	<0.001	<0.001

4. Plastic and Damage Response Determination

This section focuses on the experimental characterization and inverse calibration methodologies conducted on the AMed 18Ni300 maraging steel to determine its mechanical and tribological behaviors. It begins by introducing the compression tests on both ring and cylindrical specimens. These tests were conducted for a dual purpose: (i) a direct characterization of the material flow stress under radial strain expansion accounting for material sensitivity to strain rate and temperature; and (ii) to estimate friction coefficient at the tool-workpiece interface resorting to a combined experimental-numerical approach. Special emphasis is given to the apparatus that was developed by the authors to perform uniaxial compression tests under extreme thermo-mechanical conditions. In continuation, radial strain reduction tests are performed on both cylindrical and notched cylindrical specimens, tensile tests representing an adequate approach towards identifying the material damage behavior dependence on Lode angle and positive stress triaxialities. Double notch plane strain specimens were also tested for a complementary range of intermediate stress triaxialities at null Lode angle. The inverse analysis of the tests on double notched specimens also allowed establishing the material constitutive plastic law in a more representative state-of-stress to metal cutting. The combined experimental-numerical methodology proposed by the authors requires the inverse analysis of the experimental characterization tests. Thus, ABAQUS FEM software has been used in the full extent of mechanical and tribological characterization modelling, considering an elastoplastic approach with a von Mises (J2) yield criterion with isotropic hardening.

4.1. Initial Definition

The uniaxial compression test and the ring compression test were selected to evaluate the mechanical and tribological properties of the AMed 18Ni300 steel, when submitted to one-dimensional compression and axisymmetric expansion, at different loading rates. These are the most used tests to evaluate the flow stress and friction in metal plasticity [27,28] and can be carried out using the same experimental apparatus, allowing similar tribological conditions between the specimens and the compression platens. Under these experimental conditions and based on adequate inverse analysis procedures, it is possible to identify the individual contribution of plasticity and friction.

The compression test involves reducing the height of axisymmetric specimens, by uniaxial compression, between two flat and parallel platens (Figure 5a). Compression tests were carried out on cylindrical specimens with a cross-section of 4 mm diameter (d) and a height-to-diameter ratio, $h_0/d = 1$. In order to evaluate the mechanical response in conditions compelling to those found in metal cutting, tests were conducted for a wide range of strain rates from quasi-static up to $6000\,\mathrm{s}^{-1}$. A thin film of graphite grease was used to lubricate the compression platens. However, despite even the best interface conditioning practices, no homogeneous deformation can be attained due the impossibility to eliminate the frictional shear on both contact interfaces between compression platens and the test specimen (Figure 5b,c). The ideal mechanical response, frictionless flow curve, can be only attained by eliminating the friction contribution on the experimental load–displacement curve using post-processing techniques. To account for tribological phenomenon on the contact interfaces, ring shape specimens were also tested under compression to several pre-determined height decrements (Figure 6a).

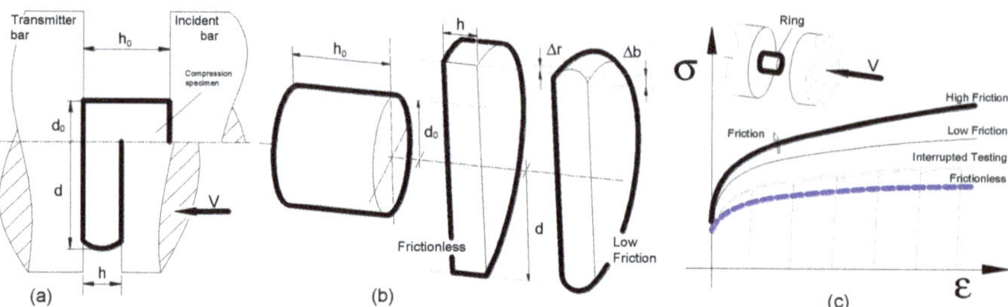

Figure 5. Basic concepts of the uniaxial compression test on cylindrical specimens: (**a**) schematic representation, kinematics and nomenclature of the compression test, (**b**) schematic representation of the friction influence on the deformed cylinder morphology (cross-section), in which the outer diameter increases as the specimen is compressed for a higher friction, (**c**) schematic representation of the typical flow curves obtained by experimental testing under different interfacial friction conditions (dry and lubricated conditions, and interrupted testing).

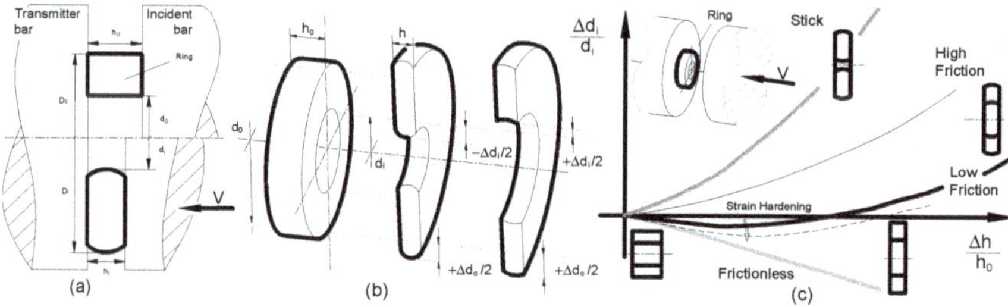

Figure 6. Basic concepts of the uniaxial compression test on ring shaped specimens: (**a**) schematic representation of the test region, showing the kinematics of the ring compression test, (**b**) schematic representation of the friction influence on the deformed ring morphology, in which both the inner and outer diameters increase as the specimen is compressed for a lower friction and the inner diameter decreases for a higher friction, and (**c**) schematic plot of the friction calibration curves obtained by finite element analysis.

As the ring height is reduced, the outer diameter expands radially outwards and the change in inner diameter depends on both compression in the axial direction and the interfacial friction between ring and compression platens. If friction is small, both the inner and the outer diameter expand. However, due to friction hindering of the radial expansion of the specimen material, the inner diameter can become smaller than the original diameter (Figure 6b), for higher values of friction [29]. In this research, the dimensions of the ring specimens were 7, 3.5, and 2.5 mm, corresponding to the outside diameter, inside diameter, and height, respectively (approx. 6:3:2 ratio). The experimental tests were performed for two different lubrication conditions: (i) lubricated contact, in which graphite grease was applied on the compression platens in order to reproduce the tribological conditions of the previously described uniaxial compressions tests, and (ii) dry contact, to study the effect of temperature and loading rate on friction coefficient under similar operative conditions as those observed in machining processes. While for quasi-static conditions it is possible to perform interrupted tests planning various compression stages (several stages and, thus a sequence of experimental measures), in high-speed tests it is not possible to apply such a practice due to the high kinetic energy of the incident bar and thus, only single stages were carried out (a single measure per experimental test). In-between compression stages, the values of specimen's inner diameter and height were measured using a digital precision caliper gauge and recorded before and after the tests. Each test/experimental measurement produces only one experimental point to be placed on the calibration curves. At least 3 tests were performed for each experimental condition to detect possible gaps in the experimental results that could be attributed to homogeneity and isotropy issues related with the AMed material production and specimens manufacturing.

Friction can be estimated by a comparison of experimental data with friction calibration curves generated in advance by numerical procedures. The calibration curves were obtained by means of finite element-based simulations using Abaqus software. Both compression platens were considered rigid; the bottom platen was considered static, whereas the upper platen motion was set to be consistent with the velocity of the experimental tests. The mechanical behavior (stress–strain data) of the materials to be used in the numerical simulation was determined by means of the compression tests performed on cylindrical specimens, as previously described. The calibration curves for quasi-static conditions are exemplified in Figure 6c and show that low values of friction give rise to increase the inner diameter of the specimens during deformation, whereas high values promote a decrease in the specimen's inner diameter.

The uniaxial compression tests were carried out in a customized split-Hopkinson pressure bar (SHPB), specially designed to provide a wide range of strain rate conditions and temperatures, schematically shown in Figure 7. This apparatus makes use of a single load cell and specific displacement transducers for all testing conditions, ranging from quasi-static, low strain rates and high strain rates tests, thus eliminating the typical calibration deviations associated with the utilization of different testing machines. The main components can be grouped into three main systems: (i) impact bench, (ii) instrumentation and data acquisition systems, and (iii) rate-based actuators.

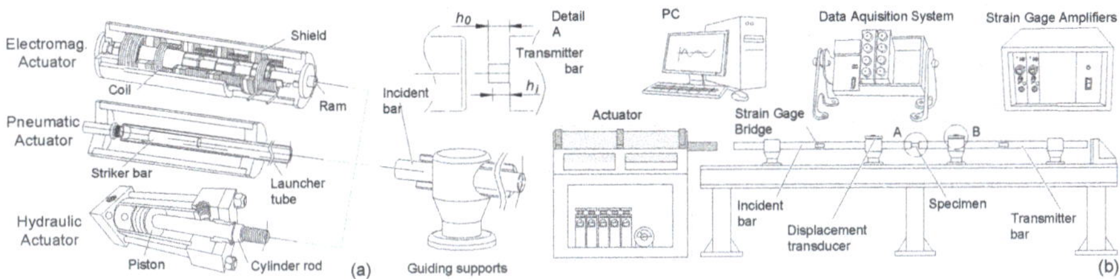

Figure 7. Schematic representation and nomenclature of the specially designed split-Hopkinson pressure bar that allows simultaneously quasi-static, low and high strain rate testing conditions by using (**a**) interchangeable hydraulic, electromagnetic and pneumatic actuators, and (**b**) assembled on the same platform as the downstream data acquisition elements; detail A shows the specimen position, while detail B shows the displacement transducer mounted on the linear guide.

The impact bench features basic structural parts, kinematics transmission systems, a thermostatic chamber, and a pair of compression platens. The compression platens were made from a powder metallurgy (PM) composite material consisting of tungsten carbide particles and a cobalt metallic binder (WC–15 wt% Co), that provides the required compromise between hardness and toughness. Thus, assuring that the rigid substrate endures the very demanding conditions of high temperatures and extreme compressive loads but also withstands the shock and vibration of impact tests. In order to limit asperity interlocking contribution to the friction mechanism, the platens' surface was polished, having shown average roughness values (Ra) in the range of 0.009 to 0.032 μm along the radial direction. Chemical adhesion between the material pair platens-specimen was controlled through the application of a TiAlSiN coating with a thickness of not less than 2.5 μm, using a PVD HiPIMS process.

An electric furnace was used to allow testing temperatures above reference room temperature (Figure 8b). This heating unit was installed directly in the impact zone, surrounding the compression platens, as seen in Figure 8c, allowing testing temperatures up to 900 °C. This furnace consists of a chamber lined with refractory material where the electric heating resistors are located, with the particularity of having a through hole in the sides to permit the compression platens' action (Figure 8b).

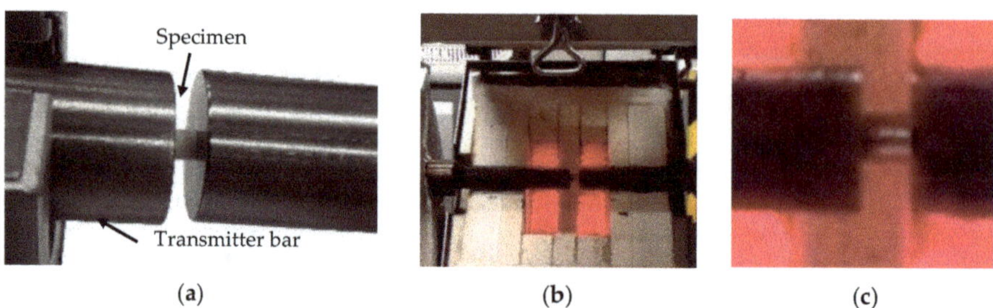

Figure 8. Uniaxial compression test under different thermal conditions; (**a**) Room temperature tests, (**b**) hot conditions using an electric furnace, and (**c**) close-up view of the compression specimen under elevated temperature.

The instantaneous strain, stress and strain-rate are calculated based on the monitorization of the three main physical parameters: time, load, and displacement; being the instant time implicit in the data acquisition systems. A multipurpose data acquisition board, National Instruments PCI-6115 combined with appropriate software (LabView) was used to monitor these parameters, allowing sampling rates of up to 10 MHz. Since the presented experimental technique deviates from the commonly used SHPB in the way that both force and displacement values are taken from direct measurements, special sensors were installed for such purpose. A full Wheatstone bridge arrangement used for load calculation and installed in the first quarter of the transmitter bar, promoting an extended time window without the interference of the reflected wave. The elastic deformation of the bar under compressive loads causes the gauges' resistance to change and thus, the load cell electrical output signal can be linearly correlated with the applied load. Yet, this voltage output is very low, so a VISHAY 2310B signal conditioning amplification system was required to amplify the electrical signal. For room temperature conditions, inductive position sensors were mounted directly between compression platens in order to promote the precision, accuracy and resolution of the experimental measurements. This experimental technique prescinds of complex data post-processing necessary to account for elastic deformation of the testing machine and allows measuring the distance between compression platens with high resolution under very high velocities and severe accelerations. For compression tests conducted at high temperature, the inductive sensors are no longer viable. To surpass this limitation, linear sliding potentiometer sensors were positioned outside the thermostatic chambers, directly on both incident and transmitter bars and thus, the performance of the sensors is not affected by even the most demanding thermal conditions. A total of 3 potentiometers were installed; 1 on the incident bar and 2 on the transmitter bar and, the distance between compression platens can be measured by correlating the differential value between potentiometers (Figure 7b).

An adequate linear actuator was selected according to the desired strain rate range. For quasi-static and low strain rate compression tests (less than 1 s^{-1}), the hydraulic ram cylinder was used (Figure 9a). Dynamic tests can be subdivided into medium strain rate tests (100 to 1000 s^{-1}), which were carried out resorting to the electromagnetic actuator (Figure 9b) and, high strain rate tests (above 1000 s^{-1}), which were performed using the pneumatic gun (Figure 9c). The electromagnetic actuator comprises several components such as the electrical circuits for charging and firing the banks of energy-storage capacitors and a series of coils that generate the electromagnetic pressure to accelerate the striker bar. The pneumatic gun consists of a pressure vessel that allows precise control of the amount of energy released during compression for a given air pressure. A pneumatic trigger valve allows the stored air volume to flow through a launcher tube, converting pneumatic energy into kinetic energy and, thus, accelerating the striker bar. The impacting velocity is limited by the mass of the striker bar and the strain rate signature of the compression tests depends on the specimen ability to dissipate the corresponding kinetic energy. Despite these fundamentals, it is difficult to achieve an adequate signal noise ratio due to the abrupt pressure impact and wave propagating through the split Hopkinson bar during test monitorization. Thus, a consumable thin nylon sheet of 1 mm was placed between the strike bar and the incident bar for better loading control. With regards to the hydraulic actuator, an electrical pump supplies the required fluid at appropriate flow rates and pressure for the quasi-static and low strain rate compression tests.

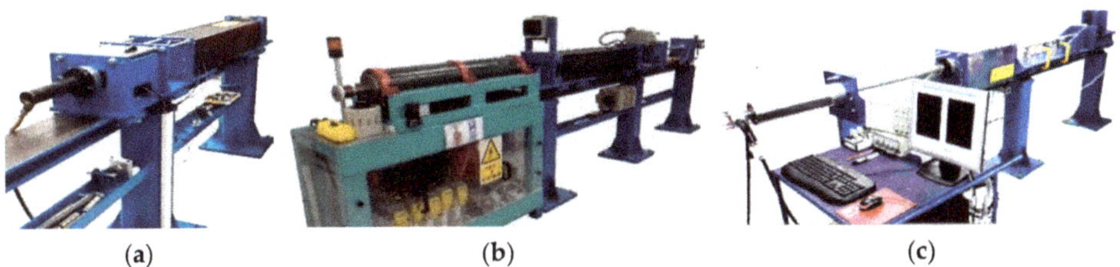

Figure 9. Specially designed Split-Hopkinson pressure bar that was utilized in the uniaxial compression tests: (**a**) in the quasi-static configuration using the hydraulic actuator, (**b**) in the medium strain rate configuration using the electromagnetic actuator, and (**c**) in the high strain rate configuration using the pneumatic gun, designed and installed at IST, University of Lisbon.

The flow curves obtained from compression tests at room temperature under different strain rate conditions are presented in Figure 10a, where three distinct stages can be observed: (i) for incipient deformation the material exhibits a typical strengthening mechanism up to strains of approximately 0.1, (ii) followed by a sudden softening behavior down to a minimum flow stress value, point after which a (iii) secondary stage of strengthening is observed. Regarding incipient deformation, an increase of 20% in maximum strength is observed from quasi-static conditions to 6000 s^{-1}. It is also worth noticing that the softening behavior seems to be more pronounced with increasing strain rates and, the instant at which the lowest flow stress was measured seems to shift in the positive direction along the strain axis. Moreover, for high strains, the rate-dependent behavior tends to similar stress values of the quasi-static compression tests. In general terms, the results show a peculiar hardening-softening behavior and markedly strain-rate sensitivity with effect on both the flow stress values and curve morphology.

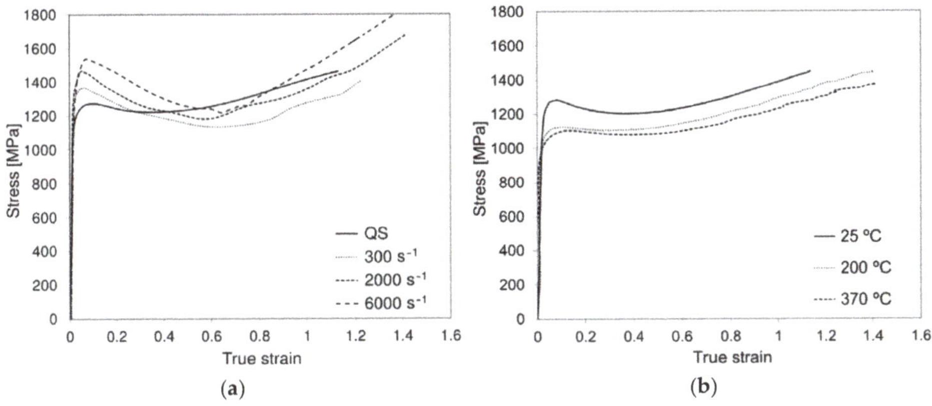

Figure 10. Experimental results of (**a**) strain-rate influence, and (**b**) temperature influence on the compressive flow curve for AMed 18Ni300 maraging steel under room temperature and lubricate conditions.

Figure 10b shows the quasi-static high temperature behavior of the AMed 18Ni300 maraging steel. Regarding the relative low temperatures of the workpiece material when submitted to cutting and the aging transformations of the maraging steel starting at 450 °C the maximum tested temperature was of 370 °C [30–32]. Apart from the already identified hardening behavior, the stress–strain response of the materials seems to present typical softening for increasing temperature in AMed metallurgical conditions and constant curve morphology.

Based on the results of the mechanical testing and after a probing calibration of the standard Johnson–Cook model (Equation (1)) for the AMed 18Ni300 maraging steel, the limitations of this equation have become clear. Despite its capacity for displaying the strain hardening and thermal softening effects of most materials, the Johnson–Cook model shows itself unable to reproduce the almost perfectly plastic behavior of the maraging steel in simple uniaxial compression, even for room temperature and quasi-static conditions. This is somewhat expected once the standard Johnson–Cook does not account for several important effects, such as the thermal-strain softening phenomena, intrinsic strain-strain rates paths, and temperature- and strain-induced phase transformation, among others. Also, in this calibration procedure the complex plastic deformation which takes place in the specimen (body) resumed as a single collection of data points, does not account for localization problems happening in the real body (e.g., fracture, adiabatic shear bands), which certainly deviate the stress–strain state from the homogenous one assumed to calculate the data originally furnished. Simpler models just Voce equation [33] and Silva equation [6] have proved to be capable to reproduce the sigmoidal stress–strain curve of the AMed 18Ni300 maraging steel. In their standard format, these athermal equations are commonly used at low strain rates with negligible temperature sensitivity and thus strain becomes nearly isothermal [34]. Its application to metal cutting should be clarified, where deformation is almost adiabatic, and heat remains where it is generated. In what follows, the authors consider that conventional metal cutting involves a cold chip formation mechanics, and the high temperature of chip is a result/output parameter of the plastic flow itself. As counterexample, the billet temperature in hot forging is an input parameter achieved prior the forging operation (e.g., by which metallurgical phase transformations may occur). This approach considers that final temperature of a formed chip or deformed cylindrical specimen should be equivalent after the plastic deformation, under comparable operative conditions (e.g., initial room temperature, strain, and strain-rate conditions). The accuracy of the model coefficients can be also improved by considering the loading signature of the process (strain-strain rate history) [6]. Thus, parameters should be obtained using experimental tests considering the signature of the machining process (strain-strain rate history). In other words, the final chip temperature can be estimated by considering the initial temperature of the uncut chip/workpiece (e.g., room temperature, external heating or cryogenic machining), heat transfer mechanisms (e.g., frictional heating, internal heating), and the inelastic heat fraction of plastic work that is converted into heat. As a result, a combined Swift-Voce law will be introduced in what follows to model the isothermal quasi-static hardening of this combined approach for the best fit to the present experimental data of AMed 18Ni300 maraging steel.

The above mentioned friction estimates method results from an experimental-numerical identification approach in which the first step corresponds to performing compressive tests on cylindrical and ring specimens under the same loading, lubrication and temperature conditions. Compression of cylinder specimens enables the estimation of trial flow stress curves that are used to simulate the corresponding ring compression test (friction calibration curves) for different friction coefficients. An example of friction calibration curves is shown in Figure 11a, where the ring inner diameter variation ($\Delta d_i/d_i$) is plotted over ring height variation ($\Delta h/h_0$). The comparison of these curves with the experimental results allows the estimation of a friction coefficient $\mu = 0.09$. The strain–stress curves used in the numerical estimative of friction calibration curves are not frictionless, even when contacting interfaces of the compression platen are lubricated. Thus, an iterative process must be applied in order to exclude the friction contribution from the trial flow stress. Furthermore, it helps to reduce the propagation of errors caused by the uncertainties of the tribomechanical calibration. This process consists in the inverse numerical simulation of the compression test using the flow stress as obtained from mechanical testing, labelled as "input" in Figure 11b, and the previously determined friction coefficient (first estimate). The resulting stress–strain curve will be an overestimate of the flow stress curve due to a double contribution of the frictional mechanism (i.e., the unknown experimental friction plus the numerically set friction), labelled as "output" in Figure 11b. By subtracting the friction overestimation, one can derive the corrected frictionless material response. Yet, an iterative process is required since the ring compression simulation is also material-dependent and new friction calibration curves should be determined, allowing a convergence towards the frictionless flow stress curve, labelled as "corrected curve" in Figure 11b, and the final identification of the friction coefficient value. It is relevant to highlight the increasing contribution of friction, for increasing strain rate and temperature, which in turn emphasizes the importance of frictionless flow stress estimation in metal cutting simulation.

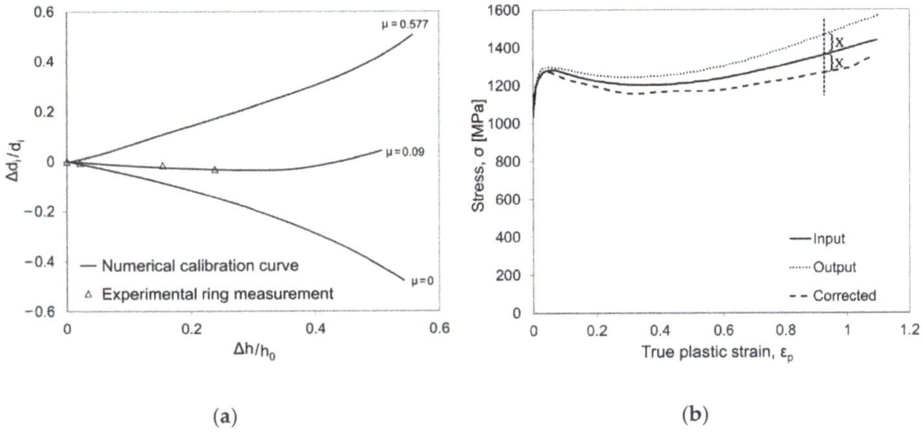

(a) (b)

Figure 11. Optimization-based procedure for identification of the effective tribo-mechanical response based on an iterative convergence between the (**a**) ring test calibration curves, and (**b**) inverse determination of the frictionless flow stress curve.

The influence of lubrication, temperature, and strain rate conditions on the friction coefficient value in tool-workpiece contact interface are shown in Figure 12. It is observed that the use of graphite lubricant (labelled as "Lub") significantly decreases friction adhesion phenomenon for the materials' interface in all tested conditions. Still with regards to lubricated conditions, a very consistent (yet modest) increasing trend is found for rising temperatures and strain rates. On the other hand, for dry conditions (labelled as "Dry"), friction coefficient tends to double when the temperature rises from room temperature to approximately 370 °C (Figure 12a). Similarly, to temperature, strain rate seems to promote an increasing friction coefficient (Figure 12b). Such results provide evidence of the friction dependence on cutting conditions where the material experiences large strains and high strain rates in the primary deformation zone and most of generated heat due to plastic work to the chip [32], which in turn slides against the tool at the secondary deformation zone.

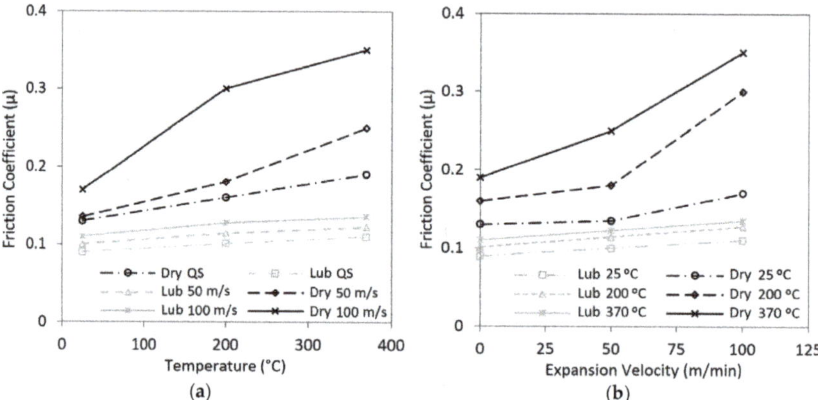

Figure 12. Experimental results of (**a**) temperature influence, and (**b**) expansion velocity of the ring radius (100 m/s is correlated with a strain rate of aprox. 6000 s^{-1}) on the friction coefficient between the AMed 18Ni300 maraging steel and the TiAlN PVD coated tools under different temperature, sliding velocity and lubricate conditions.

4.2. Compliance with Metal Cutting Conditions

Tensile tests of smooth cylindrical specimens (CT in Figure 2) were conducted in order to evaluate the stress–strain response of the AMed 18Ni300 maraging steel in radial reduction conditions (positive Lode angle parameter and stress triaxiality). Such geometrical-loading configuration allows for an assessment of the state-of-stress influence on the plastic behavior of the alloy, through direct comparison of the stress–strain curves with the ones obtained through radial expansion (compressive loading). Given that the tests were conducted for distinct notch geometries, it is also possible to access the stress triaxiality influence on ductility, enabling damage initiation model identification for a high range of stress triaxiality, which will be further discussed. The general morphology of the tensile specimens is shown in Figure 13 and the parametrized notch dimensions are described in Table 2. The theoretical stress triaxiality (η_T), assuming unaltered notch contour with deformation, has been calculated according to Bridgman [35] analytical prediction for each tensile specimen geometry. Quasi-static conditions were ensured through the imposition of a 1 mm/min pulling speed to the machine crosshead (Instron 5900R universal servo-hydraulic testing machine), which registered load through the built-in 100 kN load cell and displacement through the usage of an extensometer (MTS 632.12C-20) with a 25 mm gauge length.

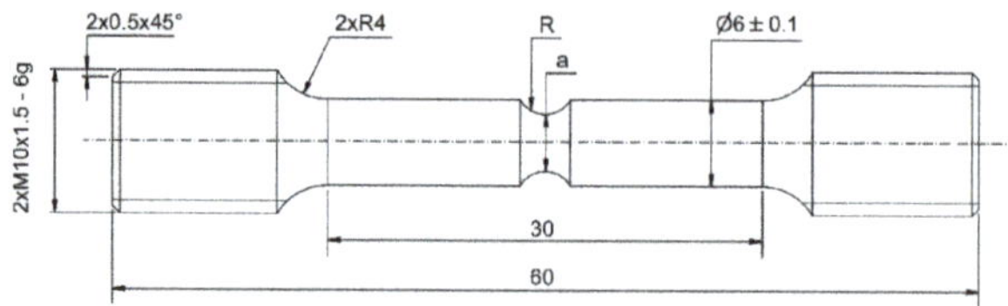

Figure 13. Tensile specimen geometry with parametrized notch dimensions.

Table 2. Tested configurations of stress triaxiality according to notch geometry.

Notch Parameter (Refer to Figure 13)	N-∞ (Smooth)	N-5 (Notched)	N-2 (Notched)
a [mm]	6	4	4
R [mm]	∞	5	2
η_T	0.33	0.52	0.74

The severe plastic deformation in metal cutting highlights the importance of its thorough characterization, especially since it is widely acknowledged that depending on the state-of-stress, the deformation behavior can significantly differ in certain materials [19]. The comparison between the true stress–strain curves of the AMed 18Ni300 maraging steel obtained from compression and unnotched tensile tests is shown in Figure 14. It is relevant to notice that due to the plastic instability, which leads to non-uniform deformation, the tensile curve is shown up to necking. Despite the very limited extent of uniform plastic deformation of the unnotched tensile tests, a clear plateau that corresponds to plastic yield is noticed, revealing a considerably distinct yield strength depending on loading direction. The observed strength differential effect between compression and tensile loading suggests the sensitivity of the tested alloy to the state-of-stress. Similar behavior has been reported for several AMed alloys [36,37], including the same maraging steel alloy [38]. However, this behavior does not seem to be exclusive of AMed metallurgical condition, given that appreciable differences between compressive and tensile yield strengths have also been noted for martensitic steels [39], including the same maraging steel alloy of conventional manufacturing (wrought) [40]. In addition, the magnitude of the tension/compression strength differential shows similar values (~10%) to that reported for the same material.

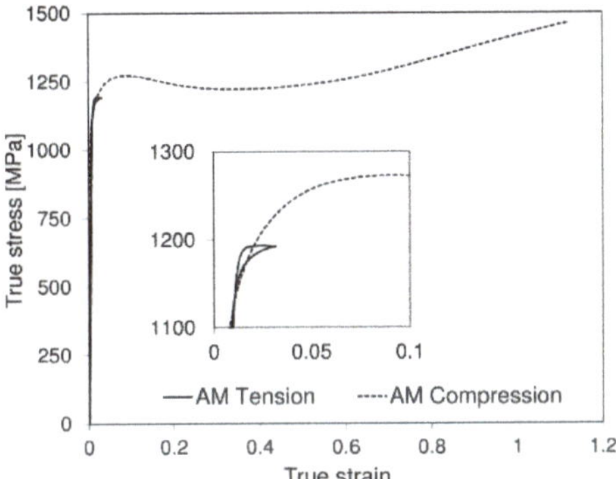

Figure 14. Mechanical response of AMed 18Ni300 maraging steel under tensile and compressive loading.

In order to accurately portray the plastic deformation during the metal cutting simulation, the mechanical response should be compliant with the plane strain conditions of the chip formation mechanism. Due to its ability to cover an intermediate range of stress triaxialities and null Lode angle parameter, the suitability of the double-notched specimen towards the calibration of plastic and damage models, for cutting applications has been demonstrated by multiple authors [1,18]. These plane strain specimens consist of symmetric double notched geometries (refer to Figure 15), where cylindrical starter cracks have been introduced to create a constant ligament length of 2 mm before loading. The length of the ligament allows to confine plastic deformation to a small region in between the cylindrical notches. The experiments consist in determining the punch shearing load–displacement evolution when the specimen is compressed between compression platens, from which critical damage and flow stress of the material under plane strain conditions may be deduced. Through the variation of the double notch configuration, it is possible to change the specimens' pressure angle, which is defined by the straight line that is colinear with the centers of the two cylindrical notches. Such modification results in a change of stress triaxiality within the same null Lode angle parameter (plane strain conditions). In the current research work, three distinct pressure angles were tested. For a pressure angle of 90°, null (or close to null) stress triaxiality due to the theoretically pure shear condition is achieved; for a pressure angle of 60° a combination of compression and shear is attained, developing intermediate negative stress triaxialities; for a pressure angle of

120°, a combination of a tensile and shear state-of-stress is achieved, yielding intermediate positive stress triaxialities.

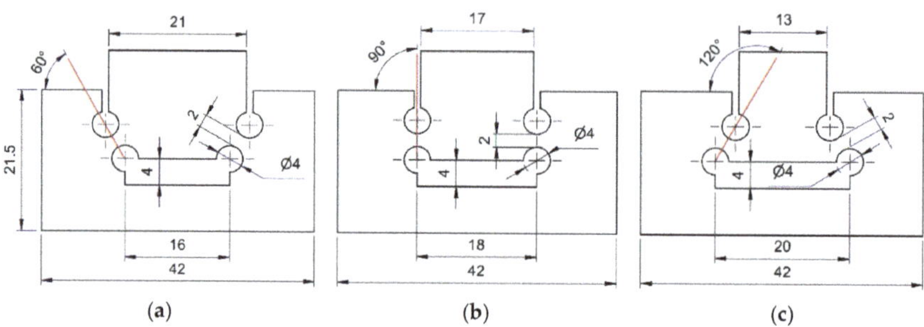

Figure 15. Double notched plane strain specimens for distinct pressure angle configurations: (**a**) 60°; (**b**) 90° and (**c**) 120°.

The tests were carried in the same universal testing machine as the tensile tests. A quasi-static compression speed of 1 mm/min was imposed to the machine crosshead and two repetitions were conducted in order to ensure experimental repeatability. Due to specimen symmetry, it was important to ensure not only that the compression plates were parallel between each other (top and bottom), but also perpendicular to crosshead displacement direction. Even though there is a major focus of these specimens towards the identification of the damage initiation and evolution model, they provide very important insight on material's plasticity and flow stress validity. Given the flaws of typical mechanical characterization methodology, say, for example, friction in compression tests and plastic instability in tensile test, added to the fact that none of both is in identical state-of-stress conditions to cutting (plane strain), the notched bar shear tests allow for a crucial inverse calibration of the plasticity models identified through direct calibration.

In this research work, a 2D plane strain approach was used for the simulation of the double notched plane strain specimens. Their symmetry enabled the modelling of half-specimen (refer to Figure 16). Four-noded elements with reduced integration (CPE4R) and an element size mesh of 0.05 mm were used. The model consists of a top die (with one degree of freedom in vertical direction) that compresses the specimen onto a bottom die (encastred). Both dies were modelled as rigid bodies.

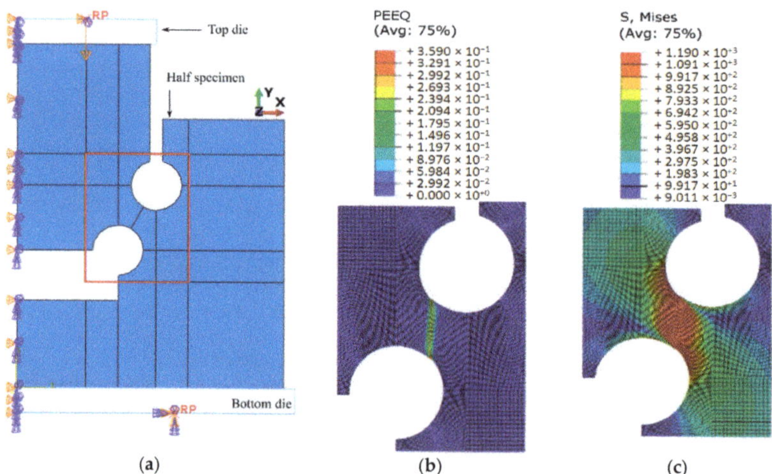

Figure 16. Numerical model for the double-notch plane strain test; (**a**) employed boundary conditions, (**b**) equivalent plastic strain, and (**c**) von Mises stress fields of the signaled notched region for a top die displacement of 0.22 mm.

The successful portrayal of the double notched specimens' plastic deformation behavior theoretically ensures a material plasticity law in a similar state-of-stress to metal cutting. Such can be achieved through the development of constitutive laws that are sensitive to the state-of-stress or, alternatively, through the inverse calibration of simpler plastic laws with the plane strain specimen. Considering the latter, it is of utmost importance to establish physically admissible upper and lower boundaries for the inverse estimation of the plane strain compliant plastic law, making the direct calibration of the tensile and compression tests a necessary initial step.

The double notched plane strain specimens have been simulated using distinct flow stresses based on the conducted characterization tests, which are shown in Figure 17a. The green curve corresponds to the (frictionless) compression test and the black dotted curve to the perfectly plastic assumption of the tensile test. An extrapolation of the tensile tests was required, given its limited extent within uniform plastic deformation. In addition, a third curve that corresponds a combined inverse calibration (based on both tensile and compression direct calibration) was built. The curve captures the tensile material response for low strain values and assumes quasi-linear negative strain hardening that is tangent to the minimum stress values obtained by compression testing. Due to allowing for an initial hardening and its saturation (or even softening) the combined Swift-Voce law (Equation (4)) has been selected to model the isothermal quasi-static hardening of this hybrid approach and the parameters are shown in Table 3. Figure 17b shows the comparison between the experimental and the numerical load–displacement results for the 60° pressure angle double notched plane strain specimens, using each of the three distinct plastic flow stress evolutions illustrated in Figure 17a. It is important to note that a damage model was not included at this stage, given the initial focus was on the assessment of plastic behavior.

$$\sigma = \alpha\left[K\left(\varepsilon_0 + \varepsilon_p\right)^n\right] + (1 - \alpha)\left[k_0 - Q\left(1 - e^{-\beta_v\,\varepsilon_p}\right)\right] \tag{4}$$

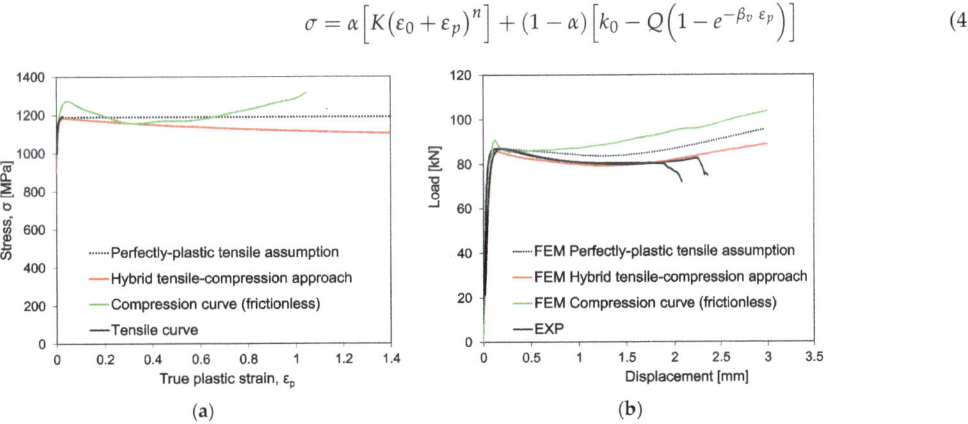

(a)　　　　　(b)

Figure 17. (**a**) Distinct flow stress curves used to represent the mechanical response of the AMed 18Ni300 maraging steel in the numerical simulation of the double-notch plane strain test; (**b**) comparison between the experimental and numerical load–displacement results, using each of the referenced flow stress curves, for the 60° pressure angle double-notched specimen.

Table 3. Swift-Voce hardening equation parameters.

K [MPa]	ε_0	n	k_0 [MPa]	Q [MPa]	β_v	α
950	1	−0.7	887.5	362.5	170	0.2

The load displacement results show very high sensitivity to the applied plastic load, which can be related with the highly heterogeneous shear strain field. It is observed that the flow stress obtained from frictionless compression testing (green curve) and perfectly plastic tensile assumption overestimate the experimental loads. On the other hand, the combined inverse calibration, inspired on both tensile and compression test data, can accurately depict the plastic behavior of the double-notched plane strain specimens.

As regards to the damage initiation model, the numerical procedure for fracture strain estimation consists in the simulation of both notched and unnotched tensile tests as well as the notched plane strain tests (using the inversely identified model) and comparing the numerical load–displacement curves with the experimentally obtained. Damage initiation was defined as the displacement at which the calibrated FEM plastic prediction and the experimental curve diverge.

For the tensile tests simulation, four-noded axisymmetric elements with reduced integration (CAX4R) were used to mesh the specimens with an element size of 0.1 mm. The adopted boundary conditions include, apart from axisymmetry, null displacement in vertical direction of the bottom nodes and a vertical displacement on the top nodes, as illustrated in Figure 18a. The numerical models were built for the extensometer gauge length, which was of 25 mm. Figure 18a additionally shows the field distributions of equivalent plastic strain and stress triaxiality of a notched specimen N-5, refer to Table 2. It is observed that both are localized in the center region of the specimen, where will be maximum. That location has, therefore, been selected for data retrieval as regards the failure strains and respective stress triaxialities, which are shown in Figure 18b, in the function of equivalent plastic strain. The results are in accordance with the predicted trend of lower ductility for higher stress triaxiality (or more pronounced notch geometry). Despite the accurate theoretical prediction of stress triaxiality (η_T) for very incipient strain values, the intense geometrical softening of the specimens (mostly as diffuse necking) results in deviation from initial notch morphology. Thus, a significant evolution of stress triaxiality up to fracture is noticed, highlighting the need for the application of this numerical methodology in order to circumvent plastic deformation localization and instability in tensile loading which hinders direct estimation of fracture strain.

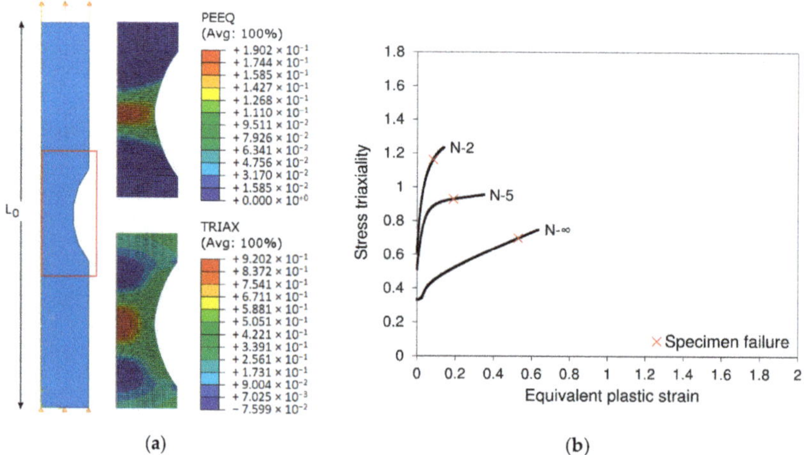

Figure 18. (**a**) Boundary conditions on the N-5 specimen geometry as well as plastic strain and stress triaxiality fields; (**b**) Estimative for stress triaxiality evolution in function of plastic strain and identification of failure point.

With reference to the double-notch plane strain tests, the determination of fracture strain in function of stress triaxiality and the equivalent plastic strain field distributions was supported by the experimental images of the deformed specimen shapes. Figure 19a illustrates the procedure for the deformed shape of the 60° pressure angle notched bar specimen at the damage onset. Relevant to highlight is the similarity between the numerical and experimental specimen deformed shape, which allows for a further validation of the inversely identified plastic law. In addition, it is possible to notice the beginning of crack propagation (signaled dashed line) which, relying on the numerical model, enables the identification of stress triaxiality (Figure 19c) and equivalent plastic strain (Figure 19b) along the actual experimental crack path. Experimental crack location is in accordance with the maximum equivalent plastic strain at the hole contour surfaces. Such has promoted the propagation of the crack from the surface to the center of the ligament and the identification of the equivalent plastic strain and stress triaxiality fracture thresholds for each notched specimen. Figure 20a shows the identified damage initiation law that was fitted to the experimentally-numerically obtained fracture strains for each distinct specimen. The reduced JC damage law (first term of Equation (2)) was fitted to the identified exponential decrease in fracture strain, for increasing stress triaxiality.

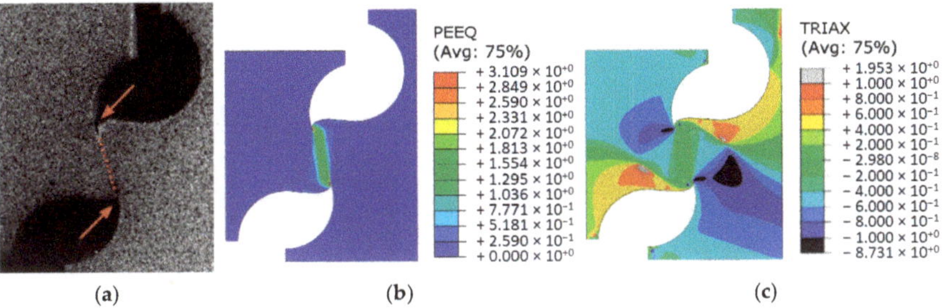

Figure 19. (**a**) Deformed shape comparison of experimental, (**b**) FEM results with plotted equivalent plastic strain (PEEQ) distribution, and (**c**) stress triaxiality distribution of 60° pressure angle double notched specimen.

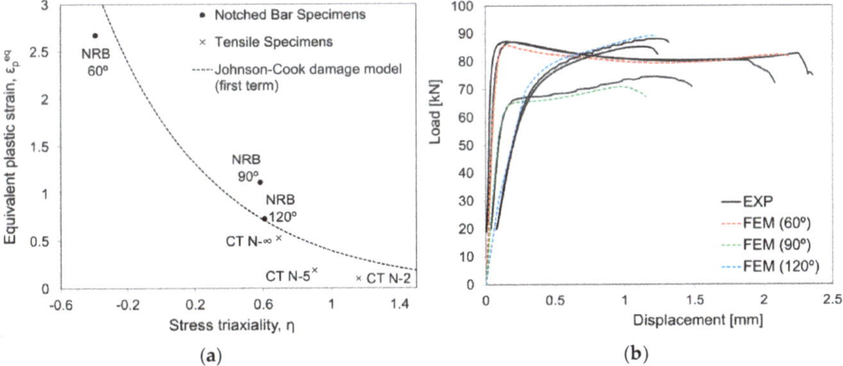

Figure 20. (**a**) Identified damage initiation law using both tensile and notched bar specimens; (**b**) Comparison between load displacement experimental curves and numerical simulation of the notched bar specimens.

With regards to material damage evolution, element degradation was defined through critical damage dissipation energy definition, as previously presented in Equation (3). It has been inversely estimated for the AMed maraging steel, through the comparison between experimental and numerical results. A critical damage dissipation energy density of $G_f = 10$ mJ/mm^3 was found. The load–displacement curves of Figure 20b present the comparison between the experimental double-notch tests and the constitutive modelling (plasticity and damage) for the AMed maraging steel. Taking into account the significantly distinct load levels and fracture strains, a good agreement seems to be found with the suggested approach and results.

With regards to the viscoplastic behavior of the maraging steel, the C parameter of the second term of Johnson Cook plasticity equation (Equation (1)) was calibrated based on the compression tests, which were conducted for increasing levels of strain rate. Despite the very challenging characterization, it is well known that fracture onset is sensitive to strain rate. In this study, that effect has been inversely identified, based on literature values on the same material. High strain rate tensile tests [41] and Charpy tests [42] are usually applied towards the identification of stain rate influence on fracture strain, which is typically modelled through the d_4 parameter in the second term of the Johnson Cook damage initiation model (refer to Equation (2)). An average value was considered ($d_4 = 0.03$), taking into account the range of literature estimation ($0.014 > d_4 > 0.05$). All material properties are summarized in Table 4. It is important to note that heat transfer, thermal conductivity, and thermal expansion parameters for this exact material batch were obtained from [43], where a detailed description of its calculation is performed.

Table 4. Material properties of AMed 18Ni300 maraging steel used in the built numerical model of orthogonal cutting simulation.

Young's modulus [GPa]	$E = 190$
Hardening Law [MPa]	$\sigma = \left[0.2 \left[950(1 + \varepsilon_p)^{-0.7} \right] + (1 - 0.2) \left[887.5 - 362.5 \left(1 - e^{170\varepsilon_p} \right) \right] \right] \times \left[1 + 0.05 ln \frac{\dot{\varepsilon}}{\dot{\varepsilon}_0} \right]$
Damage model	$\varepsilon_f = \left[-0.01 + 1.77 e^{-1.5\eta} \right] \left[1 + 0.03 ln \frac{\dot{\varepsilon}}{\dot{\varepsilon}_0} \right]$
Critical energy [mJ/mm^3]	$G_f = 10$
Friction	$\mu = 0.3$
Density [kg/m^3]	$\rho = 8000$
Thermal cond. [W m^{-1} °C^{-1}]	$\lambda = 0.0206T + 15.4$
Specific heat [J kg^{-1} °C^{-1}]	$c_p = 0.374T + 437.3$
Thermal expansion [°C^{-1}]	$\alpha_T = 1.12\text{E} - 3\,T - 4.03\text{E} - 3$

5. Validation of the Proposed Methodology

The simulation of the orthogonal cutting mechanism can be seen as a way to validate the found input data values for the plasticity, tribology, and damage laws in (relatively) closer conditions to industrial machining operations. With the plastic, fracture and friction behaviors fully depicted within the previous chapters, the current section focuses on the description of the built numerical orthogonal cutting model and the assessment of the suggested material models under varied, experimentally tested, cutting conditions.

Experimental orthogonal cutting tests were performed in order to enable the validation of the proposed methodology, through comparison with numerical modelling of the cutting process. A specially designed experimental apparatus equipped with load cell instrumentation has been employed. This type of testing machine is relevant towards the experimental representation of the orthogonal cutting due to its robustness and stiffness, that translates into the capability of sustaining high cutting loads in close velocity conditions to cutting. The linear motor is essentially an electromagnetic actuator that allows imposing-controlled velocity and energy to the kinematics transmission systems. A constant cutting velocity of 26 m/min was imposed to all orthogonal cutting tests. A scheme of the testing machine is shown in Figure 21. Load measurement was performed through a three-component piezoelectric dynamometer (Kistler 9257B). A charge amplifier (Kistler 5070A) and a data acquisition system (Advantech 4711A) enabled signal conversion and data collection, with a sampling rate of 5 kHz, which was further analysed using typical processing software. A clamping vice was built for specimen fixation and tightened to the dynamometer, which in turn was bolted to the testing machine table. All components were aligned with the ram movement. The testing machine slide allows for the definition of the uncut chip thickness, t_0, through its vertical displacement. A Mitutoyo dial indicator with a 0.001 mm resolution was clamped to the tool slide for improved t_0 definition. In order to be able to calculate the effective t_0, the height of the specimen was measured beforehand and after each cut, for the whole extent of the cutting length, in 1 mm intervals.

A Mitutoyo toolmaker's microscope (profile projector) with a 2× magnification lens was used for measuring the linear height distances along the specimens' cutting length. Two distinct tool geometries were tested, varying only the rake angle which was of 5° and 12°. The inserts' material was tungsten carbide, PVD-coated (TiAlN) with a 3 μm coating layer and their geometry is displayed in Figure 22. With regards to the cutting conditions, t_0 values ranging from 0.05 to 0.35 mm have tested. The inserts were mounted on a tool holder with a V pocket with their inverse geometry (ensuring rigidity), which in turn was fixed to the testing machine ram through a tool holder.

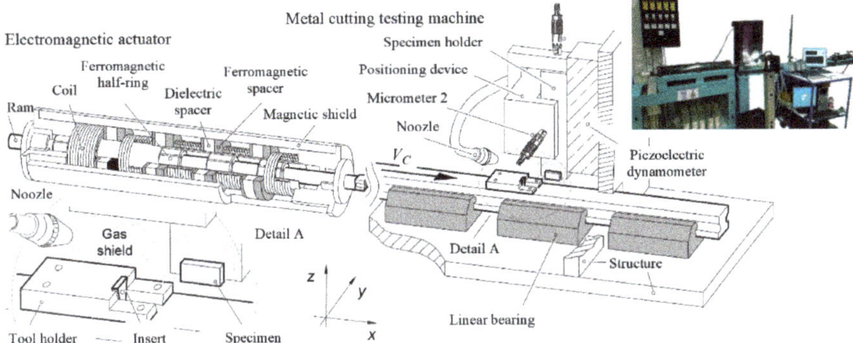

Figure 21. Schematic representation of the metal cutting apparatus with detail of the test region showing cutting tool and cutting specimen, and a picture showing the custom-built apparatus.

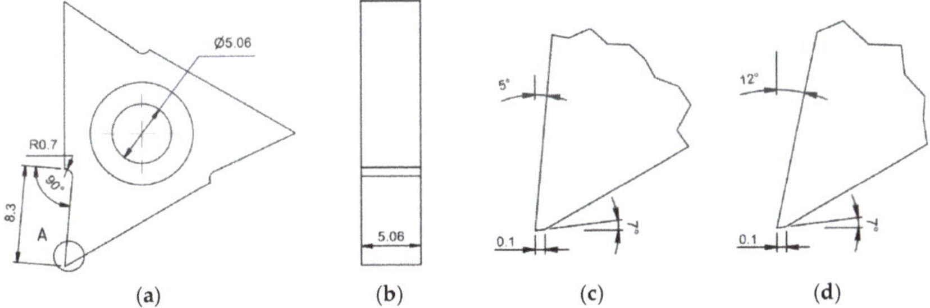

Figure 22. (**a**) Side view, (**b**) front view of the cutting inserts overall geometry employed in the orthogonal cutting; (**c**) detail A of 5° rake angle, and (**d**) 12° rake angle inserts.

In the current study, a two-dimensional orthogonal cutting numerical model was implemented, using a coupled temperature-displacement (explicit) formulation with a Lagrangian approach. Figure 23 illustrates the implemented numerical model. In order to minimize the high element distortion and eventual non-convergence, distinct aspect ratios were imposed to elements at the uncut chip thickness. Attributing a "pre-deformed" shape to the elements through its diagonal positioning allows for easier convergence of the model [44]. A distinct mesh distribution was also defined, meaning a finer mesh at the uncut chip thickness with decreasing mesh density towards the bottom of the workpiece.

Despite not significantly affecting cutting force, a finer mesh contributes to an improved realistic depiction of chip morphology, at the cost of computing time [45]. In addition, a gradual engagement of the tool was imposed with a chamfered geometry (45°) on the workpiece. An identical tool geometry to the experimental tests was employed in the numerical model, meaning that two rake angles ($\alpha_r = 5°$ and $\alpha_r = 12°$) were simulated with a fixed relief angle ($\gamma_f = 7°$). Workpiece, tool, and mesh sizes were scaled accordingly to the uncut chip thickness value and their parametrized dimensions are shown in Table 5 in accordance with the nomenclature of Figure 23. A theoretically sharp tool was considered in all tool-workpiece geometrical configurations. Four-noded plane strain elements with reduced integration and enhanced hourglass control (CPE4RT) were employed. Tool was considered rigid whilst the workpiece was considered elastoplastic. On the tool, node rotation was null for all degrees of freedom. Identically, displacement was also set to zero in yy direction and experimental cutting speed was imposed in the xx direction. It is important to note that the identified plastic model was not calibrated in function of temperature. In fact, it is assumed that plasticity-generated heat is already accounted in the flow stress obtained by compression tests, meaning that thermal softening, strain and strain-rate hardening are regarded as a coupled effect. Also, it is important to highlight, is that most of the generated heat in the primary deformation zone flows to the chip [32], meaning that at the workpiece and primary cutting zone, temperatures are not high enough to significantly influence material properties [46]. Relying on the assumption that adhesion is the prevailing friction mechanism, the identified friction values have been incorporated in the orthogonal cutting model through Coulomb's law, motivated by its simplicity and considering the general disagreement of friction modelling in the metal cutting domain. Chip separation has been applied through element deletion. When reaching the fracture strain that is determined by the damage model, elements are disregarded from the mesh. Their degradation was modelled using Equation (3), taking into consideration energy dissipation towards fracture occurrence. Failed elements are deleted from the mesh once the overall damage variable (D) reaches the value of 1. Despite mass loss, this method seemed the most appropriate for the assessment of the identified damage law. In order to promote computational efficiency, the mass of the entire model has been scaled. When used appropriately, this method can significantly decrease simulation time without jeopardizing the degree of accuracy [47].

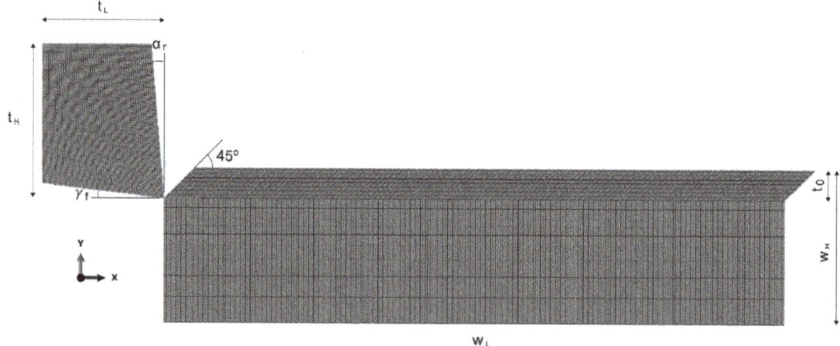

Figure 23. Finite element model used in the numerical simulation of orthogonal cutting showing workpiece geometry and parametrized dimensions.

Table 5. Parametrized workpiece, tool and mesh dimensions.

	t_H	t_L	w_H	w_L	Element Size
Dimension [mm]	$5t_0$	$4t_0$	$5t_0$	$20t_0$	$0.08t_0$

Figures 24 and 25 show he tchip formation and its evolution at four distinct cutting lengths for an uncut chip thickness of 0.1 mm and tool rake angles of 5° and 12°, respectively. Stress distribution (von Mises) is plotted which shows, as expected, maximum values at the primary deformation zone. Stress concentration is also developed at the secondary deformation zone, due to tool-chip friction. Figure 26a shows that the predicted (numerical) strain rate values at the primary deformation zone is compatible with the dynamic compression tests range. Moreover, in Figure 26b, the stress triaxiality field distribution of the chip is shown. It is interesting to note the highly negative triaxiality values at the secondary shear zone (tool-chip interface) which then evolve into positive stress triaxiality, as the chip is no longer in contact with the tool and is submitted to a bending moment, instead. Despite its tendency to bend (plastically deform) which results in a change of the stress triaxiality values at chip contour, its core still registers the mainly negative values due to the predominant compression loading of the chip formation mechanism.

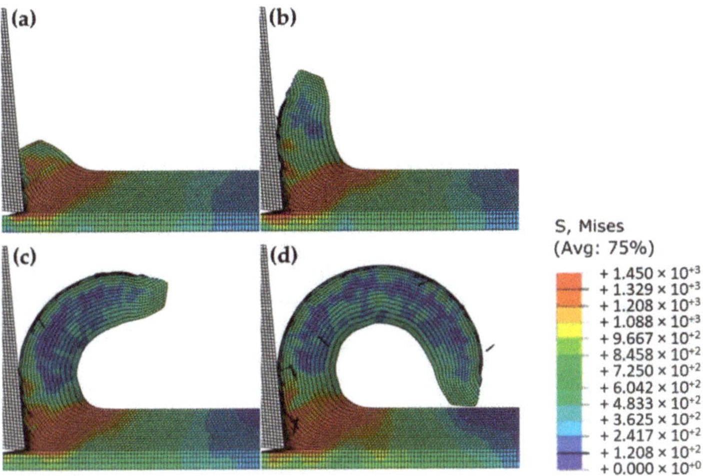

Figure 24. Numerical estimates from transient beginning to machining steady-state orthogonal cutting conditions for $\alpha_r = 5°$ rake angle, $t_0 = 0.1$ mm and cutting lengths of (**a**) 0.15, (**b**) 0.4, (**c**) 0.8 and (**d**) 1.3 mm showing the von Mises stress distribution (S).

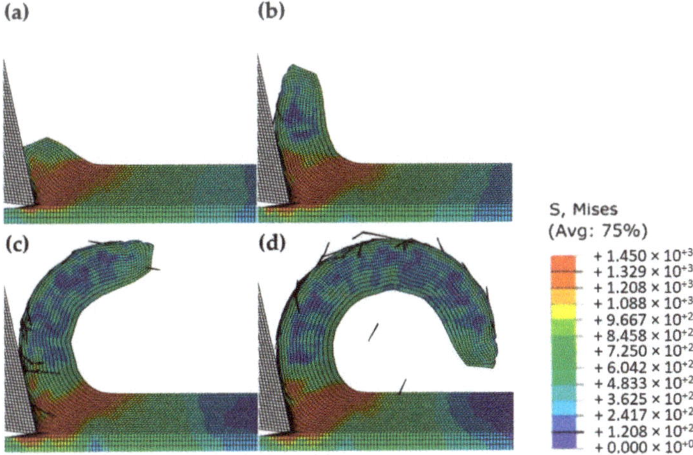

Figure 25. Numerical estimates from transient beginning to machining steady-state orthogonal cutting conditions for $\alpha_r = 12°$ rake angle, $t_0 = 0.1$ mm and cutting lengths of (**a**) 0.15, (**b**) 0.4, (**c**) 0.8 and (**d**) 1.3 mm showing the von Mises stress distribution (S).

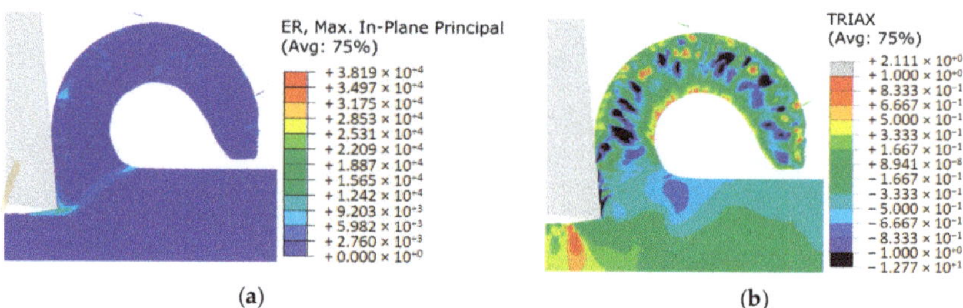

Figure 26. Computed distribution of the (**a**) strain rate and (**b**) stress triaxiality field distributions for steady-state orthogonal cutting conditions.

The study of cutting forces under the most varied machining conditions is a common approach towards cutting process improvement. It can be seen as a benchmarking parameter of cutting efficiency and cutting tool geometry suitability, especially in the development of novel geometries. Such data is typically obtained through extensive experimental testing. Numerical simulation can be highly advantageous in the virtual identification of characteristic cutting constants such as the specific cutting pressure, K_c, enabling extrapolation from the numerically validated trends and minimizing experimental machinability tests. Towards the assessment and validation of the implemented FEM models when it comes to load prediction, the experimental range of t_0 has been covered numerically for the comparison of specific cutting pressure. Figure 27 presents the numerical and experimental K_c results for the AMed 18Ni300 maraging steel in function of the selected tool geometries. The satisfactory numerical estimation for a wide range of uncut chip thickness in two distinct tool geometries enables the validation of the identified input data values and material laws, as well as the implemented orthogonal cutting numerical model. An approximate maximum error of 4% from the experimental exponential trend has been calculated.

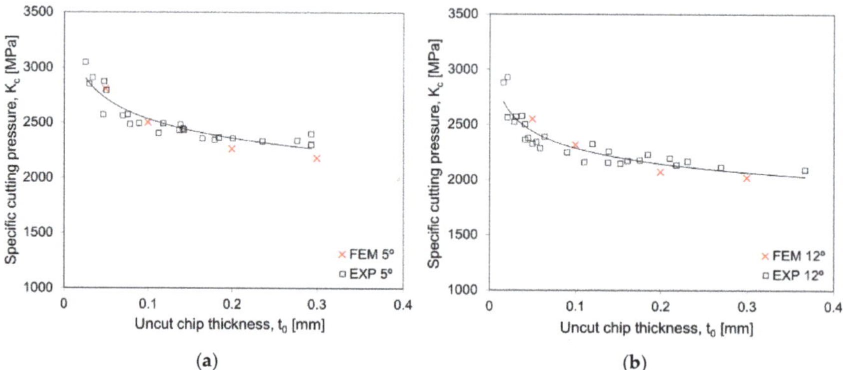

Figure 27. Comparison between experimental and numerical specific cutting pressure on the orthogonal cutting (**a**) with a 5°, and (**b**) 12° rake angle.

6. Application

The material's machinability was assessed by means of longitudinal turning tests carried out on a universal lathe EFI DU20 with 5.9 kW power under dry conditions. Given that machining operations are typically performed in additively manufactured components in order to attain better surface quality and to ensure dimensional tolerances for specific applications, the current tests focused only on finishing operations.

The cutting load is significant as an indicator of the operation stability and energy requirements of the machining process. Combined with chip morphology characterization, it allows the identification of favorable cutting scenarios in the post processing stage, envisioning an increase of the production rate and quality of the finished product. When it comes to load measurement and acquisition, the same equipment employed in the orthogonal cutting tests was used. The load cell was mounted on the lathe's tool carriage and levelled in order to ensure the alignment of the cutting tool with the center of rotation of the fixation chuck. Figure 28 shows a schematic representation and the related nomenclature of the experimental apparatus. An ISO DCGT-11T304-FS cutting insert with a 0.4 mm nose radius, chip-breaking rake surface and the same coating as the orthogonal cutting inserts and compression platens, TiAlN PVD coating, was selected to provide a representative cutting insert to the industrial turning operations and to keep the tribological conditions constant throughout the investigation. It was mounted on a tool holder providing a $-3°$ side cutting edge angle, $0°$ cutting edge inclination and $7°$ back relief angle. An unused cutting tool was selected for each segment in order to ensure equal conditions. Regarding the operative parameters, a full factorial combination of two rotational speeds (1800 and 2500 rpm), three feeds (0.05, 0.1 and 0.2 mm/rev) and three depths of cut (0.1, 0.2, and 0.3 mm) were explored. Each specimen is subjected to various cutting passes so the starting point for each cut is an already machined surface. Since the tests were performed under constant rotational speeds, the cutting speeds have evolved along the successive cutting passes due to the diameter reduction. The selected 2500 rpm resulted into cutting speeds between 118 and 77 m/min.

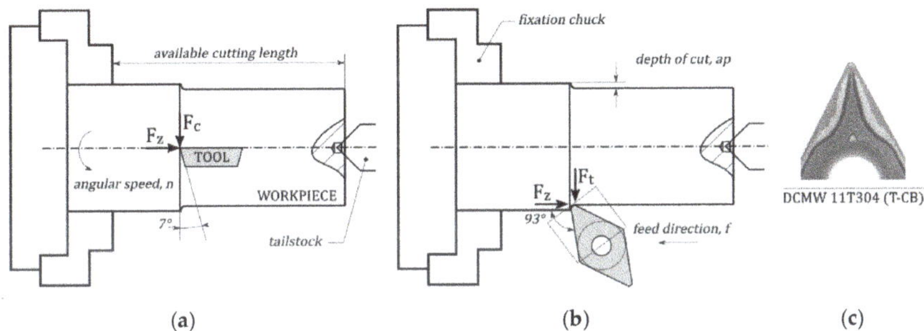

Figure 28. Scheme of the turning setup: (**a**) front view and (**b**) top view of tool-workpiece relative position; (**c**) rake surface of the used cutting insert (DCMW 11T304-FS T-CB).

Despite providing valuable insight on metal cutting mechanics, material models validation and simplified kinematics, the orthogonal cutting operation does not accurately represent 3D industrial cutting processes with intricate tool geometries, such as the turning operation. In an attempt of bringing closer the industrial processes with the numerical simulation, the same experimentally tested longitudinal turning operations were numerically modelled. Figure 29a shows the tool and workpiece schematic configuration, illustrating the geometrical cutting parameters. It is important to note that, for kinematics simplification, and similarly to other models in literature [48,49], the workpiece was defined as a rectangular extrusion with the contour of tool nose radius, focusing on a small portion (but steady-state representative) of the cutting process. The chip-breaking insert tool geometry was represented in the model through a rigid body. The workpiece was modelled with 8-node linear elements with reduced integration and hourglass control (C3D8R). In order to be able to apply two distinct structured mesh densities, the workpiece was divided in two regions, labelled as "core" and "envelope" (refer to Figure 29a). In order to ensure continuity between these two parts, a tie constraint was used, ensuring no relative motion between the nodes of each part. Chip separation is modelled through the application of the identified damage model to the whole workpiece. As regards to the tool, a finer mesh density (unstructured) was defined around nose radius. In both tool and workpiece parts, finer meshes correspond to a 0.02 mm seed size, and 0.05 mm for coarser meshes.

The workpiece base was clamped and degrees of freedom of the tool were null, except in zz direction, allowing for tool cutting speed definition which was the same as the maximum experimentally tested. With regards to material parameters, the same conditions as orthogonal cutting simulation were applied. Three distinct depths of cut a_p (0.5, 0.3 and 0.1 mm) were simulated for a fixed feed, f = 0.2 mm. The selection of such cutting parameters is related with the larger chip sections, comparatively to other experimentally tested configurations. The simulation of smaller chips is hindered by the very fine required meshes and associated computational costs.

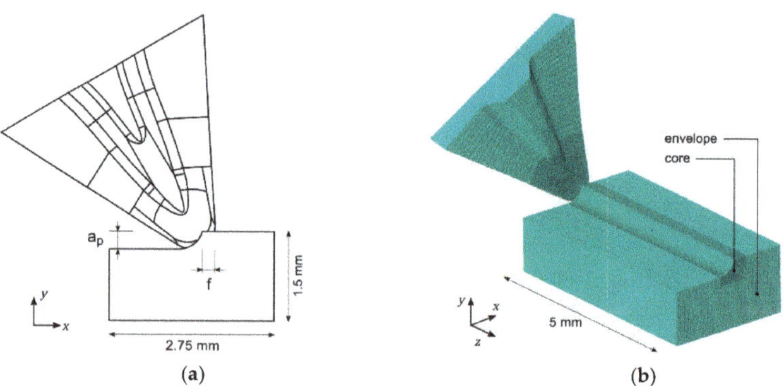

Figure 29. (**a**) Schematic representation of the turning operation, and (**b**) finite element model used in the numerical simulation of longitudinal turning operations.

Figure 30 shows the numerical final chip geometry of each simulated set of cutting parameters (for a limited cutting length of 5 mm), and the corresponding experimental chip geometry (for indefinitely long, >>5 mm, cutting length), illustrating their similar tendency to identical chip formation. This means that for minimum depth of cut (a_p = 0.1 mm) a ribbon-like chip morphology is developed, while for bigger chip sections curling seems to occur, promoting helical chip formation as the one being developed in the numerical model with a_p = 0.5 mm.

Regarding the principal cutting forces, the simulated results are compared with the experimental analogous conditions in Figure 31. A very satisfactory correlation between the experimental and numerical results is observed, showing the robustness of the identified material model towards the principal cutting force prediction. The good experimental-numerical correlation and successful simulation of the industrial turning process address the urgent need towards shifting from theoretical 2D models into more accurate and complex 3D simulations [4].

Despite the good performance of the numerical simulation for the cutting force estimates, the underestimation of the other vector components should be highlighted. In orthogonal cutting the experimental feed force (F_f) accounts for averagely 57% of the principal cutting force (F_c). Regardless of the accurate numerical prediction of the principal cutting force, the numerical feed force is underestimated (35% of the F_c). Concerning the numerical simulation of the turning process, a similar trend is observed, in which the numerical estimate of feed force is less than 35% of the principal cutting force. These results are consistent with those found in the literature [11], which are still not sufficiently understood. It is a common belief that lack of information about flow stresses and friction at the rates and temperatures experienced in practical cutting has primary responsibility for this disagreement [10]. Nevertheless, the authors of the present research, after an extensive mechanical–tribological characterization under extreme loading conditions, have the strong confidence that a lack of information about flow stresses and friction at the rates and temperatures is not the cause for the disagreement. The mechanical–tribological response of the material throughout the shear plane and over the rake surface can be adequately calibrated and modelled for the numerical simulation of the metal cutting processes. Possible reasons for mismatches between numerical estimates and experimental measurements could come from the singularities of the metal cutting mechanics located where the chip formation begins and where the chip-tool interfacial contact ends. In the first case, the size effect of the cutting-edge radius or ploughing effect [50] is pointed out as an additional contribution to the overall cutting load [51]. Some authors also claim that while the chip-rake contact interface is steady, the workpiece-clearance contact occurs under highly dynamic conditions because of inappropriate vibration (chatter) that often results in poor surface quality and feed load peaks [52]. The vibration in the direction of the cutting speed is relatively smaller, whereas in the direction of the depth of cut it is relatively higher. The vibration amplitude is promoted by the relatively low stiffness of the overall machine-tool-workpiece set and the workpiece mechanical strength that is a major obstacle, when machining difficult-to-cut materials, such as hardened steels and titanium alloys [53]. Surface free energy is also pointed out as another important contribution from the point of view of the ductile fracture mechanics [54]. Clearance surface is also of major importance since its texture scratches and rubs the machined surface masking the contribution of the primary rake face on the cutting forces amplitude. The higher the elastic recovery of the machine-tool-workpiece set, the higher the relevance of this effect.

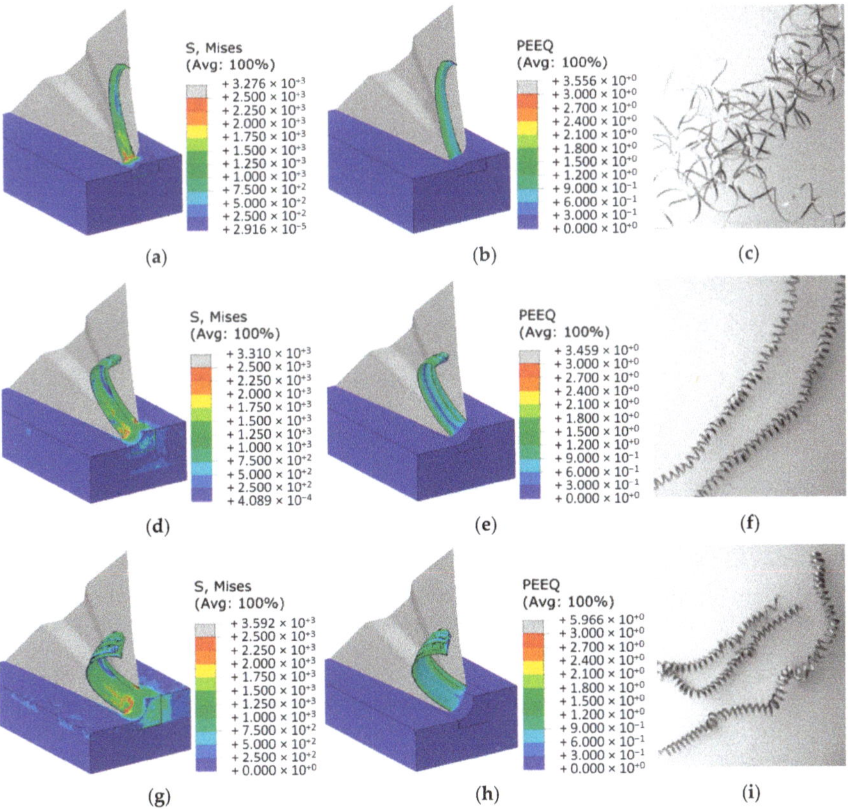

Figure 30. Numerical chip geometry showing von Mises stress (S) distribution and equivalent plastic strain (PEEQ) fields and experimental chip geometry for longitudinal turning with a fixed f of 0.2 mm and (**a–c**) for a fixed $a_p = 0.1$ mm; (**d–f**) for a fixed $a_p = 0.3$ mm, and (**g–i**) for a fixed $a_p = 0.5$ mm.

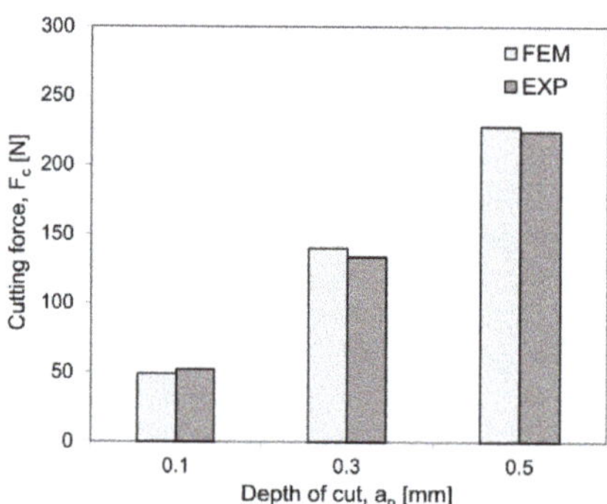

Figure 31. Comparison between the experimental measurements and numerical estimates of cutting force Fc for longitudinal turning tests.

In the second case, perhaps less known, it has been shown that pressure-welded junctions occur near the point at which the chip detaches from the cutting tool, and it has been established that there is an intimate relation between the oxide films formation and the oxygen concentration in the surrounding medium. Thus, an additional shear force parallel to the feed force should be applied to break the weld spots to allow the chip to slide over the rake surface. At these welded spots no relevant normal force to the rake surface, parallel to the cutting force, is noticed. Thus, the friction coefficient can be artificially promoted. The thinner the uncut chip thickness, the higher relevance of this effect [55].

As discussed above, the numerical underestimation of the secondary vector components of the principal cutting force seems to suggest significant limitations of the finite element codes. The effectiveness of the numerical models to predict the machining performance depends on how accurately these models can represent the actual metal cutting process, namely the severe mesh distortion around the tool cutting edge. A highly refined mesh is the first step usually taken to address the problem [56]. However, the sparse finite element mesh used in the 3D simulations of Figure 30 showed a significant computational cost of approximately 48 h, using 10 multiple processors for each numerical simulation. This high computational effort is incompatible with a practical application for the numerical simulation of metal cutting. Nevertheless, even simple 2D simulations of the orthogonal metal cutting using a highly refined mesh shown a similar underestimation of the feed force. The accuracy of the numerical methods seems not to be at stake, but rather the concept of which physico-chemical phenomena should or should not be considered in the numerical modelling of metal cutting. The availability of numerical features in the commercial codes to couple these physico-chemical phenomena is also of utmost importance to establish a realistic numerical model of metal cutting. However, despite only the cutting force can be accurately predicted, the underestimation of the secondary vector components can be minimized considering that the relative difference between the experimental measurements and the numerical estimates is about 30%. This is perhaps one of the main results of the present investigation, clarifying the influence of tribo-mechanical modelling and numerical methods on the estimates accuracy, pointing out the need to investigate the physico-chemical phenomena in the singularity zones of the chip formation mechanism.

7. Conclusions

In metal cutting modelling, the input data calibration has always relied on the uniaxial compression testing given its state-of-stress similarity and due to being the most effective way of achieving large plastic strains. As an additional benefit, uniaxial compression also allows mechanical testing in extreme mechanical conditions, resembling those in metal cutting processes. Friction models can also be calibrated by the ring uniaxial compression test over the same range of operative conditions. In this part of the study, special focus has been on identification of the individual contribution of each phenomenon involved in the tribo-mechanical mechanism. The proposed optimization-based methodology showed to be effective in subtracting the friction-dissipated energy of the experimentally determined flow stress, as well in precise determination of the friction coefficient under extreme mechanical conditions. This tribo-mechanical characterization made it possible to partly cover the deficit in the scientific literature on the material behavior over wide ranges in temperature and strain-rates for AMed 18Ni300 maraging steel.

Looking at the scientific literature, there are cases where the traditional calibration methodologies seem unable to provide accurate input data values, with negative impact on the numerical simulation results. In general, this occurs when the material plastic behavior is highly sensitive to the stress triaxiality. This sensitivity was observed in the tested AMed 18Ni300 maraging steel and is compatible with similar materials in the literature from both AMed and conventional metallurgical conditions. The input data calibrated exclusively on the basis of uniaxial compression tests provided numerical overestimates around 30% for the cutting forces. In most cases, this accuracy level is sufficient to carry out a sensitivity analysis to the process parameters, such as the study on the impact of AMed materials on the metal cutting forces and chip morphology. This research showed that where more accurate estimates are to be produced, it is helpful do have a better understanding of the plastic flow and the damage evolution associated with the manufacturing process.

An adequate constitutive model and well determined coefficients are the major key points in metal cutting modelling in order to better describe the tribo-mechanical response of the materials. The better the tribo-mechanical response is captured by the model, the better the accuracy of numerical estimates from the metal cutting simulations. However, a complete calibration of a constitutive model for general application, such as in sheet metalworking, bulk metalworking and metal cutting, is a complex an extremely time-consuming process. The proposed methodology demonstrated that the calibration procedure can be efficiently managed and simplified when the nature of the manufacturing process is considered. The almost perfectly plastic behavior of the maraging steel also exposed the inadequacy of the previously selected Johnson–Cook hardening law (first term of Equation (1)), and thus a Swift-Voce model was successfully employed even though its simplicity.

The proposed calibration methodology proved to efficiently provide accurate input data for the tribo-mechanical constitutive models by numerically reproducing all mechanical testing results successfully. Nevertheless, the simple orthogonal cutting test under controlled laboratorial conditions was also necessary for validation of the implemented numerical model for metal cutting simulation. It should be noted that the tribological condition of the ring compression test differs from the metal cutting due the existence of metallic oxides and thus a slight variation in the friction coefficient was to be expected. The friction coefficient was adjusted upwards by about 9% in order to better reproduce the experimental chip curling. Its sensitivity to strain rate and interface temperature was kept constant to that determined by the tribo-mechanical testing. Finally, the smooth cylindrical and the cylindrical-notch tests under tensile loading have been shown to be relevant for additional accuracy in the numerical prediction of the chip types and surface integrity.

The critical damage value required for chip separation is always maintained ahead of the cutting edge, but similar values are also found to occur along the rake surface of the cutting tool. This means that more damage accumulation can occurs after the material passes through out the primary shear zone. The existence of high levels of ductile damage along the rake surface of the cutting tool is not expected to promote new crack formation because this region of the chip is subjected to compressive hydrostatic stresses which inhibit crack formation. However, the chip acquires more curl as it passes through the primary deformation zone and departure from the rake surface. The material located outside the uncut chip thickness (inner side of curvature) experiences a different damage path under positive stress triaxiality. Thus, initiation and propagation of cracks from the outside the uncut chip thickness to the inside chip bulk is possible to happen and it occurs for a variable critical damage value, demanding tensile testing on easy-to-test cylindrical samples. The chip type and its curvature radius value are probably defined by an energy compromise between the plastic work, the amount of frictional work and the initiation of new surfaces/cracks. Future research at the constitutive modelling shall consider, in addition to damage mechanics, the contribution of the additional energy required to form new surfaces on metal cutting. Likewise, the establishment of new experimental methodologies to characterize the fracture toughness under similar operative conditions as those observed in machining processes.

Finally, though not less important, the double notched plane strain specimen had a central role in the inverse calibration methodology of the flow stress curve and critical damage values under plane strain conditions. This has the advantage of generating both negative and positive stress triaxiality values for the calibration of the coupled damage-plasticity constitutive model. Few double notched specimens were necessary as the previous calibrated mechanical model, based on uniaxial compression tests, served as an initial reference for the optimization-based procedure for determination of the plasticity model coefficients and damage initiation coefficients. The number of double notched specimens needed to perform the inverse calibration is of importance for industrial applications that demand a rapid effective response. The time taken in the inverse calibration procedures depends on its quantity, geometrical variants, and experimental repeatability. It is worth noting the high volume of these specimens and elevate price per kilo of the additively manufactured material to test; Furthermore, most of this material to be removed by machining. Each specimen requires a different geometry to allow different stress triaxiality values (pressure angles), thus a different manufacturing setup is required for each specimen. All this makes the double notched specimen much more expensive than the uniaxial compression cylinders. From the point of view of the authors, the double notched plane strain specimens can be improved through adoption of a simple, unique, and standard specimen geometry that allows the control of the triaxiality value by the testing machine. The adaptation to the specimen geometry should also take in account the notch geometry to enable a direct characterization of the fracture toughness values.

Author Contributions: Conceptualization, P.A.R.R. and A.M.P.d.J.; methodology T.E.F.S., P.A.R.R. and A.M.P.d.J.; software, T.E.F.S. and A.V.L.G.; validation, T.E.F.S., A.V.L.G. and P.A.R.R.; formal analysis, P.A.R.R. and A.M.P.d.J.; investigation, T.E.F.S. and A.V.L.G.; resources, P.A.R.R. and A.M.P.d.J.; data curation, A.V.L.G. and T.E.F.S.; writing—original draft preparation, T.E.F.S. and P.A.R.R.; writing—review and editing, A.M.P.d.J. and P.A.R.R.; visualization, A.V.L.G.; supervision, A.M.P.d.J. and P.A.R.R.; project administration, A.M.P.d.J. and P.A.R.R.; funding acquisition, A.M.P.d.J. and P.A.R.R. All authors have read and agreed to the published version of the manuscript.

Funding: This work has been conducted under the scope of MAMTool (PTDC/EME-EME/31307/2017) and AddStrength (PTDC/EME-EME/31307/2017) projects, funded by Programa Operacional Competitividade e Internacionalização, and Programa Operacional Regional de Lisboa funded by FEDER and National Funds (FCT). This work was supported by FCT, through IDMEC, under LAETA, project UIDB/50022/2020.

Institutional Review Board Statement: Not applicable.

Informed Consent Statement: Not applicable.

Data Availability Statement: Not applicable.

Conflicts of Interest: The authors declare no conflict of interest.

Abbreviations and Symbols

The following abbreviation and symbols are used in this manuscript:

AM	Additive manufacturing
AMed	Additively manufactured
CC	Cylinder compression
CM	Conventional manufacturing
CMed	Conventionally manufactured
CS	Cylinder shear
CT	Cylinder tensile
FE	Finite element
FEM	Finite element method
HiPIMS	High power pulsed magnetron sputtering
JC	Johnson–Cook
NCT	Notched cylinder tensile
NPS	Notched plane strain
OC	Orthogonal cutting
PM	Powder metallurgy
PVD	Physical vapor deposition
SHPB	Split-Hopkinson pressure bar
S-V	Swift-Voce
α	Material hardening parameter
α_r	Tool rake angle
α_T	Thermal expansion
β_v	Material hardening parameter
c_p	Specific heat
$\dot{\varepsilon}$	Strain rate
ε_0	Material hardening parameter
$\dot{\varepsilon}_0$	Reference strain rate
ε_f	Damage initiation strain
ε_p	Equivalent plastic strain
γ_f	Tool relief angle
η	Stress triaxiality
λ	Thermal conductivity
μ	Coulomb friction parameter
$\bar{\theta}$	Lode angle parameter
ρ	Density
σ	Equivalent plastic stress
A	Material hardening parameter
a	Notched tensile specimens effective diameter
a_p	Depth of cut
B	Material parameter
C	Material parameter
d	Diameter
d_1 to d_5	Material parameters (Johnson–Cook damage law)
E	Young's modulus
F_c	Cutting force
F_f	Feed force/longitudinal force
F_t	Penetration force/radial force
f	Feed rate
G_f	Critical damage dissipation energy
h	height
K	Material hardening parameter
K_c	Specific cutting pressure
k_0	Material hardening parameter
L	Characteristic length
m	Material hardening parameter
n	Material hardening parameter
R	Tensile specimens notch outer radius
Q	Material hardening parameter
S	Stress (von Mises)
T_0	Room temperature
t_0	Uncut chip thickness
T_m	Melting temperature

References

1. Cheng, W.; Outeiro, J.; Costes, J.-P.; M'Saoubi, R.; Karaouni, H.; Denguir, L.; Astakhov, V.; Auzenat, F. Constitutive model incorporating the strain-rate and state of stress effects for machining simulation of titanium alloy Ti6Al4V. *Procedia CIRP* **2018**, *77*, 344–347. [CrossRef]
2. Melkote, S.N.; Grzesik, W.; Outeiro, J.; Rech, J.; Schulze, V.; Attia, H.; Arrazola, P.-J.; M'Saoubi, R.; Saldana, C. Advances in material and friction data for modelling of metal machining. *CIRP Ann.* **2017**, *66*, 731–754. [CrossRef]
3. Buchkremer, S.; Wu, B.; Lung, D.; Münstermann, S.; Klocke, F.; Bleck, W. FE-simulation of machining processes with a new material model. *J. Mater. Process. Technol.* **2014**, *214*, 599–611. [CrossRef]
4. Rodríguez, J.M.; Carbonell, J.M.; Jonsén, P. Numerical Methods for the Modelling of Chip Formation. *Arch. Comput. Methods Eng.* **2020**, *27*, 387–412. [CrossRef]
5. Imbrogno, S.; Rotella, G.; Umbrello, D. On the Flow Stress Model selection for Finite Element Simulations of Machining of Ti6Al4V. *Key Eng. Mater.* **2014**, *611–612*, 1274–1281. [CrossRef]
6. Silva, C.M.A.; Rosa, P.A.R.; Martins, P.A.F. Electromagnetic Cam Driven Compression Testing Equipment. *Exp. Mech.* **2012**, *52*, 1211–1222. [CrossRef]
7. Cristino, V.A.M.; Rosa, P.A.R.; Martins, P.A.F. Revisiting the Calibration of Friction in Metal Cutting. *Tribol. Trans.* **2012**, *55*, 652–664. [CrossRef]
8. Cristino, V.A.M.; Rosa, P.A.R.; Martins, P.A.F. The Role of Interfaces in the Evaluation of Friction by Ring Compression Testing. *Exp. Tech.* **2015**, *39*, 47–56. [CrossRef]
9. Klocke, F.; Döbbeler, B.; Peng, B.; Schneider, S.A.M. Tool-based inverse determination of material model of Direct Aged Alloy 718 for FEM cutting simulation. *Procedia CIRP* **2018**, *77*, 54–57. [CrossRef]
10. Astakhov, V. On the inadequacy of the single-shear plane model of chip formation. *Int. J. Mech. Sci.* **2005**, *47*, 1649–1672. [CrossRef]
11. Bil, H.; Kilic, S.E.; Tekkaya, A.E. A comparison of orthogonal cutting data from experiments with three different finite element models. *Int. J. Mach. Tools Manuf.* **2004**, *44*, 933–944. [CrossRef]
12. Marbury, F.H. *Characterization of SLM Printed 316L Stainless Steel and Investigation of Micro Lattice Geometry*; Senior Project Report; California Polytechnic State University: San Luis Obispo, CA, USA, 2017.
13. Leça, T.C.; Silva, T.E.F.; de Jesus, A.M.P.; Neto, R.L.; Alves, J.L.; Pereira, J.P. Influence of multiple scan fields on the processing of 316L stainless steel using laser powder bed fusion. *Proc. Inst. Mech. Eng. Part L J. Mater. Des. Appl.* **2020**, *235*, 19–41. [CrossRef]
14. Bai, Y.; Chaudhari, A.; Wang, H. Investigation on the microstructure and machinability of ASTM A131 steel manufactured by directed energy deposition. *J. Mater. Process. Technol.* **2020**, *276*, 116410. [CrossRef]
15. Johnson, G.R.; Cook, W.H. Fracture characteristics of three metals subjected to various strains, strain rates, temperatures and pressures. *Eng. Fract. Mech.* **1985**, *21*, 31–48. [CrossRef]
16. Zheng, X. Extreme mechanics. *Theor. App. Mech. Lett.* **2020**, *10*, 1–7. [CrossRef]
17. Gambirasio, L.; Rizzi, E. On the calibration strategies of the Johnson–Cook strength model: Discussion and applications to experimental data. *Mater. Sci. Eng. A* **2014**, *610*, 370–413. [CrossRef]
18. Abushawashi, Y.; Xiao, X.; Astakhov, V. A novel approach for determining material constitutive parameters for a wide range of triaxiality under plane strain loading conditions. *Int. J. Mech. Sci.* **2013**, *74*, 133–142. [CrossRef]
19. Bai, Y.; Wierzbicki, T. A new model of metal plasticity and fracture with pressure and Lode dependence. *Int. J. Plast.* **2008**, *24*, 1071–1096. [CrossRef]
20. Moakhar, S.; Hentati, H.; Barkallah, M.; Louati, J.; Haddar, M. Influence of geometry on stress state in bulk characterization tests. *C. R. Mécanique* **2019**, *347*, 930–943. [CrossRef]
21. Abushawashi, Y.; Xiao, X.; Astakhov, V. Practical applications of the "energy–triaxiality" state relationship in metal cutting. *Mach. Sci. Technol.* **2017**, *21*, 1–18. [CrossRef]
22. Buchkremer, S.; Klocke, F.; Lung, D. Finite-element-analysis of the relationship between chip geometry and stress triaxiality distribution in the chip breakage location of metal cutting operations. *Simul. Model. Pract. Theory* **2015**, *55*, 10–26. [CrossRef]
23. Zhang, M.; Sun, C.-N.; Zhang, X.; Goh, P.C.; Wei, J.; Hardacre, D.; Li, H. Fatigue and fracture behaviour of laser powder bed fusion stainless steel 316L: Influence of processing parameters. *Mater. Sci. Eng. A* **2017**, *703*, 251–261. [CrossRef]
24. Lang, F.H.; Kenyon, N. Welding of maraging steels. In *WRC Bulletin*; Welding Research Council: New York, NY, USA, 1971; Volume 159.
25. Król, M.; Snopiński, P.; Czech, A. The phase transitions in selective laser-melted 18-NI (300-grade) maraging steel. *J. Therm. Anal. Calorim.* **2020**, *142*, 1011–1018. [CrossRef]
26. U.S. Department of Defense. *MIL-S-46850D—Steel: Bar, Plate, Sheet, Strip, Forgings, and Extrusions, 18 Percent Nickel Alloy, Maraging, 200 KSI, 250 KSI, 300 KSI, and 350 KSI, High Quality*; U.S. Department of Defense: Washington, DC, USA, 1991.
27. Zhang, D.; Zhang, X.-M.; Nie, G.-C.; Yang, Z.-Y.; Ding, H. Characterization of material strain and thermal softening effects in the cutting process. *Int. J. Mach. Tools Manuf.* **2021**, *160*, 103672. [CrossRef]
28. Rao, K.P.; Sivaram, K. A review of ring-compression testing and applicability of the calibration curves. *J. Mater. Process. Technol.* **1993**, *37*, 295–318. [CrossRef]
29. Sofuoglu, H.; Rasty, J. On the measurement of friction coefficient utilizing the ring compression test. *Tribol. Int.* **1999**, *32*, 327–335. [CrossRef]

30. Zorev, N.N. Inter-relationship between shear processes occurring along tool face and shear plane in metal cutting. In *International Research in Production Engineering*; American Society of Mechanical Engineers (ASME): New York, NY, USA, 1963; Volume 49, pp. 143–152.
31. Astakhov, V. *Metal Cutting Mechanics*; CRC Press: Boca Raton, FL, USA, 1998.
32. Shaw, M.C. *Metal Cutting Principles*, 2nd ed.; Oxford University Press: New York, NY, USA, 2005.
33. Sung, J.H.; Kim, J.H.; Wagoner, R.H. A plastic constitutive equation incorporating strain, strain-rate, and temperature. *Int. J. Plast.* **2010**, *26*, 1746–1771. [CrossRef]
34. Vaz, M., Jr. On the numerical simulation of machining processes. *J. Braz. Soc. Mech. Sci.* **2000**, *22*, 179–188. [CrossRef]
35. Bridgman, P.W. *Studies in Large Plastic Flow and Fracture*; McGraw-Hill: New York, NY, USA, 1952.
36. Chen, W.; Voisin, T.; Zhang, Y.; Florien, J.-B.; Spadaccini, C.M.; McDowell, D.L.; Zhu, T.; Wang, Y.M. Microscale residual stresses in additively manufactured stainless steel. *Nat. Commun.* **2019**, *10*, 4338. [CrossRef] [PubMed]
37. Joseph, J.; Stanford, N.; Hodgson, P.; Fabijanic, D.M. Tension/compression asymmetry in additive manufactured face centered cubic high entropy alloy. *Int. J. Plast.* **2017**, *129*, 30–34. [CrossRef]
38. Cyr, E.; Lloyd, A.; Mohammadi, M. Tension-compression asymmetry of additively manufactured Maraging steel. *J. Manuf. Process.* **2018**, *35*, 289–294. [CrossRef]
39. Drucker, D.C. Plasticity theory strength-differential(SD) phenomenon, and volume expansion in metals and plastics. *Metall. Trans.* **1973**, *4*, 667–673. [CrossRef]
40. Spitzig, W.A.; Sober, R.J.; Richmond, O. The effect of hydrostatic pressure on the deformation behavior of maraging and HY-80 steels and its implications for plasticity theory. *Metall. Trans. A* **1976**, *7*, 1703–1710. [CrossRef]
41. Fu, H.; Wang, X.; Xie, L.; Hu, X.; Umer, U.; Rehman, A.U.; Abidi, M.H.; Ragab, A.E. Dynamic behaviors and microstructure evolution of Iron–nickel based ultra-high strength steel by SHPB testing. *Metals* **2020**, *10*, 62. [CrossRef]
42. Madhusudhan, D.; Chand, S.; Ganesh, S.; Saibhargavi, U. Modeling and simulation of Charpy impact test of maraging steel 300 using Abaqus. In *IOP Conference Series: Materials Science and Engineering 330, Proceedings of International Conference on Recent Advances in Materials, Mechanical and Civil Engineering, Hyderabad, India, 1–2 June 2017*; IOP Publishing: Bristol, UK. [CrossRef]
43. Silva, T. Machinability of Maraging Steel Manufactured by Laser Powder Bed Fusion. Ph.D. Thesis, Faculdade de Engenharia da Universidade do Porto, Porto, Portugal, 2021.
44. Rosa, P.; Kolednik, O.; Martins, P.; Atkins, A. The transient beginning to machining and the transition to steady-state cutting. *Int. J. Mach. Tools Manuf.* **2007**, *47*, 1904–1915. [CrossRef]
45. Haddag, B.; Atlati, S.; Nouari, M.; Znasni, M. Finite element formulation effect in three-dimensional modeling of a chip formation during machining. *Int. J. Mater. Form.* **2010**, *3*, 527–530. [CrossRef]
46. Astakhov, V.P. *Tribology of Metal Cutting*; Tribology and Interface Engineering Series; Elsevier: Amsterdam, The Netherlands, 2006; Volume 52, pp. 1–425.
47. Smith, M. *ABAQUS/Standard User's Manual, Version 6.9*; Dassault Systèmes Simulia Corp: Providence, RI, USA, 2009.
48. Silva, T.F.; Soares, R.; Jesus, A.; Rosa, P.; Reis, A. Simulation Studies of Turning of Aluminium Cast Alloy Using PCD Tools. *Procedia CIRP* **2017**, *58*, 555–560. [CrossRef]
49. Liu, G.; Huang, C.; Su, R.; Özel, T.; Liu, Y.; Xu, L. 3D FEM simulation of the turning process of stainless steel 17-4PH with differently texturized cutting tools. *Int. J. Mech. Sci.* **2019**, *155*, 417–429. [CrossRef]
50. Sahoo, P.; Patra, K.; Szalay, T.; Dyakonov, A.A. Determination of minimum uncut chip thickness and size effects in micro-milling of P-20 die steel using surface quality and process signal parameters. *Int. J. Adv. Manuf. Technol.* **2020**, *106*, 4675–4691. [CrossRef]
51. Vipindas, K.; Anand, K.N.; Mathew, J. Effect of cutting edge radius on micro end milling: Force analysis, surface roughness, and chip formation. *Int. J. Adv. Manuf. Technol.* **2018**, *97*, 711–722. [CrossRef]
52. Munoa, J.; Beudaert, X.; Dombovari, Z.; Altintas, Y.; Budak, E.; Brecher, C.; Stepan, G. Chatter suppression techniques in metal cutting. *CIRP Ann.* **2016**, *65*, 785–808. [CrossRef]
53. Singh, K.K.; Kartik, V.; Singh, R. Modeling dynamic stability in high-speed micromilling of Ti–6Al–4V via velocity and chip load dependent cutting coefficients. *Int. J. Mach. Tools Manuf.* **2015**, *96*, 56–66. [CrossRef]
54. Rosa, P.A.R.; Martins, P.A.F.; Atkins, A.G. Ductile Fracture Mechanics and Chip Separation in Cutting. *Int. J. Mach. Mach. Mater.* **2007**, *3*, 335–346. [CrossRef]
55. Rosa, P.A.R.; Gregorio, A.V.L.; Davim, J.P. The Role of Oxygen in Orthogonal Machining of Metals. In *Measurement in Machining and Tribology*; Springer: Cham, Switzerland, 2019; pp. 49–88. [CrossRef]
56. Outeiro, J.C.; Umbrello, D.; M'Saoubi, R.; Jawahir, I.S. Evaluation of Present Numerical Models for Predicting Metal Cutting Performance and Residual Stresses. *Mach. Sci. Technol.* **2015**, *19*, 183–216. [CrossRef]

MDPI
St. Alban-Anlage 66
4052 Basel
Switzerland
Tel. +41 61 683 77 34
Fax +41 61 302 89 18
www.mdpi.com

Journal of Manufacturing and Materials Processing Editorial Office
E-mail: jmmp@mdpi.com
www.mdpi.com/journal/jmmp